Interactive
Computer Systems

Videotex and Multimedia

Interactive Computer Systems

Videotex and Multimedia

Antone F. Alber

Bradley University
Peoria, Illinois

Springer Science+Business Media, LLC

Library of Congress Cataloging-in-Publication Data

Alber, Antone F.
   Interactive computer systems : videotex and multimedia / Antone F.
Alber.
      p.   cm.
   Includes bibliographical references and index.
   ISBN 978-1-4613-6251-7   ISBN 978-1-4615-2888-3 (eBook)
   DOI 10.1007/978-1-4615-2888-3
   1. Interactive computer systems.   2. Videotex systems.
3. Multimedia systems.   I. Title.
QA.9.I58A43   1993
650'.0285'66--dc20   93-3393
                                                              CIP

ISBN 978-1-4613-6251-7

© 1993 Springer Science+Business Media New York
Originally published by Plenum Press in 1993
Softcover reprint of the hardcover 1st edition 1993

To Mary Beth,
my wife and friend

Preface

A generation ago, the raw material of a computer was data, the material from which information is derived—facts, concepts, and instructions. Today the raw material is information, data that have been transformed into a meaningful and useful form. A vast gulf separates the way data and information are handled and used.

Data are compact and often consist of numbers and individual elements that can be conveniently handled in printed reports or in columns on a display screen. Normally data must undergo a transformation to be useful. Previously that transformation occurred in the viewer's mind or on pads of paper. Today, the viewer works directly on the information by interacting with the computer. Unfortunately information is much more demanding of computer and network resources than data and therein lies the rub. Converting data into information by introducing text in the form of sentences and paragraphs, along with color, graphics, picture-quality images, audio, and even motion video consumes many more computer resources than just data alone. Compounding the problem is the unhappy prospect of individuals who work in a data-processing environment thrust into the roles of graphic illustrators, script writers, or electronic journalists—roles for which many individuals in data processing are ill suited.

This book explores the realm of interactive computer systems, the varied media forms that are possible, and access to this information, often at great distances. Treatment of the topic does not focus on technical details; concepts are fully explained, so that anyone with even a very limited knowledge of computer technology can understand them and recognize the possibilities that exist for intelligently using interactive computer systems based on videotex and multi-media applications.

Chapter 1 discusses the changing business environment and the rise of the computer culture. This sets the stage for examining how some organizations use computers to realize operational and strategic goals. Chapter 2 expands on several points raised in Chapter 1 and introduces a model for studying operational and strategic applications of systems that have been enhanced by varied media forms.

Chapters 3 and 4 present the host hardware and the user's equipment. Of particular interest are such new media forms as CD-ROMs and their derivatives, CD-I, CDTV, and so forth. A section about multimedia information storage and some of the inherent difficulties will interest the reader. Chapter 4 describes end-user equipment ranging from dumb terminals to computer telephones and video-phones. Public access terminals, also known as kiosks, which are springing up in airports, shopping malls and hotel lobbies, are discussed.

Chapters 5–7 form a trilogy devoted to the links between where information resides and those who access it. Chapter 5 explains storage and transmission requirements of multimedia applications. Chapter 6 presents some of the communication services that an organization has at its disposal externally and internally. Chapter 7 outlines events leading to the breakup of AT&T and the development of Open Network Architecture (ONA) for business and residential applications.

Chapter 8, devoted to the ethereal subject of software, explains the special set of functions that differentiates a software program presenting information from one presenting data. Since many organizations will end up buying the software they need and thus be vulnerable to the supplier, a discussion of escrowing software and a software bill of rights are included.

Chapters 9–11 focus on producing the contents. A strategy for selling the system to users, managers, and the data-processing department is proposed along with ideas for nurturing expectations without ignoring the economic realities of implementing enhanced systems. A 10-step methodology shows how to plan and design the contents in the context of developing an electronic version of an organization's annual report. Incorporated within the methodology are tips and procedures ranging from designing menus to picking colors and writing a storyboard.

Chapters 12–15 describe how organizations have actually implemented the types of systems described in the book. Chapter 12 reports on the results of an extensive mail survey that shed light on why organizations are deploying enhanced systems, the approaches used, and the problems to avoid. Ten case studies of organizations that have implemented projects are presented in Chapters 13–15. In each case, technology, objectives, communication links, and how it was designed are explained in the context of the material presented in the preceding chapters. These cases reveal how organizations are using information technology in innovative ways to further their goals.

Antone (Joe) F. Alber

Peoria, Illinois

Acknowledgments

This book was a labor of love and a lot of work. It benefited from the contributions of many people who shared their time, talents, and resources to help me write it. For those who find the information useful, please remember that the following people played a role.

Special thanks to these individuals who took time from their busy schedules to let me visit them to write the 10 cases appearing in Chapters 13–15: Gerry J. Barker, *Fort Worth Star-Telegram* (StarText); Helen G. Bradley, American Airlines SABRE Travel Information Network (EAASY SABRE); Fred L. Craig and Sue Davis, Chevrolet-Pontiac-Canada Group (Town Crier); Larry R. Dale, North American Automotive Sales Operations, Ford Motor Company; Don R. Forsyth and Eileen M. Smith, Ford Motor Credit Company; William Leonard and Lucy Keister, National Library of Medicine (HARPP); Brenda J. Madding, Clark Distribution Services, Inc. (ClarkNet II); Ann Mintz, the Franklin Institute Science Museum (Unisystem); Mike Noar and Roman R. Gackowski, Merck & Company (Heartfelt Advice); Prabal Roy and Connie S. Tarburton, E. I. Du Pont De Nemours and Company (Corporate Homes for Sale).

There were many individuals who participated in the mail survey discussed throughout the book, especially in Chapter 12; to maintain the anonymity promised, their names cannot be listed, but thanks are due nevertheless. There were several individuals who made material contributions to the survey effort whom I can thank publicly; they are Gary H. Arlen, Arlen Communications, Inc.; Geary Barnes; Arthur Esch, NuMedia Corporation; Kevin Gazzara, Intel Corporation; Michael Gordon, ByVideo; Joel R. Lapointe, Essense Systems; Peter Levin, Cadillac Motor Car; Tom Morgan, InterMedia Development Corporation; Leslie Ney, AT&T; Doug Peter, St. Clair Videotex Design; P. A. (Peter) Richardson,

Tayson Systems, Inc.; Elwood Schlesinger, Pontiac; Roland J. Sharette, J. Walter Thompson; Hilary B. Thomas, Interactive Telecommunications Services, Inc.; Peter Winters, Newspaper Advertising Bureau, Inc.; Reinhard Ziegler, Andersen Consulting.

There were a variety of people at Bradley University who provided assistance. Lynne Kowalske did many of the illustrations. The following secretaries lent a hand at one time or another: Sharon Dew, Jewel Gray, Ruth Morris, and Mary Lu Reither. These graduate students assisted with library research: Mohumad Fayad, Mike Haag, and Brandon Velde.

Thanks also to Fred Takahashi of Panasonic and Leslie Stanbrough of EyeTel Communications, Inc., for tracking down some difficult to obtain information.

Special thanks to Bradley University and Caterpillar Corporation. In the normal course of each day, the university directly and indirectly underwrites the expense and provides the support that enables faculty members to engage in these sorts of activities. Each summer Caterpillar Corporation provides fellowships to selected faculty members at Bradley for research and writing in a wide assortment of fields. I was the fortunate recipient of a summer fellowship. Many thanks to both organizations for their financial support and encouragement.

I extend my gratitude to all of you and apologize to anyone whose help I neglected to mention.

Contents

Chapter 1

The Business Scene

This book is about easy-to-use, interactive computer systems that possess some or all of the following features: text, graphics, color, picture quality displays, sound, animation, and motion video. The focus is on business or organizational applications rather than consumer applications, but much of the information is applicable to either environment. These types of computer systems are distinguished from the traditional systems found in data processing departments by their information rather than their data orientation, by their stylized presentations and ease of use, and by their connectivity to other systems in and outside the organization. Computer systems with such features are known by a variety of names: videotex, multimedia, hypermedia, telemedia, and mixed media. Because these terms may mean different things to different people, the expression enhanced system is often used in this book. Hopefully this term is neutral enough to avoid misunderstanding yet specific enough to convey the idea that the computer systems being discussed represent an innovative use of information technology. Some of the names associated with enhanced systems surround the perimeter of Figure 1.1, while the interior lists acronyms of the communications services that make connectivity and the two-way flow of information possible. These terms are expanded on in later chapters, and there is a list of acronyms and a glossary in the appendices.

In Chapter 1, we examine the changes occurring in the business environment and the increasing importance of information and computer resources.

1

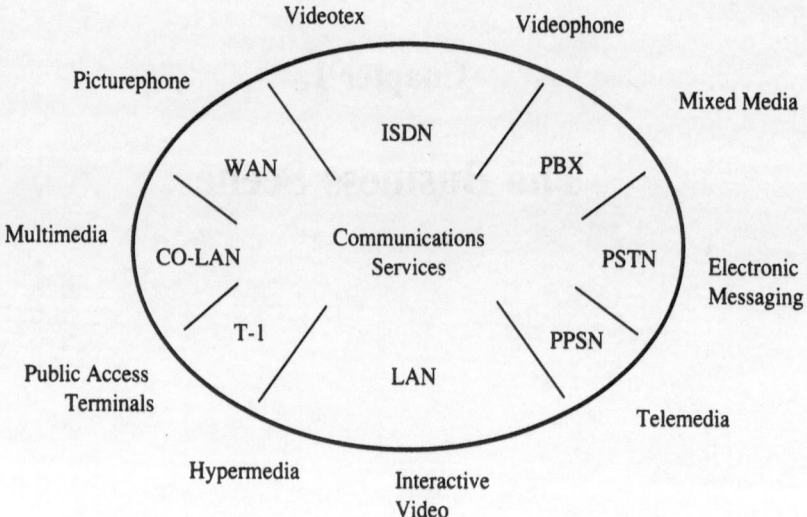

Figure 1.1. A potpourri of terms associated with enhanced systems.

1.1. THE CHANGING BUSINESS ENVIRONMENT

After the 1980 census findings were released, many books were published that chronicled the changes occurring in the U.S. economy; *Megatrends* is probably the best known example. When researchers eventually finish studying the results of the 1990 census, they will identify some new trends, such as a work force that is both aging and changing radically. They will also find deregulation and privatization rampant in this country and abroad. They will see a global economy in which new forces are emerging to challenge the perennial economic powers and a world dominated by a single superpower with an unchallenged military machine and a sick economy. Amid the chaos produced by rapid change and political realignments will be an ever-increasing demand by organizations for ways to use information technology more efficiently and more effectively than the competition.

1.1.1. The Graying Society

In a free and competitive society, structural changes in the economy can have a profound and long lasting effect on how a company manages its business internally and externally. One of the more visible demographic changes is humorously referred to as the graying of America. Between 1986 and 2000, the projected population increase in persons aged 45 and over will be 30%, compared to an increase of 2% for persons aged 18–44. To put this into perspective, the percentage of the population over 65 will increase from the current 12% to 22% in 2030. The three principal reasons for the aging of the U.S. population are: (1) Life expectancy at birth is increasing; (2) baby boomers are moving into middle age; and (3) birth rates in the 1960s and 1970s were low. As this segment of society moves into the ranks of senior citizens it will bring new values, new expectations, and new demands for information and services.

The graying society will impact the work force and thus employers. The average age of workers is expected to jump from the current age of 32 to 40 by the year 2000. There will be shortages of qualified workers in jobs traditionally filled by young workers. Labor costs (per hour worked) will increase due to longevity factors, and pension costs will increase because people live longer. Technological obsolescence is more likely as employee skills and job knowledge fall behind job requirements. The half-life of information workers, which is already estimated to be as short as five years, may be reduced further due to the unwillingness to provide training for older workers. An aging work force will create new challenges for employers. New policies and management practices must be designed to use the skills of older workers effectively. Part of the solution will lie in developing innovative management practices, such as job sharing, job redesign, and rehiring retirees. But part of the solution resides in effectively using information technologies.

Multimedia training facilities that are computer driven and self-pacing will accommodate the need for new and more extensive in-house training. Organizations will welcome such innovations as telecommuting to encourage some senior citizens to continue working, or to accommodate those who are less mobile, particularly in northern climates during the harsh winter months. Information technology will enable adaptive strategies to accommodate handicapped and physical infirmities that are more prevalent among senior citizens. These will include large-screen print displays and audio output for the visually and hearing impaired, image capturing and retrieval systems for the less mobile, electronic mail and computer-conferencing systems for those working at home or working part-time, and other systems tailored to meet the needs of the older worker. Whether organizations are able to employ information technology as an adaptive strategy will depend in some part on the willingness of older workers. Unfortunately studies have shown that older individuals are less likely to adopt a technology, independent of other demographic and attitudinal factors.

1.1.2. The Changing Work Force

Not only is society in general growing older in the United States, society's cultural and racial mixtures are undergoing significant change. In the next century, racial and ethnic groups will outnumber whites for the first time. By the middle of the next century, the average American will trace his or her ancestry to Asia, Africa, Arabia, or almost anywhere else but white Europe. Referred to by William Henry as the "browning of America," this will have a profound impact on everything from politics and education to personal values and ideas. This ethnic change will be acutely felt at the work site. According to the Bureau of Labor

Statistics, the work force will become increasing minority and female. Minorities, principally blacks, Hispanics, and Asians will account for 57% of labor force growth between 1986 and 2000; factoring in white women sends the figure to 90% of labor force growth. Thus the era in which the typical worker is white and male is rapidly coming to an end.

Between the late 1980s and the year 2000, more than 21 million new jobs will be created, mostly in the service sector. By their very nature, many of these jobs will require improved quantitative skills, language skills, and reasoning ability. Since much of the growth forecast for society as a whole and the work force in particular will be composed of relatively less educated members of minorities, the expanding labor pool will not be well matched to the jobs being created.

Part of the solution to this problem lies in the judicious use of information technology which will facilitate training, improve communications, and provide flexibility through automation. Organizations will have to invest in training, education, and employee involvement, and they will also have to accommodate the needs and deficiencies of the workers.

Training with technology has an especially promising future. According to Minneapolis-based Lakewood Research, corporate America spent an estimated $43.2 billion on employee training in 1991. As the need for training intensifies, so does the cost unless new and more efficient strategies are found. Technology-based training is one possible strategy: It costs only 15%–25% as much as traditional classroom instruction, and it enjoys economies of scope. As the number of students increases, the cost per student declines, because training can go to the student rather than require the student to go to the training. In the Guglielmo reference at the end of the chapter, it is reported that at IBM, it costs approximately $350/day with travel to hold a workshop at a central site. It costs $150/day for an on-site workshop and $75 for a self-paced and technology-based course.

In 1992 Bell South reported it expected to save $5 million in travel expenses alone over a 4-year span by using a multimedia training program. This program reduced a 7-day course taught in Atlanta to a 7-hour course offered in the employee's office. NCR has adopted multimedia on the shop floor. Instead of time consuming briefings and manual checklists, employees view a monitor that guides them through the process of assembling products.

Another aspect of the training and educational opportunities offered by technology is learning on demand. This is made possible by the ability to access vast amounts of information either on-line or from mass storage devices anytime of the day or night. A single CD-ROM disc can deliver several shelves worth of books to the desktop electronically. One popular CD-ROM contains over 950 works of drama, poetry, religion, history, medicine, science, and more, complete with illustrations. Mass storage coupled with commercial on-line information sources means tomorrow's student can be freed from the time and place constraints that often determine when and where learning will take place.

1.1.3. Deregulation

Government deregulation of the transportation, financial, and communication industries has created fundamental changes in the way businesses in these industries compete. The probusiness climate of the 1980s has encouraged business expansion, job creation, and entrepreneurship; and the relaxation of antitrust policies has led to many corporate mergers and takeovers that would have been unlikely a few years earlier. Information technology has become the fuel that powers many of these renaissance businesses. For example, American Airlines makes more from its computer reservation system then it does from flying passengers, and financial services companies like Merrill Lynch offer a portfolio of services that a few years ago were available only from banks, insurance companies, or savings and loan institutions. These companies are able to do this because of some of the most sophisticated information systems found anywhere. Perhaps the largest wave, which has just barely begun to form, will be created by the Baby Bells as they launch the information services they fought so hard for. Deregulation provides the opportunity, but information technology makes it possible.

1.1.4. Accent on Information

A major occupational change in the labor force has been the shift in the importance of information processing. In 1860 information handlers, consisting of clerical workers, professional and technical workers, managers, and administrators, made up less than 10% of the total labor force, compared to workers in agriculture, who made up more than 40%. In 1980 nearly one out of two workers was employed in the information sector and only 2% in agriculture. The shift from brawn to brain has created new opportunities for women, and it has altered the lifestyle of a major portion of the population. In the 1980s more women than men were enrolling in college, and the number of women employed in jobs categorized as professional reached 49%; by 1989 21% of the population had completed college as compared to 8% in 1960. As these college graduates advance in their careers, they will increasingly work in the expanding information and service sectors of the economy, and not in the shrinking industrial and agricultural sectors.

Two components of information processing hold particular significance for business in this decade: on-line databases, referred to in this book as infobases, and content-based software. On-line services for information, entertainment, and transactions grew by 90% between 1986 and 1990, and they are expected to grow 67% more over the next 5 years according to a study by Simba Information, Inc. Content-based software consists of dictionaries and other reference works, marketing information, video libraries, and text databases, such as the white and yellow pages. Content-based software is licensed in a form suitable for electronic retrieval using such media as CD-ROM and magnetic disks. Some people in the industry believe that the market may exceed the applications software industry in size in a few years.

1.1.5. The Global Economy

The fifth major structural change that has had a fundamental impact on how business functions is the emergence of a global economy and increased foreign competition. In 1991 the foreign trade deficit was $66.2 billion, and competition continued to intensify in areas once the province of U.S. manufacturers, especially in Japan, which accounted for $43.4 billion of the deficit. In the 1960s, competition was in light manufacturing and soft goods such as clothing and shoes; in the 1970s, it was in cars and steel; by the 1980s, competition had extended to the high-tech industries of computers and integrated circuit chips. Interestingly much of the deterioration in the high-tech trade balance is due to international U.S. companies seeking low wages and tax advantages by locating factories in Asia and Mexico. A global economy has a twofold impact on the importance of information: Firms operating domestically and overseas need to know exchange rates, shipping schedules, inventory levels, and other information related to the flow of goods; and there is an increased need for easy-to-use, companywide information and communication systems capable of operating at locations where there is little or no data-processing support available.

A major factor affecting the global economy is the political upheaval in Eastern Europe and what was formerly the Soviet Union. Coupled with the unraveling of countries in that region of the world and the reunification of Germany are the disappearance of geopolitical economic alliances. Comecon, the Soviet-led block that united Communist economies around the world was dissolved in July 1991, the same month the Warsaw Pack was terminated. Trade restrictions by the United States and other industrialized nations are being rescinded. The economic unification of the European Economic Community in 1993, and cooperation forged among the coalition nations that successfully waged the Gulf War are creating new trading opportunities. As an illustration, in 1990 the installed base of personal computers (PCs) in Eastern Europe numbered between 1 and 2 million, compared with 50 million in the United States. (Unfortunately U.S. companies had been largely prevented from entering this market.) Today manufacturers are addressing this new market vigorously. To a lesser extent economic opportunities like this are blossoming in many other regions of the world.

1.1.6. Trends in Handling Information

Structural changes to the economy like those just mentioned are increasing the importance of information and how it is handled in business. Figure 1.2 shows five recognizable trends.

Emphasis on development or acquisition of information rather than data

Decisions based on knowledge rather than intuition and guesswork

Rapid growth of computer supported environments that is supplanting traditional paper and pencil modes of work

Assimilation of technology and willingness to try new labor-saving products

Recognition of the importance of timely information

Figure 1.2. Trends in handling information in business.

There is an emphasis on developing or acquiring information rather than data; the raw material from which information is derived. Data consist of facts, concepts, and instructions suitable for communication, interpretation, or processing by human or automatic means; whereas information is data transformed into a meaningful and useful form. The computer era was spawned by the Second World War, when it become necessary to process ballistic equations and compute the trajectories of naval artillery. Originally the focus was on computer hardware; later emphasis was placed on software, then communications. Now the shift is toward information use. The increasing complexity of business, the development of new markets, the movement toward a freer and less government-controlled economy, a changing labor force, and competition on a global scale create a need for accurate and timely information rather than computer printouts a foot thick, containing disorganized and unrelated facts and figures.

Decisions are based on knowledge rather than intuition and guess work. The presence of affordable and readily available information gives decision makers the opportunity to research and analyze questions before taking action. This contrasts with the *modus operandi* of management as recently as 20 years ago, when decisions based on experience and intuition were the basis for action. Such phrases as market research, operations research, business plan, economic analysis, and feasibility study are part of management's lexicon today.

There is a rapid growth in the numbers of computer-supported personal workbenches, and these are replacing paper-shuffling pencil-pushing jobs. According to a Dunn & Bradstreet survey of 5000 companies in 1983 and again in 1985, the number of firms using personal computers increased sixfold. The sample was selected to represent the full range of businesses. Banks, utilities, accounting firms, insurance companies, educational institutions, just to mention a few types of organizations, provided terminals and microcomputers or arranged affordable financing so every employee had access to a unit. These statistics indicate the rapid shift that occurred from data center computing to end-user computing in the early 1980s. Most of these users were not technically sophisticated; they were managerial, professional, secretarial, and clerical employees interested in applications and results rather than the technology of computing. They knew enough to recognize what was needed but not available in the way of computer support from their organization.

Technological change is being assimilated with surprising ease by business. The principal motivation for accepting new technology is its relative cost compared to the cost of labor. In 1970 80% of the total cost of technology was attributed to terminal costs and 20% attributed to labor costs, according to the American Productivity Center. By 1983 the cost of technology was approximately 20% in most clerical operations, and labor was approximately 80% (a clerical worker at a single work station). The cost trade-offs are even more pronounced when expensive professionals and managers are involved. One measure of technology growth is the ratio of workers to terminals; in some companies, it is a one to one ratio now.

Technology has not always advanced at the rapid pace of the past 30 years. The dial telephone was introduced in 1896, and it was unchallenged until the introduction of the touch-tone telephone in 1963; in the late 1970s the cordless telephone appeared; in the early 1980s, cellular mobile phones and computer phones appeared. In 1961 the first commercially available integrated circuit (the chip) was announced by Fairchild Semiconductor for $32; by 1971 the price of a chip had fallen to $1.27, and its first consumer application was in a Texas Instruments' pocket calculator in the same year. That company sold five million calculators the next year; and in 1976 Keuffel & Esser Company, the premier slide rule manufacturer went out of business. The first electronic computer, the ENIAC, was dedicated in 1946, but the first really popular computer, the IBM 1401, was not announced until 1959. Since then the installed computing capacity has doubled approximately every 3 years in the United States. Once technology produces a cost-effective product with a clearly identified benefit, business is quick to adopt it.

For information to be useful in decision making, it must be accurate, contain sufficient detail, and be delivered in a timely manner. Print-based information served the needs of business well for many years, and for some applications, it will be more than adequate for years to come. But when a company is faced with new competitors, changes in the laws and regulations that control its industry, or other factors that breed instability in the marketplace, the timeliness of information becomes crucial. Often by the time a corporate document is researched, compiled, printed, and physically distributed, the contents are obsolete; if research is based on trade journals, another 2 months to 1 year must be added to the age of the information. In the case of a book, the time from when information is first recorded in a draft to when it appears in print can easily range from 6 months to 2 years. Because of the unavoidable delays associated with paper-based methods of distributing information, many companies are turning to electronic forms of information distribution and communication.

1.2. RISE OF THE COMPUTER CULTURE

Between 1986 and 1991 the revenues of the world's top 100 information services companies grew from $174 billion to $290 billion; IBM's 1991 revenues alone were $63 billion. To put this into perspective, if IBM were a country, it would be among the top three dozen countries in the world in size measured by gross national product.

The introduction of the microcomputer has made it extremely difficult to determine accurately the number of computers in business today. Many large companies have lost control of computer acquisitions because departments can easily buy microcomputers possessing the capacity of large mainframes of just a few years ago with their discretionary funds. Unfortunately this often leads to incompatible machines that can not share software nor communicate with one another.

Estimates of how many workers will have PCs in the 1990s vary, but most forecasters agree it will be a major part of the workforce. Research companies like Booz, Allen & Hamilton, Inc., and Dataquest, Inc., agree that more than nine white-collar workers in ten will sit at computer work stations in the 1990s. According to a January 1990 survey by COMTEC, 39.3% of work sites now have at least one PC installed, compared to 30.5% in January 1988; similar findings were reported for small businesses by BIS CAP International, and 45% of businesses with 5 or fewer employees had a PC in 1990. This figure increases to 75% for businesses with between 20–99 employees. All of these figures ignore people who have a computer at home. The Cadogan reference reports that 61% of the personal computers purchased in 1990 for the home were designated for work-related tasks. Coupled with the future spread of PCs will be such features as multimedia to facilitate the human interface. A survey published in the July 1991 edition of *Datamation* reported that 20% of those surveyed had implemented or planned to implement multimedia in the near future and another 29% were currently in the process of evaluating the technology. The diffusion of information technology at the work site and the intensity of the changes underway are increasing. The few years left in this millennium will pose challenges and opportunities unparalleled in human history.

The major reason that companies are willing to expand computer power so rapidly is that it continues to be one of the best buys around. Computing costs are declining by better than 20% a year; as a result of this decline, it becomes practical to expand the reach of the computer to include an ever-increasing list of applications. In many offices, such tasks as typing, editing, filing, retrieving, billing, and copying are done by a secretary, but a microprocessor will gradually assume these tasks because it can handle words and numbers faster and often more cheaply. In one case, productivity measured in terms of letters or documents produced per hour increased by 300% when a microprocessor was used. In another study, it was reported that an office worker with a word processor can produce standardized letters 400 times faster than with a conventional typewriter.

On the factory floor, robots are producing an industrial renaissance. It has been predicted that after 1990, it will become technically possible to replace all manufacturing operatives in the automotive, electrical equipment, machinery, and fabricated metals industries; and the American Society of Manufacturing Engineers forecasts that by 1995, 50% of automobile assembly will be done by robots. In retailing computers are being used in stores and shopping malls to answer shoppers questions and advertise sales. In education computer-assisted instruction (CAI) has supplemented the instructor in some courses. CAI allows students to proceed at their own pace; CAI does not permit the students to advance to new material until mastery of the present material has been demonstrated on a computer-controlled test.

Rapid strides are occurring in the marriage of computer technology and the human psyche. The ultimate personification of the technology is through the creation of an information environment known as virtual reality. A computer-simulated environment in which devices attached to the head, hands, or other parts of the body provide feedback to a computer enabling a computer to reproduce and control the perceived environment. Artificial environments created by flight simulators are widely accepted; in time playing sports or living adventurously in a virtual world may be just as acceptable and much more common.

Computers equipped with communication devices link individuals with a wealth of information resources. The on-line services industry has enjoyed double digit growth since its inception in the late 1970s and the trend continues. In 1991 the 54 leading on-line services reported a 19% growth from the year before. There were 5.2 million customers accessing everything from legal information to credit reports and sports scores. Table 1.1 lists the top ten services in terms of customers at the close of 1991.

Table 1.1.   Estimated Number of Customers for the Top Ten Information Services[a]

Company	Service	Focus	Customers[b]
Prodigy	Prodigy	General interest	1,350,000
CompuServe	CompuServe	General interest	903,000
Dow Jones	DJ News/Retrieval	Financial	330,000
Mead Data Central	LEXIS/NEXIS	Science/technical/professional	314,683
GE Information Services	GEnie	General interest	267,000
AT&T	Easylink	General interest	245,000
British Telecom/Tymnet	Dialcom	General interest	236,000
Reuters	Monitor	Financial	205,000
TRW	Credit Data Services	Credit reporting	106,000
American Airlines	SABRE	Airline information and reservations	96,845

[a]Consumer Services Dominate Online Growth, *Information Industry Bulletin* **8**(3) (1992), pp. 1–4.
[b]These statistics are as of December 31, 1991.

The pervasiveness of computers in homes, businesses and schools is creating a computer culture; as shown by how people choose to entertain themselves at home and how they perform their job and behave at work. One of many possible examples is the impact of E-mail on the human psyche. Studies have shown that when groups of people are networked, it took longer to make a decision. On the other hand, the process is much more democratic because individuals with higher status in the organization are perceived by their subordinates as less powerful than on a person-to-person basis. The electronic medium sometimes encourages "flaming" or impassioned self-expression. People with a poor self-image have reported feeling more lively and confident when communicating electronically rather than in person; others who are small in stature or have soft voices report they no longer have to struggle to be taken seriously.

The impact of the computer culture is most evident every day in how people speak. A person speaks of being programmed, receiving input, interfacing with coworkers, simulating an outcome, having the decision hardwired, producing output, and working out the bugs. The computer has become so much a part of our culture that we use these terms without reflecting on their origin. Computer technology also has its dark side; in his book *Information Anxiety*, Richard Wurman describes the undesirable effect of information anxiety, bewilderment caused by an excess of information. In a society that publishes a thousand books a day and doubles its printed knowledge every 8 years, the frustration caused by the glut of information is growing. But despite this dark side, information is taking on added importance as an asset rather than an expense.

1.3. THE INFORMATION INVESTMENT

Knowledge is power, and the amount of power a person or organization possesses depends on the amount of knowledge held. Information is the foundation of knowledge and the basic ingredient of decision making. Figure 1.3 shows the characteristics that distinguish information from other company resources. For

Expendable

Compressible

Substitutable

Transportable

Diffusible

Sharable

Figure 1.3. Characteristics of information.

example, information is expandable; it is not consumed when used, and the more people who access and use it, the more potential value it has. Information is also compressible; it can be summarized to provide a quick overview, or can be expressed as a formula or theorem to state a concept precisely. Information is substitutable; it can replace capital and labor by enabling people to make more timely and better decisions than is possible with "guesstimates." Information is transportable; it can be stored on a piece of paper and physically transported or represented as a set of magnetic spots and electrical pulses transported at nearly the speed of light. Information is diffusible; it can be copied at very low cost and distributed easily. The more it is diffused, the more potential value it has. And last information is shareable: If I have a pencil and give it to you, then you possess it; but if information is shared, we both have and can use it at the same time.

These characteristics make information unique among resources, but most companies do not view information this way; most think of information as simply an expense. To an accountant, an expense is any expenditure whose benefits do not extend beyond the present. Except for very limited forms of information, such as a tornado alert or a flood warning, this is an inappropriate view of information. In fact, information is really an asset. To an accountant, an asset benefits a future period. This definition is consistent with how information normally functions in a business. Thus information represents an investment, and it should be treated as one: It should be acquired for a specific purpose and expected to produce a return.

1.3.1. Form and Focus

Investing in information should be a conscious business decision borne of the need to know more and operate differently. The information investment can be examined according to its form and focus, as shown in Figure 1.4. Form refers to

	Focus	
Form	Internal (Employees)	External (Customers/Public)
Operational	Better information for decision making and Cost-effective communications	Cost-effective communications and Revenue generation
Strategic	Improved organizational climate and Ability to perform opportunity analysis	Improved client relations Product differentiation New business opportunities

Figure 1.4. Objectives of the information investment.

whether the information is used for operational or strategic purposes and focus to whether it is directed at internal or external applications. The following section provides some examples of how companies are investing in information, and Chapter 2 carries this paradigm a step further by discussing how companies can invest in enhanced systems using the structure in Figure 1.4.

1.3.2. How Some Companies Invest in Information

The operational use of information is intended to help an organization function more efficiently and thus presumably more economically by providing better information for decision making and cost-effective communications. In its most obvious form, the operational use of it involves substituting information and information-handling technologies for human labor. The rationale for this substitution is straightforward: Computer-processing costs have been decreasing by more than 20% a year; on the other hand, personnel costs have been increasing by 7% a year. Between the mid-1970s and mid-1980s, people became 20 times more expensive than the cost of computer-processing power; over a 20-year period, people became 400 times more expensive.

The astute organization leverages this trade off between information-handling technology and its human resources. A $3500 microcomputer that is 8% of a manager's salary in 1987, may be only 4% of that manager's salary in 1992, but based on current trends, the microcomputer will be several times more powerful. This trade-off will encourage the continued substitution of information technology for human labor whenever efficiency and economy will result.

Systems that produce operational information for internal uses are generally known as management information systems (MIS). An MIS is an organized way of providing information to the appropriate level of management for planning, organizing, staffing, and controlling the organization more efficiently. The MIS is the system that produces the payroll reports, accounts receivable statements, employee benefits analysis, and the inventory stockout sheets. The MIS has become the *sine qua non* of most organizations; turn it off, and the organization stops functioning.

Many organizations are fashioning operational applications out of enhanced systems for internal use. Anheuser-Busch maintains its extensive collection of corporate memorabilia in a photo database containing written records and pictures of tens of thousands of such artifacts as bottles, cans, coasters, key chains, and pocket knives. These and many other items are photographed in color, and the pictures are then scanned into a photo database. Written material is keyed into the system. Once captured in memory, the contents can be easily researched for advertising purposes, copyright claims, speeches, and even for maintaining and preserving the collection.

An evolutionary step for MIS is its application outside of a business organization. Like many other companies, Northern Telecom has implemented an on-line catalog containing its extensive documentation offerings. Access is free, but users must obtain an identification code and password. Access may be made from virtually any type of work station or terminal having dial-up capability at speeds from 0.3–2.4 Kbits/sec. Information may be accessed by product, topic, document type, or document number, and it is usually available within 48 hours of the document's release. This service greatly simplifies identifying technical material, and it is a boon to Northern Telecom customers.

Clark Service Company has taken this application one more step with a text and graphics system called Clark Net II. The system is an upgraded version of an earlier on-line service that enables dealers to inquire about the status of inventory and previously shipped orders, transmit mail electronically, and view classified advertisements. The system is described more fully in Chapter 13.

An operational system can quickly become strategic in importance; the SABRE system is such an example. A case in Chapter 14 outlines the history of the SABRE system and its most recent strategic incarnation, EAASY SABRE. The SABRE system evolved from an operational tool for managing passenger reservations to a competitive weapon for capturing market share and thwarting attempts by rival airlines to undercut American Airline fares. Competitors underestimated and overlooked its potential to their chagrin. One of the biggest losers was People's Express, the Volkswagen Beetle of its day in the airline industry. People's Express had been built around the theme "every employee an owner" and rock bottom fares. It was not unionized, which contributed to low costs, and its seat pricing made it one of the 10 largest airlines in the world almost overnight. Unfortunately American Airlines discovered two-tier pricing and it used computerized "yield pricing" to match People's low fares for seats that would otherwise be unoccupied. Within a brief 2 years, People's Airline was out of business; in the words of a senior manager, it had been "obliterated by the chip."

Other airlines concerned with the competitive threat posed by American's SABRE system and United's Apollo reservation system brought suit against both airlines, charging them with monopolizing the computerized airline reservation industry. The suit was settled in 1989 in favor of American and United. The die is cast; the plane has left the gate, and a number of airlines have missed their ride.

What American Airlines did is not unique. Organizations interconnect their computers with suppliers and customers throughout the world. Digital Equipment Corporation has hundreds of such gateways. It views its international internetworking capability as a strategic advantage for customer service and product development. The Clark Distribution Services case in Chapter 13 describes how that company realizes both operational and strategic advantages from its network.

These strategic uses of information can be divided into strategies that are internal and affect the organizational climate and how opportunities are analyzed and those strategies that are external and enable organizations to improve client relations, differentiate products, or pursue new business opportunities.

A classic example of the first use of information occurred in the bruising sport of football: The Dallas Cowboys refined the process of player identification and selection to an art by strategically analyzing data. Following the creation of the franchise in 1960, a project was begun to define the attributes that made a good player on a position-by-position basis. Although it is not possible to determine how much of a role this system played in the team's early success, there is no dispute that the approach is superior to the haphazard approach used by some teams. Eventually other teams applied computer technology similarly.

An example of the second use of information occurred in the Owens-Corning Fiberglass Corporation, which turned information developed through in-house research and development into a marketing tool. The company developed a substantial body of information on the energy efficiency of different house designs when conducting research to develop new insulation materials. In order to boost sales of its insulation products, the company later developed a computer program to use data from its research to come up with energy-efficiency ratings for new designs. Owens-Corning then offered builders free evaluations of their designs if they agreed to purchase insulation from Owens-Corning and meet a minimum standard of energy efficiency.

The most publicized and glamorous strategic application of information technology occurs when it is used as a competitive weapon. The classic example is American Hospital Supply, which manufactured and marketed health care products. In 1976 the company introduced ASAP, the analytical systems automated purchasing system, which allowed customers to place direct orders for over 100,000 products. By 1984 over 4200 customers were accessing ASAP through the electronic order entry system, which saved hospitals money by simplifying the ordering process and shortened order lead time. More importantly for American, it connected them directly to the customer base by creating an exclusive communication link and effectively thwarted competition. According to competitors at the time, it was almost impossible for a hospital to avoid doing business with American once an ASAP terminal was installed, and it was difficult for another supplier to sell its products once American was entrenched.

Another classic electronic order entry network that locked in customers was developed by McKesson Corporation, a major drug distributor. Customers used a hand-held order entry device. The order entry procedure was triggered by scanning a product bar-coded label on a shelf or by entering a product identification number from a keyboard. Information was transmitted by telephone to a data center. It was estimated that the order entry procedure reduced the time a typical drugstore spent on ordering by 50–75%, reduced inventory requirements significantly for the customer and McKesson, cut paperwork, and enabled McKesson to offer such service enhancements to customers as Economost, a service that provided a detailed report on fast- and slow-selling items.

The hotel industry is putting a slightly different twist on the use of information technology. Upper-tier hotels catering to the business trade are allowing guests to check out of the hotel from their rooms. The guest retrieves the bill using a keypad, displays it on the television set, and after reviewing it, either approves payment or contacts the desk to resolve a problem. This process speeds up checking out; avoids the need to stand in line; allows the guest to pick a convenient time to carry out the process, since it may be sometimes done an hour or more before vacating the room; and avoids paperwork for the hotel. It is a good example of a high-tech application taking on a chore dreaded by most travelers.

Effectively using the information an organization already possesses can lead to new and previously untapped markets. Appliance manufacturers and retailers routinely send out notices to customers about maintenance service contracts that are available for their products. By monitoring each customer's purchases, Sears created a powerful marketing tool that can boost maintenance revenues and customer goodwill.

The infrastructure a company must create to process its information resources can often lead to new business opportunities. Sears uses its excess computer capacity to provide credit authorization, transaction-processing, and retail remittance-processing services to other companies. J. C. Penney sells time on its high-speed data network. Merrill Lynch recoups some of its investment in satellite dishes and private microwave channels by reselling services to local businesses in certain areas. Perhaps the most visible example is Holiday Inn, which offers satellite teleconferencing using communications facilities that were initially installed to support a national data network to speed and simplify room bookings.

1.3.3. The Strategic Use of Information

The perception of information as a strategic resource is a recent phenomenon. Strategic management involves formulating, implementing, and evaluating actions that enables an organization to achieve its objectives. Only in the past few years have companies learned how to use information and information technology strategically.

In the 1950s and late 1960s, computers were viewed as a centralized repository of information and service. The machine was attended by white-coated technicians and fed a diet of data delivered in mass and processed in batches. The 1970s ushered in the era of time sharing; a mode of operation in which the computer was partitioned to support multiple users. Time sharing enabled the cost of computer resources to be shared, and it enveloped a larger circle of users. The introduction of personal computers in 1975 and their widespread adoption during the 1980s irrevocably joined computing technology with the end-user at work and home. The 1990s is consolidating the position of networking and media integration, complete with touch screens, written and voice input, and massive storage capabilities.

In the 1950s and 1960s especially, the computer controlled the person— machine dynamics. Because of its limited ability, the human element in the equation was forced to work on the machine's terms supplying data and receiving in exchange reams of paper containing the results of computations. Each decade has brought forth a less structured relationship between the human and machine. Initially the rigid structure was capable of supporting only the operational needs of the business. It handled the routine day-to-day computational requirements. As the user's perceptions changed and computer technology progressed beyond its formative years, management turned to tactical applications involving planning and control. Now companies are designing information systems to support or shape competitive strategy. They are using computers for strategic purposes. This evolution is due to engineers who have advanced the technology and in the process changed the economics of computing. No better example exists of how far the technology has evolved than the computer chip.

The overall trend in the number of transistors that can be squeezed onto a chip continues to rise and the cost to decrease. This is sometimes expressed as Moore's law, which states that approximately every 2 years, the number of transistors on a chip doubles (See Figure 1.5.) and the cost falls 50%. (Moore is the president of Intel, a manufacturer of computer chips.) By the year 2000, it is predicted that there will be 100 million transistors per chip running at 250 MH_z and performing 2000 million instructions per second.

The widespread availability of computing power extends a person's ability to

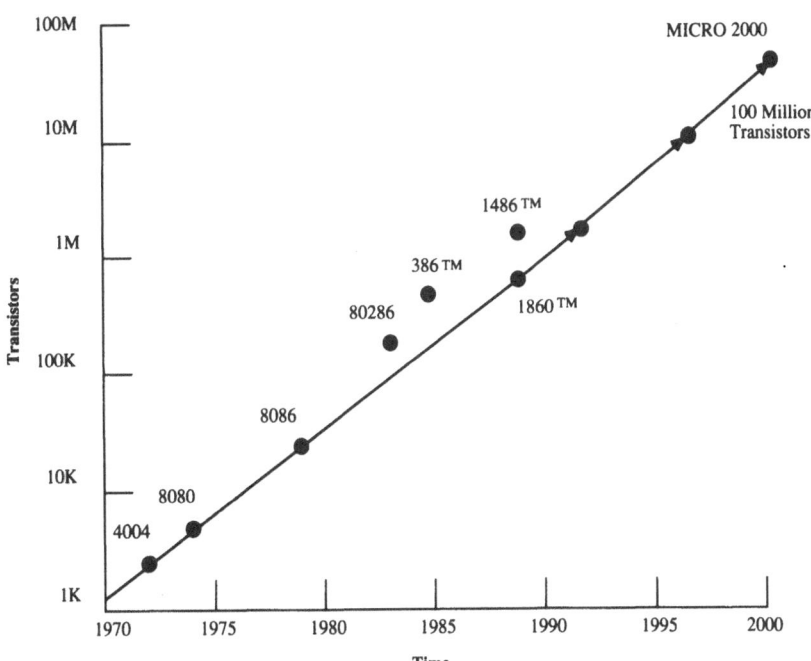

Figure 1.5. Overall trends in the evolution of the transistor: Moore's Law. (Courtesy of Intel Corp.)

deal with, and respond to, business variables. Past generations of managers were primarily limited by their span of control. In a labor-intensive environment, a manager's horizons were limited by how many people could be effectively managed. In an information-intensive environment, a manager's horizons are limited by how effectively information can be analyzed, processed, and evaluated. The declining cost of computer power means that financial spreadsheets, market studies, forecasts, engineering analyses, and similar activities that formerly required a staff of people can now be done by one person using a microcomputer or computer terminal. Answering what-if questions, accessing external databases, and doing competitive analyses using the same equipment is a natural progression. Creative thinking and all that it makes possible replace supervising personnel and enable managers to respond to the changing business environment in new and creative ways.

Despite these factors, most organizations do not use information in a strategic sense. There are several reasons why: Senior management is often ignorant of information technology and its potential uses; poor communication may exist between the information system group and the rest of the company; there may be strong and determined resistance to change among the information systems personnel and other key people in the organization; and last, a mechanism to direct corporate resources toward strategic goals may be missing. Whatever the reasons, an organization that fails to consider the strategic application of its information resources may be ignoring one of its most important assets.

Organizations that do build strategic systems with their information assets have a common element in their corporate culture. According to an article in the November 1989 issue of *PC Week*, the winners had three things in their favor: (1) a vision of how information resources could be used competitively and the ability to sell the vision to management; (2) strong bonding between the systems and business units of the organization to ensure the strategic systems solved real business needs; and (3) commitment to continue funding worthwhile systems during difficult economic times.

1.4. SUMMARY

There is a digital consciousness developing throughout society. It is visible in the general public by the proliferation of microcomputers and in business by the increased reliance placed on computer resources. This digital consciousness is occurring at a time when the business environment is undergoing significant change.

One of the most profound factors affecting the business scene is the shift in emphasis among users from data processing to communication and information use. This shift is producing a whole genre identified by a new vocabulary that personalizes automation: Friends interface and people program themselves. Even companies have begun to look at information differently—it is being viewed as a resource, a commodity that can be bought or developed and managed for the benefit of the organization.

REFERENCES

L. Allen, Who Are End Users?, *Computerland* **18**(47) (1984), pp. ID19–ID20, ID22, ID24, ID26.
American Association of Retired Persons, *(The) Aging Work Force*, Washington, DC (1990).
American Association of Retired Persons, *America's Changing Work Force*, Washington, DC (1988).
D. Anderson, Case Studies and Implementations of LDI Arrangement, *Data Communications* **15**(2) (1986), pp. 173–174, 177–178, 181–182.
D. Bunnell, Toward Transcendental Multimedia, *New Media* (May 1992), p. 3.
S. Burke, International PC Markets Take up the Slack in Domestic Sales, *PC Week* **7**(27) (1990), p. 120.
J. Canning, Multimedia Extensions Breathe Life into the PC, *Info World* **13**(42) (1991), pp. S90, S95, S100.
W. J. Canogan, Fiber: The Tie That Binds, *Lightwave* **8**(13) (1991), pp. 20, 22–24.
W. Childs, Users Want Multimedia Applications, *Datamation* **37**(14) (1991), p. 94.
Consumer Services Dominate Online Growth, *Information Industry Bulletin* **8**(3) (1992), pp. 1–4.
M. Cooper, *Expanding the Information Age for the 1990s: A Pragmatic Consumer Analysis*, American Association of Retired Persons and the Consumer Federation of America, American Association of Retired Persons, Washington, DC (1990).
D. Coursey, PCs: Revolution in '80s, Evolution in '90s, *MIS Week* **11**(1) (1990), p. 34, 37.
B. Davis, The Impact of Microcomputers on Organizational Behavior, in *Managing Organizational Behavior* (John R. Schermerhorn, James G. Hunt, and Richard N. Osborn, eds.), 2d ed., Wiley, New York (1985).
L. Day-Copeland, Growth Rate of PC-Installed Base Slows As U.S. Micro Market Matures, *PC Week* **7**(36) (1990), p. 158.
Ford Says EDI Saves Millions, *Software News* **7**(1) (1987), p. 16.
A. P. Garvin and H. Bermont, *How to Win with Information or Lose without It*, Bermont Books, Washington, DC (1980).
C. Guglielmo, Cheaper, Better, Snazzier, *New Media* **2**(3) (1992), pp. 19–21.
C. Harlan, Information As a Resource, *The Futurist* **16**(6) (1982), pp. 34–39.
W. A. Henry, Beyond the Melting Pot, *Time* **135**(15) (1990), pp. 28–31.
Introducing the Compass Directory, Marketing Bulletin 50078.03/03.88, Northern Telecom, Nashville, TN, (1) (1988).
B. Jacobson, The Ultimate User Interface, *Byte* **17**(4) (1992), pp. 175–176, 178, 180, 182.
M. Karr and J. McQuillan, Strategic Models: Where Information Means Profit, *Infosystems* **32**(2) (1985), pp. 88–89.
Office Automation and the Workplace: Standards for the 80s, *Productivity Brief*, no. 36, American Productivity Center, Houston (1984).
On-Line Services Soar, *Datamation* **37**(13) (1991), p. 24.
P. J. Pane, Perspectives, *Info World* **13**(10) (1991), pp. 47, 51.
T. R. Reid, "The chip," *Computerworld* **19**(11) (1985), pp. 32–33.
S. Salamone, Third Parties, EDI Shape Enterprise Internetworks, *Network World* **7**(2) (1990), pp. 1, 27–28.
R. L. Scheier, How Winners Build Strategic Systems, *PC Week* **6**(45) (1989), pp. 125, 127.
R. L. Scheier, "Obliterated by the Chip": The Crushing of People's Express, *PC Week* **6**(45) (1989), pp. 125, 131.
P. E. Schindler, How to Convert Information into a Competitive Weapon, *Information Week* (073) (1986), p. 26.

D. M. Simons, "Romancing the Telephone," *Information Times* (Fall 1984), pp. 43–53.

Small Businesses Go for PCs in a Big Way, *Personal Computing* **14**(6) (1990), p. 38.

P. Sniger, Office Systems Struggle for User Acceptance, *The Interpreter* **17**(15) (1984), pp. 89–92.

L. Sproull, and S. Kiesler, Computers, Networks, and Work, *Scientific American* **265**(3) (1991), pp. 116–123.

M. Stoll, Brewery Keeps Its Memorabilia on Tap, *PC Week* **5**(1) (1988), pp. 39, 44.

P. A. Strassmann, *Information Payoff*, Free Press, New York (1985).

K. Sullivan, Multimedia's Impact Blunted by Expense, *PC Week* **9**(11) (1992), p. 20.

L. G. Tesler, Networked Computing in the 1990s, *Scientific American* **265**(3) (1991), pp. 86–93.

S. Turkle, *The Second Self: Computers and the Human Spirit*, Simon and Schuster, New York (1984).

A. Warfield, Societies and Organizations in Transition, in *Managing Organizational Behavior*, (John R. Schermerhorn, James G. Hunt, and Richard N. Osborn, eds.), 2d ed., Wiley, New York (1985).

C. Wiseman, *Strategy and Computers*, Dow Jones-Irwin, Homewood, IL (1985).

C. Wiseman, *Strategic Information Systems*, Irwin, Homewood, IL (1988).

R. Wurman, *Information Anxiety*, Bantam Doubleday Dell, Inc., New York (1989).

Chapter 2

Enhanced Systems

An enhanced system was described in the previous chapter as an easy-to-use, interactive computer system that incorporates one of more of the following: text, graphics, color, picture quality displays, sound, animation, and motion video. Enhanced systems are evolutionary rather than revolutionary. They are another step in the continuing trend to place computer resources in the hands of the end-user, but enhanced systems are considerably different than their forebears— systems found in a traditional data-processing environment. Understanding these differences is the first step in learning how enhanced systems can improve day-to-day operations and provide a competitive advantage.

Chapter 2 begins with a discussion of the general features of enhanced systems. To plan for and manage these types of systems it is essential to understand their basic features. This information is followed by a description of how systems with these features can be employed in an organization. This provides the reader with the foundation for generating his or her own applications. The chapter ends with a brief discussion of how enhanced systems will impact data processing—the part of the organization that will have to change the most in the era that is beginning to dawn.

2.1. GENERAL FEATURES

The four general features characterizing an enhanced system are (1) functional simplicity; (2) the ability to integrate different equipment, types of media, and transmission services; (3) graphical user interfaces and visualization; and (4) a focus on information and transaction-oriented applications (See Table 2.1).

21

Table 2.1.   Ideal Set of General Features
Characterizing an Enhanced System

Feature	Components
Functional simplicity	User directed
	Training/ease of use
	Documentation/self-documenting
	Loading
	Access methodologies
	Control
	System directed
Integrative	Hardware
	Software
	Communications
	Media
	Security
Image based	Graphical user interface
	Graphics, pictures, motion video
Informational and transactional	Design
	Creation
	Edit and maintain
	Presentation
	Authorization
	Security
	Billing

2.1.1. Functional Simplicity

Functional simplicity consist of those features that make the system easy to use and those features that enable the system operator to implement new applications or modify existing ones. Easy-to-use is a relative term, and it is used here for a system that requires no formal training. Such a system is self-documenting to the extent that once a user has logged on, the system automatically provides all the necessary instruction. Even the process of logging on may be reduced to turning on a terminal and if it is a video unit, providing the appropriate response to automatically displayed questions. The finest example this author has ever seen is the Heartfelt Advice system described in Chapter 15.

If the system is on a networked computer, access to the system should be available from multiple locations at whatever times the user determines is appropriate, and a sufficiently large number of simultaneous users should be able to access the system. Depending on the application, different ways of accessing the information base or performing a transaction may be necessary. Three common methods of access are by menu, direct access, and through key words. Some systems may use all three forms of access.

A menu-based approach is the most common, and it is sometimes referred to as treeing. A list of options is presented to the user, and the user selects one of the options by entering a prompt number or letter located beside that choice or by touching the screen in a specific location in systems supporting that kind of user input. Direct access may be possible by using a number or word related to the information to be retrieved or the transaction to be performed. The numbers or words are sometimes stored in paper-based directories, or they are stored electronically on the system itself.

Keyword-based techniques are more sophisticated, and they involve keyword access or keyword search. In the former, a keyword is stored in the system— perhaps in a special index or with a specific piece of information. Retrieving the keyword automatically retrieves the sought-after information. In a keyword search, the contents of the information base are searched for occurrences of the keyword. Boolean operators (AND, OR) may be used to combine terms in a keyword search. The common thread that unites all three methods of access is ease of use. Whether the system interface is a video terminal or a touch-tone telephone, very little prior training is needed and instructions are available at any point on how to proceed to the next step.

Enhanced systems use standardized interfaces and clearly defined functions. One of the problems encountered in data-processing systems and early versions of enhanced systems is their inability to support equipment from different vendors. An analogy can be found in the early history of the telephone industry when it was often necessary to have a different telephone set for each person with whom you wished to communicate. The same problem existed in the early 1980s when most public videotex services of that era required a special terminal capable of accessing that system and that system alone. As adherence to international standards increases, incompatible equipment is becoming less of a problem.

A function is a procedure initiated by an uninterrupted series of one or more key strokes, pulse tones, or other forms of input. An automatic log-on or the display of a new screen or frame of information when a particular key is depressed are examples of functions. Traditionally different systems have adopted different function names to achieve the same results; for example, on some systems, the command to repeat the immediately preceding frame is the key labeled back; on other systems, it is the symbol < and on still others, an arrow pointing to the left. Even more annoying, however, is when a function meaning changes for different applications on the same system.

Related to the issue of clearly defined and consistent functions is the set of rules for information retrieval and transaction processing. Even among systems with consistent functions, the method of conducting a search for information, for example, often differs. This situation requires the user to become familiar with different ways of performing the same application. Ideally functional simplicity will enable an enhanced system automatically to translate its operating procedures so that they are compatible with the procedures of the system with which the user wishes to communicate.

An enhanced system permits selective control at the point where the interface with the system occurs. This requires a two-way flow of communication and an input-output device. This feature excludes teletext and other broadcasting-based technologies from the realm of enhanced services, as discussed in this book. An enhanced system also pays considerable attention to behavioral issues affecting the user population, for example:

Text and graphics are often merged in systems with visual forms of output, although graphics is by no means a requirement.

Upper and lower case text are available and color, flashing, extra large fonts, and standard screen widths of 40 columns are common features.

System response times, especially in transaction-based systems, such as banking and conferencing, are sufficiently brief to prevent user frustration.

Transient indicators, such as a flashing symbol in the corner of the screen, inform the user that the system is still working and has not forgotten or lost the user's request.

On the system side, functional simplicity enables a new application to be implemented rapidly and economically once the infrastructure of an enhanced system exists. No major software development is necessary. The principal tasks instead are to prepare the functional specifications for the application and acquire or develop the information base. As an example, a simple hotel-booking system was implemented by one company using both a public videotex service and a time-sharing bureau. The videotex solution required about one-third of the effort and elapsed time compared to running the same application with the time-sharing approach.

2.1.2. Integrative

Another distinctive feature of an enhanced system is its ability to integrate different technologies. These technologies may be manifested in the hardware and software to be used or in the communications services that will support the applications. From the user's perspective, integration begins at the terminal; thus the terminal should be capable of supporting the user's data-processing needs as well as word processing, electronic messaging, information retrieval, telephone management, and the many other data and voice needs that are becoming increasingly important in business.

As the sophistication of the terminal increases and the user becomes increasingly dependent on its availability, an additional rather than a reduced burden is placed on the computer resources supporting the system. Fault-tolerant and nonstop computers become a requirement for centralized systems rather than a luxury and optical-based storage media become important due to the increased need for capacity as image-capturing techniques finally begin to make the paperless office a possibility.

Internal company communication must support both data and voice needs, and the integration of the two are a common feature of enhanced systems. These internal communication services must be reliable and capable of providing the high bit rates required to support business applications. Media integration is expected to be the primary desktop technology of the 1990s; some enthusiasts predict that the market will double each year through the middle of the decade. Combining text, audio, graphics, and video on a PC will be the most common implementation, but mainframes sporting X Windows terminals will share the limelight. X Windows technology is discussed in Chapter 4.

External gateways to public packet switching services must be available for electronic document interchange (EDI) and to enable business users to access such databases as the Dow Jones News/Retrieval Service, on-line banking services, nationwide electronic message facilities, and other transaction-based applications that are emerging. The backbone of these external communication services will be the integrated services digital networks (ISDN) that are being implemented by the regional Bell operating companies and long-distance carriers. The interconnectivity that will take place will increase the need for data security and conformance to international standards for data processing and communications.

2.1.3. Images and Graphics

Graphical user interfaces, such as Microsoft Corporation's Windows, and imagery will increasingly supplement text on the screen. A graphical user interface or GUI (pronounced gooey), uses similar elements, such as icons, scroll bars, and dialogue boxes to provide a common denominator for different application software. A GUI attempts to simulate how a person works and what a typical desktop looks like. In a GUI environment, users can move from application to application much as they move among tasks at their desk during the course of a day.

Computers can produce an overpowering and endless menagerie of facts and figures in the form of spreadsheets, tables of numbers, searchable databases and so forth. Understanding and assimilating these data can be simplified through graphs and diagrams. The proliferation of inexpensive CD-ROM drivers and to a lesser extent the availability of laserdiscs provide a platform for picture databases and reference material containing pictures. Even motion video is available on the desktop by adding boards to a PC for under $1,000.

2.1.4. Informational and Transactional

An enhanced system normally encompasses within its scope the organization's data processing activities; however, it extends beyond the traditional role of a data-processing system because it deals in information as well as data. Data are facts, concepts, and instructions suitable for communication, interpretation, or processing by human or automatic means. Information is data that have been transformed into a meaningful and useful form. Data are the raw material of information. The transformation may be done initially and the information stored in logical units, sometimes referred to as a page. Because of its size, a page may be equivalent to one or more screens if viewed on a visual display unit. Another term often used is frame. A frame is a physical unit of storage, equivalent to a single screen of information. The terms page and frame are important in discussing enhanced systems because they describe how the user perceives information is created, stored, modified, transmitted, presented, and sometimes even paid for.

Information may also be stored as data in some systems. When a particular frame is requested, the appropriate data are retrieved, and the frame is created by special software and presented. Regardless of when a frame is created, the information it contains is considerably different from that found in the typical computer database. Contents of a conventional database are often numerical and contain little value until assembled into a meaningful form. A database is usually processed in an arithmetic sense, and not simply updated. When its contents are presented, these are often columns and rows of numbers without any attempt to present the statistics through color and graphics; it is up to the user to infer the meaning of the data.

In enhanced systems, frames of information are displayed and such trans-actions as document interchange are accomplished using preformatted templates. One frame may be sufficient to answer a question or complete a transaction; if not, additional frames can be linked together. But whether presented singularly or collectively, the information stands as a complete expression.

Enhanced systems give a perceived structure to the information base that may in fact not really exist. The menu access described earlier can create the illusion of an inverted tree, and while the frames may not actually be organized in this way, the link connecting them enable a user to proceed through the contents of the information base in a way that fits the psyche.

Information must be current to be of value; consequently it must be updated periodically. Information must also be accessible when and where it is needed. If users are not centrally located, information must be available around the clock and often accessible through the existing telephone network. Transactional services must be capable of sustaining many short sessions of a few minutes and provid-ing a rapid response; a session of 4–8 minutes is typical with a response of 2 seconds or less. There must be appropriate levels of security provided, especially for financial services and provisions made for detecting transmission errors due to signal interference. Also any system that charges users real money or sells goods and services must provide a scheme for validating authorized users, ensuring that a complete audit trail is created, and giving customers a detailed billing for charges rendered.

Enhanced systems present information in a nonlinear fashion, which differs from the conventional way of presenting information in print, on videos, and through sound. Information presented through these traditional media forms is cast in a straight, one-dimensional, beginning-to-end format. An enhanced system provides a fully interactive and intelligent way of storing, retrieving, cross-referencing and presenting information dynamically, thereby creating a better tool for many applications, such as training, than many of the alternatives available.

2.2. ENHANCED BUSINESS APPLICATIONS

In Chapter 1 we saw that a company can focus its investment in computer resources internally and externally to satisfy operational and strategic objectives. In most of the examples provided, management used business acumen and existing data-processing resources. Many of the features differentiating an enhanced system from the traditional approach to data processing were not discussed. As companies invest in such features and add them to their computer resources, a number of new and exciting applications emerge that lead to reduced costs, increased market share, new products, and more effective management of corporate resources. In the following sections, these applications are categorized according to their form and focus, and some specific examples are listed; these examples just scratch the surface of what is possible.

Before looking at these applications, two important points should be made: First whether these enhancements can be justified as opposed to the more traditional data-processing solution depends on the characteristics of the application, the ultimate system user, the equipment available, the environment in which the application must strive to succeed, and finally the resources and time that can be invested. Second divisions between these neatly boxed applications sometimes blur when viewed from different perspectives. For example, a financial institution considering the strategic decision of offering an electronic banking service to small businesses looks at the trade-offs much differently than a small business weighing the operational decision of subscribing to an electronic banking service. Perspective is everything!

2.2.1. Internal and Operational Applications

Enhanced systems provide new opportunities for distributing and gathering information and educating and training employees. Organizations spend enormous sums of money typesetting, printing, storing, distributing, and updating information crucial to the daily operation of the business. According to a study by Arthur Andersen, U.S. businesses produce 21 trillion paper documents a day. The cost to organizations of individuals making decisions based on inaccurate or obsolete information is incalculable. Electronically distributing information can reduce or eliminate inaccurate and obsolete information and the problems that it creates, and it can do it economically. Although an individual may infrequently have to retrieve information, the aggregate demands of all the employees in the organization can make information retrieval a frequent occurrence and therefore important to the organization's overall productivity. The proliferation of equipment and the development of new storage media, such as laser discs make electronic distribution an economical alternative to paper-based systems and the availability of equipment considerably broadens the range of access locations.

Between 1980 and 1985, millions of microcomputers were introduced into the corporate world, and intricate wiring schemes became available to attach them to the organization's database. According to one estimate, there was one computer installed in the United States for every 160 Americans in 1980; 5 years later, it was one computer for every 10 Americans, and in 1990 it was about one computer for every four Americans. The proliferation of microcomputers and the availability of dumb terminals in many businesses means that accessing information stored electronically is no longer difficult.

While a great deal of information already resides in the corporate computer system, these systems usually require special training to retrieve the information. This training is often costly and time consuming and therefore impractical to make available to everyone who needs it; however with the type of software described in Chapter 8, training can be made available through simple and easy-to-use menu-prompted requests. Examples of applications that can be performed with this software are listed in Table 2.2. The majority are listed under Information Distribution. In this application, the typical user has a spontaneous need for information on personnel matters, or he/she is seeking specific information about the business itself and its products.

More novel uses of computer resources involve gathering information from unskilled users and allowing sales personnel to order products when away from the office. The rapid changes of computer technology have left sales and marketing functions behind, since most of the effort has been to provide systems for financial and technical users. Marketing managers frequently receive very little information, and if they have sales personnel who travel extensively, available information is often outdated.

In many organizations, sales personnel fill out forms to report their activities; at the end of the week or month, these forms are collected, checked, returned for queries, and then analyzed and summarized. It would be faster and more economical to have sales personnel to input data into a computerized system. In addition to the actual order, information about call reports, competitive activity, expenses, forecasts, customer feedback, orders for promotional material, and specific product information could also be included. Portable microcomputers, briefcase-sized terminals, and even pulse-tone telephones can serve as input devices, and some hotel chains now rent equipment for the traveler who prefers to travel light. Similar procedures could be established for dozens of other data collection needs.

Enhanced systems are an effective training medium because they are easy to use and require no training. The high-quality graphics, color, and two-way flow of information enables complex concepts to be presented in an easy-to-understand

Table 2.2   Internal and Operational Applications

Application	Example
Information distibution	Personnel
	Policies and procedures
	Job postings and promotions
	Lobby, phone, and plant directories
	Benefits
	Organization charts
	General
	Supply catalogs
	Airline, train, bus schedules
	First aid
	Library card catalog
	Communications
	Bulletin boards
	Trade show calendar
	Newsletter
	Weather
	Safety bulletins
	Business and product related
	Product catalogs
	Pricing information
	Customer order status
	Engineering specifications
	Budget and financial
	Product specifications
	Service guides
	Product news
Information gathering	Financial and sales performance
	Survey and opinion polling
	Inventory status
Education and training	Interactive instruction
	Registration management
	Course schedules
	Training materials

manner. An additional advantage is the ability to deliver training to geographically dispersed employees. As in most computer-based educational systems, the learning pace can be varied to suit the learner, and testing can be automatically performed, with immediate feedback provided.

A number of peripheral activities crucial to a successful training effort can also be supported; such as distributing course schedules and training materials electronically; on-line registration, which enables students to add, wait list, and drop courses from any location. Creating and updating material are simplified, since these can be done quickly and in fewer steps than with paper-based information.

2.2.2. External and Operational Applications

Investing in the organization's information resources can produce new and cost-effective ways of communicating with customers and the general public (see Table 2.3); it also has the potential of producing new sources of revenue. Three areas where this change is most likely to occur are in transaction processing, electronic messaging, and information distribution. Electronic document interchange is a method of transaction processing that will revolutionize business practices in the late 1990s; EDI refers to any type of communications among computers. It could involve transferring information in prescribed formats among inhouse computers; drawing from or transmitting information to, external databases owned by customers or vendors; or integrating in-house and external data sources on a real-time basis. The American National Standards Institute (ANSI) and several trade associations have developed standards for the electronic interchange of ordering, invoicing, billing adjustments, payments, and other common business transactions. When implemented EDI will reduce from weeks to minutes the time it takes to place an order or pay a bill.

Small businesses especially will benefit from the improved management of financial resources that enhanced systems make possible. On-line budgeting, loans, income tax preparation, investment management and financial planning, and arranging lines of credit can all be done in minutes on enhanced systems. AT&T, IBM, and many other companies market hardware and software that support such business applications, and Chemical Bank, Chase, Citicorp, Bank of America, and many others offer enhanced services on a commercial basis.

Another application with the potential for containing costs is teleservicing, which is also known as remote sensing and load management. Teleservicing can be divided into monitoring, metering, and control functions. Telemonitoring is

Table 2.3.   External and Operational Applications

Application	Example
Transaction processing	Electronic document interchange
	Financial resource management
	Teleservices
	Reservation management
Electronic messaging	Mail
	Interactive facsimile
	Teleconferencing
	Photophone
	Videophone
	Audiotext/voice
Information distribution	Public access terminals
	Multilisting services

most often used for security services and fire detectors. Telemetering eliminates the costly and error prone process of visiting every customer each billing
period to read and record electric, gas, and water usage. Gathering meter data
electronically is also advantageous because it is then in a form that the computer
system can use to prepare the bill automatically. Telecontrol enables a utility to
manage demand for its services better and also helps individual customers limit the
total consumption of services during periods each day when rates are highest.

There are several types of electronic messaging available to any organization
that wishes to acquire the necessary hardware. Electronic mail and facsimile
do not even require the recipient of the message to have equipment, since there are
companies with offices throughout the country that will receive a message by
computer, print it out, then either mail or personally deliver it. Another form of
messaging called interactive fax enables anyone to call an automated information
source, page through the contents using a push-button telephone, then retrieve a
printout immediately by fax without ever speaking to anyone else. Several forms
of video messaging are readily available. Teleconfering is rather costly and not
widely used, but the photophone, a microcomputer and video camera contained in
one small unit measuring approximately 1 by 1.5 feet is a reasonably priced
alternative. It can send and receive black and white still pictures over regular, dial-
up telephone lines in as little as 7 seconds. Interactive audio messaging, such as the
976 services offered by the telcos are another example. Depending on the form in
which these appear, they are commonly referred to as voice messaging and
audiotext.

An organization's ability to distribute information electronically to its customers and the general public is severely constrained by the hardware these two
groups possess. The problem is compounded by the difficulty of convincing either
group that the benefits of receiving information from the organization is worth the
cost of acquiring the necessary hardware and software. Consequently the external
distribution of information in color with graphics is largely limited to in-house
audiences and special groups, such as subscribers to commercial on-line services
at this time. However as more systems become available, it will supplement paper-
based catalogs and direct mailing.

In the mid 1980s, public access terminals began appearing in shopping malls,
building complexes, airports, hotel lobbies, and at conferences and exhibitions.
Through attractive and colorful displays of text and graphics, public access
terminals can provide a wide range of information drawn from centrally located
information bases or from data bases stored in the terminal itself. The storage
media can range from floppy disks to laser discs offering sophisticated and high-
resolution displays. In lobbies such terminals provide the name, location, phone
number, and mail address of everyone in a building; in these terminals banks
report current interest rates and other information about available services; and at
trade shows exhibitors use them to demonstrate products and services.

Closed user groups and multilisting real estate services are other examples of enhanced system applications. Several multilisting services transmit narrative along with floor plans electronically to their members. Systems for the real estate market that are capable of distributing photographic-quality pictures have been demonstrated commercially by several equipment vendors.

2.2.3. Internal and Strategic Applications

Enhanced systems can produce significant changes in an organization's climate and its ability to assess and react to new developments in the marketplace. Investments in enhanced systems can affect the long-term economic health and competitive position of the organization. (see Table 2.4).

Enhanced systems facilitate the introduction of new technologies and the integration of these technologies within the organization; for example, the graphic and communication features of an enhanced system can provide the infrastructure to unite a number of new technologies that are being introduced as stand-alone and independently functioning entities. The North American Presentation Level Protocol Syntax (NAPLPS),* a graphic standard developed in 1983, which is resolution independent, (not committed to a terminal of a specific resolution) and thus can function as the common denominator for an organization that wishes to combine the high-resolution graphics of computer-aided design in manufacturing with the color requirements of electronic audiovisual presentations and the low-resolution demands of information distribution. The interconnectivity of an enhanced system enables all of the machines supporting these applications to share the same host computer and communicate with each other to support electronic messaging and access external computers and users through gateways.

Another illustration of system integration involves the joint use of NAPLPS and teleconferencing. With this capability, participants can share a common visual space, creating text and graphics that are instantly viewed and can be modified by remotely located participants. In a sense, such a system constitutes an electronic blackboard for teleconferencing.

Audiovisual presentations with color and graphics can be created and shown live and on-line for briefings, sales meetings, training sessions, and motivation programs. The displays can be shown on a television or monitor for small groups and projected on a screen for larger gatherings. Simple animation and synchronized sound complete the package. The electronic audiovisual medium gives the speaker complete control, since displays can be controlled by a small keypad and any display stored in the system can be retrieved without moving through a series of slides, as is necessary with a carousel tray. Displays can be stored in a stand-alone unit or retrieved from a computer information base located elsewhere. Once created displays can be used for other purposes and copied and distributed to

*The NAPLPS includes another standard, the American Standard Code for Information Interchange (ASCII), which is the most widely used code for text in the United States.

Table 2.4. Internal and Strategic Applications

Application	Example
Facilitator and integrator	Text and graphics
	Teleconferencing
	Photophone
	Videophone
	Electronic audio visual
Telecommuting	Information distribution
	Information gathering
	Data processing
Research and analysis	On-line database industry
	Artificial intelligence and expert systems

other locations in a matter of seconds. Displays can be enlarged or reduced, rotated, added to, color coded differently, and deleted seconds before a presentation begins.

The office of the future may not really be an office. An enhanced system that enables a technical neophyte to become adept at using its features and to access it through gateways from remote locations will have a profound effect on the structure of the organization. Working at home and maintaining contact with the office by microcomputer or computer terminal is known by a variety of terms; several of the more popular ones are telecommuting and teleworking.

From a strategic standpoint, there are good reasons for a company to support telecommuting. It reportedly increases productivity between 10–20% and expands the available labor pool by making it possible for people to work who are handicapped or have young children. Since fewer desks, office equipment, and space are needed, the organization's overhead is reduced. And for employees in favor of working at home, telecommuting increases morale. But there are tradeoffs. Since employees and managers are at different locations, it is difficult to supervise employees, so new forms of control must be developed. There may be a reduction in corporate identity because employees associate their work with home rather than the organization. And most importantly, it is necessary to develop and maintain the computer resources necessary to provide the infrastructure for making telecommuting a feasible alternative to going to the office every day.

Gateways from corporate computer systems to external databases enable anyone with the proper authorization, equipment, and ability to extend his or her research horizons significantly. It is possible today to search the contents of dozens of daily newspapers, examine and download census data, read over 150 newsletters, search the abstracts of several thousand journals, and tap into the news wires of every state in the union. The potential is enormous. Once acquired, data can be downloaded to host computers or micros and with commercially available expert systems, dissected and analyzed.

2.2.4. External and Strategic Applications

Strategic investments being made today in information technologies will alter the marketplace drastically, and companies that fail to adjust will disappear from the scene much like the manufacturers of slide rules and the publishers of evening newspapers. New techniques will emerge to change client relations, enable organizations to differentiate their products, and create new business opportunities (see Table 2.5).

Business-to-business electronic services will revolutionize customer service. The availability of a link between a business and its customers means that information could be shared about: ordering and shipping, lead-time and availability, service and warranties, as well as product specifications, product bulletins, parts lists, and assembly instruction and help. Initially equipment and standards that will enable business-to-business services to function will not be widely used; therefore these electronic applications will flourish in high-profit niche markets and between organizations that are highly dependent on each other for goods and services. Electronic services will be used to marry organizations to one another strategically. However once the capability widely exists, it will then be necessary to engage in the electronic distribution of information because the competition is doing so.

Marketing personnel can use enhanced computer systems as an additional product distribution medium. Some of the applications that can be supported are distributing product and parts catalogs, price lists, promotional literature , product use and training information, and competitive comparisons. This frequently updated information can be supplemented with an on-line ordering capability. An added feature of an electronic marketing support system is the fact that it can operate 24 hours a day. Even small businesses can implement these applications using a PC and inexpensive software.

Once in place, these electronic services can be accessed by the general public

Table 2.5. External and Strategic Applications

Application	Example
Information distribution	Customer service
	Product promotions
	Public relations
Electronic selling	Order entry
	Retailing
	Financial services
	Public access terminals
Commercial service provider	America OnLine
	CompuServe
	Prodigy

and the news media. Enhanced systems can provide company news, information about the company's support for community projects, newsletters, stockholder information, quarterly and annual reports, and similar types of public relations and investment information. The steps for developing such a system are discussed in detail in Chapters 10 and 11.

Distributing information electronically may build good customer relations, but it does not necessarily produce revenue. Selling goods and services electronically does. An enhanced system can provide color and graphics to reinforce the sales message; it can operate around the clock. Its message is consistent, and it can be changed to reflect advertising and special promotions.

Electronic selling had its genesis in the order entry systems initiated by such companies as American Hospital Supply and McKesson Corporation, described in Chapter 1. More and more companies are establishing gateways to their computer systems as the benefits of doing so become more widely recognized. However electronic selling has evolved beyond these dedicated and single-application systems: There are electronic malls operating on nationwide videotex services, such as CompuServe and Prodigy, along with many city and regional services that host a wide variety of stores. A shopper can visit Sears or Neiman-Marcus, buy rattlesnake meat from Texas or macadamia nuts directly from Hawaii, order financial publications from Dow Jones or the latest rock video from Waldenbooks.

The financial service industry, especially banks, are very heavily involved in electronic selling. Many banks offer electronic banking services to residential and small business customers. These services are aimed initially at attracting new customers, since in a highly competitive industry like banking where the total number of customers is relatively stable, a bank increases its customer base by taking customers away from someone else. Electronic banking enables an institution to differentiate its service. Customers can obtain account information, such as balances on hand and lists of transactions that have occurred, check interest rates, transfer funds, analyze their portfolios, pay bills, access budgeting software, and merge financial information from various accounts, and develop budgets. Electronic banking systems can operate 24 hours a day as a convenience to customers and provide new services.

In the short run, these services are costly to implement and difficult to sell. Some banks have promoted the service by selling or renting special terminals or software for microcomputers. However in the long run, electronic banking will enable financial institutions to reduce their expenditures for brick and mortar and operate with fewer employees while at the same time adding new services that can be sold to customers.

Enhanced services have made possible the development of new self-service marketing channels using public access terminals (PATs). The PATs are appearing in hotel and company lobbies, airports, shopping centers, at travel rest stops, and in large stores. As a marketing aid, they can carry advertisements, disperse coupons, quiz users, and offer guidance in making purchase decisions; in some cases, they can accept orders for goods and services. These units help speed up the buying process, and they can eliminate many of the steps that traditionally take place between an initial request for information and delivery of the good. From a cost standpoint, they can eliminate the need for branch stores, and when used in-house, they can reduce the number of employees needed to provide the same service. One unit located in a sales department and operating from 9:00 A.M. to 9:00 P.M. can eliminate three sales clerks. Furthermore if the unit is tied to the store's computer, it can eliminate much of the paperwork involved in selling and billing the customer. In addition the unit can automatically update the inventory. Laser discs and sound can be a strong enhancement in systems of this type, and if gateways are provided between host and public data networks, portions of the same information base can support electronic retailing applications.

In January 1992, Sears, Roebuck announced it was going to install 6000 automated kiosks to reap these types of benefits. The units will give customers access to information on inventory and credit card accounts and also take catalog orders. By placing these functions directly in the hands of the customers, Sears can move employees from the backoffice to the sales floor to generate additional revenue, thus permit office space to be converted into merchandising space. Sears also expects to eliminate several thousand jobs.

A commercial service provider is any entity responsible for operating an enhanced system that may be accessed by the general public or closed user groups. Leading candidates for this role are the telephone companies, cable television operators, public data network providers, retailers, and banks with excess computing capacity, publishers, and broadcasters. Most major U.S. cities have one or more small services today that supports information retrieval. These services represent strategic thrusts designed to penetrate new markets and position firms for competing in the next century.

2.3. IMPACT ON DATA PROCESSING

There are few areas in any organization that have undergone as much change in as short a time as data processing. Driven by advances in technology, change has become endemic to the industry, and this is unlikely to change in the near future. Changes have been tolerated because they have produced their share of benefits albeit a few problems also.

2.3.1. Generic Benefits

The benefits from operating an enhanced system will not all accrue to just the users or the organization overall. Data processing will directly benefit from the inherent ease of use, reduced media transformations, incremental growth features, reduced shadow functions (shadow functions are explained below), and shared control.

Ease of use is usually thought of as a user benefit, but there are two important benefits for data processing. First it reduces training requirements. Data processing is often the unit that must pay for and provide system training. Menus, pop-up screens, icons, color, graphics, and other aids that simplify system use mean personnel in data processing can spend less time holding the user's hand. Second ease of use facilitates the willingness of prospective users to evaluate and accept the system. Thus data processing has less selling to do, and users have less learning to do in order for the system to be deployed successfully.

Media transformation involves converting information from one medium to another, for example, from paper to computer disk to someone's paper calendar. Such transformations take time and energy and introduce errors during the conversion process. The more media that are automated and interconnected within the same system, the fewer transformations needed.

Another important benefit from enhanced systems is the ability to implement new applications without making major revisions to the existing system. Conformance to industry standards makes it possible to interconnect systems that otherwise might have to operate as separate entities. Integrating data, voice, and video in a communications network coupled with multifunctional terminals supporting data, voice, and video eliminates the need for redundant systems and permits modular growth as new applications are conceived. A multifunctional terminal is illustrated in Figure 4.5c.

Shadow functions are time-consuming activities that are part of every job but do not contribute to productivity. They accompany a person throughout the day, largely unnoticed until the total time they consume is recognized. For example, attending a meeting may involve travel time, waiting time, introductions, and polite conversation. The actual productive time spent at the meeting may be only a small portion of the total time consumed. An enhanced system reduces nonproductive time spent by users, but it also affects the productivity of data-processing personnel.

Since systems are designed to support a wide range of users without significant software modification or installing special equipment, personnel are not needed to perform these activities. Well-documented and easy-to-use systems reduce the number of questions from users. Conformance to standards reduces the problem of interfacing equipment from different vendors operating with different software. In short fewer problems for users translates into fewer requests for help.

Another factor that is often forgotten is that data processing is the biggest source of paper in most organizations. Computers have a tremendous capacity for generating vast quantities of paper to print, store, log-out, and distribute. This activity often consumes more labor than any other single task, and it is a slow and inefficient way of distributing information. Fortunately it can be automated out of existence and is!

Control is shared between the user and data processing. The primacy of the user is assured by the intelligence and functionality placed in the terminals, especially if the unit is a microcomputer or a computerphone. But data processing also maintains a degree of control by managing the communications services that connect the terminal to other units, by creating and managing the information base, and perhaps by authorizing access to external services through a gateway. This shared control blends the control that normally belongs to users of stand-alone microcomputers and to the manager of a centralized data-processing facility. Shared control forces both parties to engage in a dialogue and work toward creating and operating a system that will serve both their needs.

2.3.2. Problems and Hurdles

Planning, designing, and implementing applications that offer enhanced features is not without its difficulties. The traditional approach to providing computer support is different, and these differences will create a variety of problems for personnel in data processing. Problems and hurdles for the data-processing manager to overcome include the following:

New way of thinking
Measuring the investment
System integration
Infobase creation
Changing organizational relationships
Deceptively simple system

2.3.2.1. New Way of Thinking

When first introduced to a system that has all the bells and whistles users often find desirable, computer technicians have difficulty taking the system seriously. Menus and color graphics are slow; software that makes a system easy to use consumes a disproportionate share of the computer overhead; and incorporating audio and video is viewed as overkill. The problem is how these features are viewed. Enhanced systems are an extension, and not a replacement for traditional data processing. As computers shift from data analysis to a vehicle for presenting information and processing transactions, from working with numbers to working with words and ideas, these features become appropriate. Just as it is an anachronism to call a train an iron horse, or a car a horseless carriage, it is inappropriate to view a computer system today in terms of how it was used 10 years ago. It is especially difficult to determine in advance the extent to which the intended beneficiaries of these enhancements will find them valuable and actually use them.

A difficult transition occurred in one company when innovation encountered intransigence. In this particular company, the MIS department was responsible for approving computer purchases. When a request was made by the training unit for a multimedia computer with a touch screen, the request was modified to conform to MIS policy, which was limited to a regular monitor with a keyboard and mouse. When the equipment arrived, the trainer inquired about the changes that had been made to the original request. The trainer explained that the touch screen was essential for the type of training intended which was literacy training. The department head, unyielding, told the trainer the best method was to teach the trainees to type first and then to read.

2.3.2.2. Measuring the Investment

If information is an investment, how is its value supposed to be determined; after all, if it is an investment, does it not belong on the balance sheet with other assets? The answer is yes if its value can be determined. But sometimes the true value can not be quantified in dollars and cents, and management should not delay making decisions by trying to determine the price tag. There are cost-based and appraisal-based techniques for determining the value of information. Some techniques for determining the value of information are:

Cost-based valuation
Appraisal-based valuation
    Current market value
    Income generation
    Replacement cost

Cost-based techniques are modeled on methods used in the construction and hard-goods manufacturing industries. Information obtained inhouse or purchased information is valued at its original cost, then depreciated over time. The life of the information is estimated, and a depreciation schedule established on the basis of how quickly the information ages. Unfortunately this technique has several shortcomings. Some information is extremely short lived; for example, the value of certain financial news may be measured in minutes. Other information may gain value as additional facts become known. These temporal features wreak havoc with cost-based techniques for some kinds of information.

There are three appraisal-based techniques for calculating value: current market value, income generation, and replacement cost. Market value works well if the information can be sold outright to another company. An example would be the exclusive rights to data collected in a survey of a business market. Income generation is appropriate if the information can be packaged and sold to other companies and consumers on a nonexclusive basis. Credit files and on-line data bases containing abstracts of periodicals are examples. Replacement cost is based on the current cost to replace the information, which may differ considerably from the original cost.

These techniques demonstrate some ways of attaching a value to information; but information often has a value that cannot be quantified. Information that is timely, accurate, and accessible is worth a lot to a manager, but it may be impractical to try to determine its value.

It may also not be possible to speak about improving productivity in a quantifiable sense. The common way of measuring productivity is to divide output by input. Output and input are usually physical in nature or represented in a monetary form. Unfortunately this method cannot be applied to knowledge workers. For a manager, the output may be frames of information and the input the cost of his or her time, a prorated charge for computer resources consumed, and other costs equally difficult to determine. Productivity was designed to measure the tangibles found in a manufacturing environment; the concept is difficult to apply to the intangibles in a knowledge-based environment. Productivity tends to focus on outputs, rather than outcomes and impact.

If a performance-based measure is needed, effectiveness may work; because it transcends a single person's sphere of operation. Effectiveness tends to be a social concept, so it applies to groups. If a person's actions as a result of information processed have a desirable outcome or impact on the actions of others, whether customers or fellow employees, the person is effective. This means that effectiveness is not a ratio; instead it is a relative measurement best expressed as an inequality. In his book *Information Payoff*, Strassmann defines effectiveness "as people cooperating to produce a result in which the value produced exceeds production costs, so that the product may be sold for a profit." He expresses it as Customer Benefits > Prices > Costs.

In whatever mold effectiveness is cast, it is a better measure than productivity for expressing the impact of an enhanced information system. If data-processing personnel are called on to justify their investment in information resources, they should not shirk from the challenge. There are techniques available for such an assessment.

2.3.2.3. System Integration

The data-processing manager faced with the task of integrating enhanced services with an existing system has several major problems. The first is what to do with existing programs and data files. Should they be superseded by the new system or maintained because of their relation to a historical database or for users who like them in their current form? What new access methods and programs should be created? How should security be ensured? Who should be authorized to update the information if it is maintained on-line? If terminals, especially microcomputer-based terminals, are acquired by users rather than data processing, how is hardware and software compatibility assured. These questions do not have easy answers, but the data-processing manager is fortunate in one respect: The user will not judge the system on the basis of how such questions are resolved. As far as the end-user is concerned, the nature of the information, how it is presented, and how responses must be formulated are the criteria governing acceptability.

2.3.2.4. Infobase Creation

Color, graphics, animation, interactive video, audio, and similar types of enhancements lead to appealing and informative displays, but the effort involved in creating them is unlike anything people in data processing have had to deal with in the past. Alphanumeric text is and will remain the predominant method for conveying information. Provided the wording is correct and properly crafted, there is no ambiguity about the message. There are ways of presenting text to improve the reader's comprehension, but the basic process involved in collecting data, converting it into a digital format, and storing it in computer memory is well understood by data processing. The skills involved in providing enhancements are not native to the data-processing environment. Graphics and color require artistic skills and integrating these elements with the text requires preparing storyboards and scripts, skills not found in data processing. Interactive video combines television production and training delivered through the computer medium. Audio and audiotext are an outgrowth of activities in the telecommunications industry. Chapters 9–11 deal with designing and producing the information base and many of the skills that are involved. If data processing is to retain its role within the organization and at the same time meet the needs of its users, management must make sure these skills exist.

2.3.2.5. Organizational Relationships

The information industry has been enamored with "big concepts" throughout its short life. The 1960s was the era of the large mainframe in the glass-encased showroom. The 1970s saw the introduction of the centralized database managed by large and expensive software packages. The proliferation of microcomputers and minicomputers distributed processing power to the desktop in the 1980s. In the era now dawning, the user will reign supreme. Anyone, regardless of background or level of technical competence, can use a computer. Traditionally computer systems were developed to process large volumes of specific raw data. The results were then provided in a highly structural format to a small and relatively sophisticated set of managers. This will continue, but now everyone or anyone becomes a user.

Today information is collected and stored for selective retrieval without prior knowledge of specific users, their needs, or habits, and without prespecified fixed outputs. Considerably more importance is placed on display presentation than before. Users have less patience, and they will not tolerate delays. Data processing becomes less a source of computing power and more a service center responding to user needs and ensuring connectivity for everyone in the system.

2.3.2.6. Deceptively Simple Systems

The easier to use the system, the more demanding users become. Enhanced systems can be deceptively simple. One of the trade-offs to make accessing information as simple as possible, is the user may have to forego the ability to make complex searches or operate the terminal as a stand-alone processing unit. The information base, which sometimes has the gloss of a slick magazine, is considerably more complex to design and create than the numerical databases of the 1970s and 1980s. Color must be carefully chosen, and graphics must be used sparingly to complement the text. The contents must be indexed thoughtfully, and if keyword searches are allowed, those keywords must be selected carefully so searchers are not misled.

Connectivity, which allows the user casually to access a computer on the other side of the country as easily as accessing a computer on the other side of the room, can be difficult to achieve, especially if the system's hardware components are produced by different vendors. Standards are not always adhered to, and when they are, there is always the risk that two different sites will create the infobase using different standards, or they will have communications protocols not found elsewhere in the system.

Potential problems and difficulty overcoming them may seem endless. The data-processing manager has been successful if users remain oblivious to the problems that have to be overcome to implement and maintain the system.

2.4. SUMMARY

An enhanced system is a logical extension to time sharing but with several important differences: It is designed for the unsophisticated user, and it integrates different technologies and communication services into a single system with extensive information retrieval and transaction-processing capabilities.

The possible applications can be examined according to whether they take place within or outside the organization and are operational or strategic in nature. Enhanced systems can impact the data-processing department positively and negatively. Data processing will benefit from system simplicity, reduced media transformations, potential modular growth, elimination of wasted time, and user involvement. But all that glitters is not gold: There are many problems and hurdles to overcome to use an enhanced system effectively, not the least of which is the need to develop a new way of thinking about how computers and information resources can be applied in a business environment.

REFERENCES

A. F. Alber, *Videotex/Teletext: Principles and Practices*, McGraw-Hill, New York (1985).
An Executive Guide for Understanding and Implementing Business Videotex, Videotex Industry Association, Rosslyn, VA (1986).
D. Anderson, Case Studies and Implementations of LDI Arrangements, *Data Communications* 15(2) (1986), pp. 173–174, 177–178, 181–182.
W. L. Anderson, If Information Is Really an Asset, Put It on the Balance Sheet, *Information Week* (056) (1986), p. 68.
A. Brigish, Videotex Fits the Electronics Industry, *Electronic Business* (November 1984).

R. Buday, The Strategic Use of Information: Seizing the Competitive Edge, *Information Week* (067) (1986), pp. 26–60.

R. S. Buday, 1987 Budgets: Tightening the Belt, *Information Week* (097) (1986), pp. 13–15, 17–18.

R. Ellson, Visualization at Work, *Academic Computing* 4(6) (1990), pp. 26–28, 54–55.

R. Graham, Videotex in the Business World, *Viewdata 1982*, *Proceedings*, Online Conference, Northwood Hills, UK (1982), pp. 169–180.

B. Ives and G. P. Learmonth, The Information System As a Competitive Weapon, *Communication of the ACM* 27(12) (1984), pp. 1193–1201.

B. O'Connell, Media Integration, *DEC Professional* 10(7) (1991), pp. 38–40, 42, 44, 46–47.

B. O'Keefe, Adopting Multimedia on a Global Scale, *Instruction Delivery Systems* 5(5) (1991), pp. 6–11.

G. Phippard, Videotex in an MIS Environment, *Videotex World* 2(3) (1986), pp. 36–40.

A. Radding, GUIs on Every Desktop, *Digital News* (April 16, 1990), pp. 1, 31–32, 34, 47.

J. Schwartz, PC Point-of-Sale Network and Customer Information Kiosks Are in Store for Sears, *Communications Week* (386) (1992), pp. 27–28.

D. M. Simmons, Implementing Corporate Videotex, *Videotex World* 2(3) (1986), pp. 4–7.

C. W. Steinfield and L. Caby, Strategic Organizational Applications of Videotex among Varying Network Infrastructure, *First International Research Seminar in Service Management*, France (1990), pp. 657–675.

P. A. Strassman, *Information Payoff*, Free Press, New York (1985).

Videotex in Top Gear, *Telecom France* (7) (1984), pp. 6–10.

Videotex and the World of Business, Telidon Marketing, Dept. of External Affairs, Ottawa, Ontario, Canada, (1984).

T. Woods, Image Conscious, *Business Computer Systems* (October 1984), pp. 60–63.

Chapter 3

Computer Options

Computers have evolved from centralized and dedicated machines requiring highly trained operators to distributed units that can be and often are run by children. When computers were introduced for general business applications in 1952, each user wrote his or her own program, loaded it into the machine, and set the switches that enabled the computer to operate. By the mid-1960s, a support group had emerged, and it was no longer necessary for the user to travel to the machine.

Systems analysts, programmers, and machine operators provided support functions for the user who was often connected to the main computer by a terminal and telephone line. In the 1970s, smaller computers began to appear in departments and plants to service a local group of users, but the users were still connected to the machine in a starlike configuration using telephone lines or a point-to-point cable.

In the 1980s, microcomputers placed computing power squarely on top of the user's desk in a footprint that measured 1 square foot, only 0.00067 the footprint of ENIAC, the first electronic computer. Each step in the development of computer technology has moved the computer closer to the user. In the case of microcomputers, this has resulted in moving the machine to within arm's reach and in the case of large mainframe computers, by making access to the system uncomplicated and convenient.

In the 1990s, the introduction of new computer hardware and software is occurring at a breath taking pace; trying to keep track of the changes is a job in itself. Rather than trying to keep up, it makes more sense to approach the issue from the other side and to develop an understanding of the computer features that are suitable for an enhanced environment. To this end, Chapter 3 begins by reviewing these computer features. In Chapter 3, when the term enhanced is used, it is in connection with the whole system. The term multimedia applies to storage techniques and the contents stored. It is especially important for the reader to understand the rudiments of optical storage, since that is the principle way of saving multimedia information. Unfortunately since information requires so much space, the rules for managing the information base change, which creates another challenge for the unwary; both of these points are explained in Chapter 3.

Organizations approach the implementation of these systems in a variety of ways. Although the machine the viewer looks at may be a Personal Computer, it is often connected to a mainframe or minicomputer lurking in the background, which is supplying some of the content. Several possible configurations a manager might pursue for configuring a system are explained in the chapter and tied to the cases in Chapters 13–15. Chapter 3 ends with some important terminology and a review of that damnable subject, comparing computers based on performance.

3.1. SYSTEM CHARACTERISTICS

Networked computers that support enhanced services fall within the domain of on-line transaction processing (OLTP) systems. Since the mid-1960s on-line transaction processing has assumed a steadily growing role in business data processing. It is currently estimated that perhaps as many as half of all new mainframe and medium-scale computers are intended exclusively or partially for on-line applications.

On-line transaction processing is considerably different from simple time-sharing. The latter creates the illusion that the user has sole control of the computer through such features as multiprogramming and multitasking, when in reality the machine in being shared by many others (see Table 3.1). On-line transaction processing is more complicated, because it contains the basic elements of multi-programming and multitasking systems, but in addition, it allows concurrent, on-line common database sharing by multiple users. The earliest large-scale systems were used by airlines for reservations; these were later expanded to include such functions as crew and equipment scheduling; these were followed by reservation systems in the hotel and car rental industries and funds transfers in the financial services industries.

On-line transaction processing differs from traditional data processing in another way. Profit-motivated firms anticipate increased profits from investing in a computer resource. Nonprofit and not-for-profit organizations expect to operate more effectively. In both cases, computers are expected to reduce the cost component by improving efficiency. On-line systems are often intended to in-crease revenue. In the banking and travel industries, this is done by using automatic teller machines to attract new customers and reservation systems to facilitate the sale of seats or hotel rooms.

Table 3.1. Terminology—Multi Terms Lead to Multi Confusion

Term	Explanation
Multicomputer	Computer with multiprocessors for simultaneous use; the processors may share memory and other system components.
Multiprocessing	Simultaneous execution of one or more programs by two or more processors (see parallel processing).
Multiprogramming	Concurrent execution of two or more programs by allowing the programs to share computer resources; *e.g.*, to execute one program while a second is outputting a result to a terminal.
Multitasking	Partitioning a job that has been submitted to the computer into several cooperating tasks that can be executed simultaneously.
Parallel processing	Simultaneous execution of two or more sequences of instructions, or one sequence of instructions operating on two or more sets of data, by a computer having multiple arithmetic and/or logic units.

On-line systems introduce more risk into the organization. When a conventional computer stops functioning, it is inconvenient and work is often delayed. When an on-line system that is an integral part of an organization's minute-by-minute operation stops, orders can not be taken, customers can not be served, and business comes to a standstill.

3.1.1. Integration

An enhanced system that is networked is often an on-line transaction-processing system with additional features the most important of which is the ability to integrate different technologies (see Table 3.2). Voice-, data-, and visual-based applications may all be present on the same system, and the latter may involve graphical forms of imagery, photographic-quality still pictures, and interactive video.

Combining these dissimilar technologies is considerably more difficult than most people realize. A variety of equipment must be supported. Terminals suitable for displaying text may not be suited for graphics. Extensive storage is needed for photographic-quality and motion displays; this storage is often provided by attaching such peripheral equipment as a CD-ROM or videodisc players to the user's equipment. Audio and video each require special processing chip sets. Such differences make enhanced systems more complex to design and build than conventional data-processing systems that produce accounting reports and maintain inventory records in a pure text-oriented format.

Until quite recently, system designers and integrators specially configured equipment to support enhanced features; mini- and microcomputer systems were jerryrigged configurations built from components selected from a variety of manufacturers. At least this was true until May 1991, when a group of microcom-

Table 3.2. Characteristics of an Enhanced System
That Is Networked

Feature	Characteristic
Integrate heterogeneous technologies	Voice
	Data
	Visual
Connectivity	Response time
	Simultaneous users and loading ratios
Storage and information handling	Large information base
	Real-time access
	Special storage devices
Communications environment	Data transmission requirements
	Communications services
	Gateways
Operational readiness	Available and reliable
	Consistency
	Modular
	Secure

puter vendors led by Microsoft Corporation issued the Multimedia PC (MPC) Specification, Version 1.0. The impetus created by this specification will spur vendors of larger systems to undertake similar activities.

A summary of the specification appears in Table 3.3. The MPC hardware specification is a minimalist's approach to a multimedia system: It specifies the minimum capabilities needed to sustain a multimedia environment. It represents a specification needed by an industry aggressively seeking common platforms on which to develop a market.

3.1.2. Connectivity and Communications

Another characteristic of enhanced systems is their degree of connectivity. This term refers to the ability to interconnect equipment through a communications network. Enhanced systems typically support a large number of simultaneous users with an acceptable response time. For a given equipment configuration, the number of simultaneous users depends on the response time deemed acceptable for the applications involved. There are several ways of measuring response time, but in the discussion that follows, response time represents the total time elapsed between a user's inquiry and the appearance of the response on the screen. Response time equals the sum of the hardware time needed to satisfy the request and the communication time requirements.

Hardware time depends on the nature of the application. A simple information query involving text requires very little actual processing and consumes few resources. On the other hand, a transaction involving graphics and text may require

Table 3.3.   Partial Description of the Minimum MPC
Hardware Specification, Version 1.0

Hardware	Description
CPU	Minimum requirement: 80386SX or compatible microprocessor
	Two MB of extended memory
	3.5 inch high density (1.44-MB) floppy drive
	30-MB hard drive
Optical storage	CD-ROM drive
	Sustained 150 K/sec-transfer rate without consuming more than 40% of the CPU bandwidth
	Average seek time of 1 sec or less
	10,000 hours MTBF
Audio	CD-ROM drive with CD-DA outputs and front-panel volume control
	8-bit (16 bit recommended) digital-to-analog converter (DAC)
	8-bit (16 bit recommended) analog-to-digital converter (ADC)
	Internal synthesizer hardware with multivoice, multitimbral capabilities, six simultaneous melody notes plus two simultaneous percussive notes
	Internal mixing capabilities to combine input from three (recommended four) sources and present the output in stereo
Video	VGA-compatible display adapter and color monitor
	Minimum 16 colors, recommended 256 colors
User input	Standard 101 key IBM-style keyboard
	Two-button mouse
Input/output	Asynchronous serial port programmable up to 9600 baud
	Bidirectional parallel port with interrupt capability
	MIDI port
	IBM-style analog or digital joystick port

considerable time to process in its final format, and the transaction may consume considerably more system resources. Hardware time is really a combination of the time to access and retrieve the information, usually from disk storage, and the time spent by the central processing unit (CPU) to perform whatever calculations may be necessary. In most instances, access time and retrieval time take longer than the actual processing time.

Communication time refers to the length of time it takes the transmission to travel back and forth between the user and the computer. Communication time is a function of the type of link, transmission speed, protocol, number of characters transmitted, and the time spent at intermediate routing nodes. In most applications, user input requires an insignificant amount of time compared to the system's response time.

The longest delay is created by the user. It may take only a few seconds to scan a menu frame, then make a decision, but reading a frame of information may consume many minutes. A calculation of response time and the number of simultaneous users that can be theoretically supported are shown in Figure 3.1. In actual practice, an extensive analysis of the type and length of transactions must be

Case: A manager wants to retrieve a frame of information displaying a graph of the aged accounts receivable. Since the address of the frame is not known, it will be necessary to proceed through a set of four menu frames to locate the information.

Response Time (secs):

Hardware time

The menu frames and the information are located on disk storage. The time to access either the menus or information has three components; seek time, latency time, and transfer time. The seek time involves the movement of the read/write heads to the proper track on the disk. Latency is the average time for the information being sought to revolve under the read/write heads. If it has just passed by, the disk must make a full revolution; if it is just behind the read/write head, access will be immediate. Transfer time is the movement of the information from the disk to the CPU.

Processing time for the menus will be insignificant. However, the time requirements to produce the graph could be extensive. If the information already exists in a tabular format it only has to be translated to a graph format. But, if every accounts receivable record must be examined and the results tabulated, the resources and time expended could be considerable. It is assumed that the information exists and only needs to be translated. The time requirements in seconds for each menu and the information return are:

	Menu Retrieval	Information Retrieval
Seek	0.030	0.030
Latency	0.080	0.080
Transfer	0.000	0.001
Processing	0.000	0.100
Total	0.110	0.211

Communications time

The menu frames each contain approximately 38 characters. The graph is much more detailed and including colors, special symbols, and the text has the equivalent of 1500 characters. If the desired response time is not to exceed 2 seconds the line speed must be at least 6,708 bits/sec. This is calculated by subtracting the hardware time of 0.211 seconds from 2.0 seconds and determining how fast the line must operate to transfer 1,500 characters. If only the menu frames were being transmitted the line speed would only have to operate at 170 bits/sec.

User Delay (secs)

Menu Retrieval	Information Retrieval
Study Time 10	Study Time 300

Figure 3.1. Calculating response time and examining factors to consider in determining the number of simultaneous users and the loading ratio.

Cycle Time (secs)

	Menu Retrieval	Information Retrieval
Response	2	2
Study time	10	300
Total	12	302

Simultaneous Users
If the computer can satisfy 20 requests per second, the theoretical maximum number of simultaneous users is:

Menu Retrieval	Information Retrieval
240 (12 × 20)	6040 (302 × 20)

Loading ratio 10:1

Menu Retrieval	Information Retrieval
2400 (10 × 240)	60,400 (10 × 6040)

Figure 3.1. (*Continued*)

done and an average cycle time calculated; then it is possible to determine the number of simultaneous users that can be supported for a given configuration. Since every potential user will not access the system simultaneously, a statistical analysis is performed to determine the loading ratio possible for a given level of service. If the ratio is properly set, a predictable and presumably acceptable number of users are likely to receive a busy signal during peak activity periods.

Figure 3.1 illustrates how sensitive the number of simultaneous users and the loading ratios are to the type of application. Although at first it seems improbable, the longer the user takes to assimilate what is on the screen, the greater the number of simultaneous users and loading ratios possible. A discussion of how the Prodigy network satisfies user requests is presented in the Ford Motor Company case in Chapter 14. Figure 14.7 illustrates the network architecture.

3.1.3. Storage and Information Handling

An important system feature is the ability to store and quickly retrieve large quantities of information. Computers supporting enhanced applications do very little actual data processing; most of the activity involves unrelated and random requests to provide information or support a transaction, such as electronically transmitting a document. These applications require a real-time addition, deletion, and modification of the data and management of the communications process; actual processing time is minimal.

Information retrieval applications in a typical videotex system may be handled in two ways. Information may be stored and maintained in discrete units called a frame, which corresponds to a single display screen. The amount of storage required for a frame depends largely on the type of display. A frame of text material might consume several hundred bytes; on the other hand, a photographic-quality image may require many thousands of bytes.

Information also may be stored as raw data in application files. Data are retrieved, modified if necessary, and displayed in predesigned templates. This is the approach often used for business applications. It trades the time and expense of creating and storing frames of information for the delay involved in retrieving data and doing whatever processing is needed to make the data fit the preformatted template.

Enhanced applications frequently involve transactions processing. A transaction is a collection of related actions whose completion identifies a discrete unit of work. The work may be transferring a document or money, transmitting an electronic message, or monitoring and controlling a heating system. Such transactions as the last one mentioned do not require an extensive database, but these applications are in the minority. Most transaction-based applications require large on-line databases because major delays will occur if data are not immediately accessible. Data availability is especially important in applications where customers are being serviced.

3.1.3.1. Media Forms

Special storage needs are required for some applications; for example, voice message systems require considerably more space to store a digitized message than is required for the equivalent text version. A minute of uncompressed telephone-quality sound requires a half MB of storage space, and 1 minute of CD-quality music consumes 5 MB. Fortunately popular compression algorithms can reduce storage requirements by as much as one-sixth for voice and one-third for music.

Photographic-quality imagery and interactive video are usually stored on optical discs, which can be classified according to whether they are digital or analog based. There are three types of digital technologies discussed here—read-only, write-once, and erasable media. It is convenient to think of these technologies as a publishing media (CD-ROM) or a data-processing media (WORM and magneto-optical). Each is explained in more detail in the following paragraphs.

The technology for CD storage, introduced in 1982, was originally developed for recording digital music. Since its introduction, CD technology has been extended for recording text and images. The application determines which form of the technology is used, and the differences lie in how digital information is encoded, decoded, formatted, and so forth. Three CD formats are discussed here: The first addition to the CD family of formats is called CD-ROM (Compact Disc-

Read Only Memory); it was introduced in 1985 and intended principally for storing text, but it is frequently used today for recording images as well. The first commercial uses were a bibliographic product called Bibliofile and the Academic American Encyclopedia. The standard defining CD-ROM technology is called ISO 9660; this standard is a derivative of an earlier specification called High Sierra, but it differs slightly. Most of the titles in production today conform to ISO 9660. There are two disc sizes—a 4.75-inch diameter size and a smaller format, called mini-CD, which is 3.5 inches in diameter. An interesting part of CD-ROM folklore is the genesis of the larger size disc. It seems that an audio CD had to be 4.75 in. to permit an uninterrupted performance of Beethoven's *Ninth Symphony* to be heard, which lasts about 75 minutes. When the compact disc format was adapted for data storage, the size of the disc was not changed.

The CD-ROM discs can be manufactured at the same facilities that produce audio discs; a front-end computer is required to insert error codes, manage formatting, and perform other tasks unique to computer data. Consequently the cost of producing CD-ROM discs is fiercely competitive with other media forms. Many production houses have a minimum order of 200 discs. Fortunately small quantities can be made on special equipment costing under $8000. This equipment may require anywhere from 30 minutes to 2 hours to make each disc which limits the equipment to archiving data or supplying small quantities of discs for in-house users or a few customers.

A CD-ROM reader resembles an audio player except for an error correction chip replaces the digital-to-analog conversion chip. Also the servo mechanism supports direct access of the contents rather than the linear access characteristic of music. An interface board attaches the driver to the PC and special driver software completes the connection.

A CD-ROM can be read by relatively inexpensive hardware under the control of machines as small as a microcomputer. There are many advantages to the medium: Magnetic fields do not affect the contents, head crashes are an impossibility, and environmental hazards, such as water and dust do not endanger the integrity of the contents. Discs can be handled and wiped off if they become soiled. They are portable, and they can be mailed without concern; discs last a minimum of 10 years. Storage capacities currently range between 550–660 MB, depending on the data format. A single CD holds the equivalent of about 200,000 pages of text or the contents of 1,200 5.25-inch high-density, single-sided magnetic diskettes or as many as 860 3.5-inch high-density, single-sided magnetic diskettes.

The major disadvantage of all optical storage technologies is the time required to access data when compared to magnetic hard-drive storage. The delay is due to the weight of the optical head, with its laser, lenses, and mirrors. Access time is between two and six times slower than for high-performance magnetic hard disks. A popular misconception about CD-ROMs is that they will last forever; in fact CD rot is a probability due to a variety of factors, including oxidation, contaminates in the manufacturing process, improper handling, and such environmental conditions as heat and humidity. Vendors typically warrant CDs for about 10 years.

Discs are created by burning tiny indentations approximately 1-μ wide (a human hair is 50 μ) on a plastic-coated surface with a laser. The presence of a pit represents a digital 1, and a flat surface, which is called a land, represents 0; the combination of 1s and 0s are then used to code the material being stored.

The major cost of creating a CD-ROM is the expense of compiling or developing the contents of the disc. The cost of producing the master disc (a die for pressing the plastic replicas) from a tape starts at about $1200. Most organizations prepare the contents but then have an outside firm specializing in producing discs make the master copy. Once a master exists, it can be duplicated for under $1.50 per copy.

There are three steps in producing a master: data conversion, premastering, and mastering. In the data conversion step, the contents are organized and formatted; the premastering adds the error codes and indices. The master is then created and used to stamp the pits in the production discs.

Two other forms of CD technology are CD-I (Compact Disc-Interactive) and CD-ROM XA (eXtended Architecture). A CD-I drive resembles a CD audio player; it has limited memory, a mouse, and a remote control unit. The unit is attached to a stereo system for broadcasting audio and to a television set for displaying images. It is capable of full-motion video, with a capacity of 7000 photographic-quality pictures. The CD-ROM XA combines the storage of a CD-ROM with the dynamic qualities of CD-I. The extended architecture supports full-motion video and sound in a relatively inexpensive platform.

A novel application within the CD industry occurred in 1992, when Kodak introduced the Photo CD System. This is a new technology that digitizes and stores film negatives on a CD-ROM to be viewed later on a television set or computer monitor. A consumer simply takes the film to one of approximately 200 processing centers where the contents are transferred to the CD for under $20 per roll of 24-exposure film. The capacity of the CD permits multiple rolls of film to be stored, and interfaces to popular software permit clipping and pasting the images for a variety of applications.

The write-once read-many-times (WORM) optical disc provides slightly more flexibility than a CD-ROM. The discs are typically 5.25 in. in diameter, although large sizes exist. Data are recorded on tracks and sectors similar to a floppy disk, thereby making access faster than a CD-ROM. Unfortunately data cannot be mastered on a Worm disc as efficiently as on a CD-ROM, and the equipment is considerably more expensive. Recording takes place by creating a series of blisters on the surface of the disc with a high-powered laser. Then the disc is placed in a WORM drive and a low-powered laser produces just enough heat to break the prerecorded blisters. During the read operations, a laser illuminates the disc's surface. The burst blisters provide a higher contrast than the adjoining area, and they are interpreted as either a 1 or 0. The ability to write on the discs enables updates to be made on unused portions of the disc. When the disc is read, new portions containing updates are accessed in place of the former contents.

The third form of optical disc is magneto-optics; the surface, called the magneto-optic layer, is subjected to a steady magnetic field. To write on the surface, a laser heats selected locations that correspond to bit positions. The heat inverts the polarity of the magneto-optic laser at these positions. Reading is done with a polarized laser light. The positions that did not have their polarity changed rotate the polarized light in one direction; the positions with the inverted polarity rotate the polarized light in the opposite direction. A 1 or 0 is read based on the direction. Like the other two forms of optical storage, this media can hold extensive quantities of information, but unlike the other forms, data can be erased by reorienting all the particles to zero. Theoretically present technology permits rewriting on the surface up to 10 million times, and data will last for between 10–20 years. Erasable systems are not alternatives for electronic publishing, since the contents can be modified and it is difficult to reproduce the contents in large quantities.

Interactive videodiscs represent the analog side of the family. The analog signal conforms to a standard known as National Television Standards Committee (NTSC), the same standard used by the television industry in North America. There are two methods for recording on, and playing back, video from a videodisc: constant angular velocity (CAV) and constant linear velocity (CLV). Both systems store images as frames. A frame corresponds to a single screen of information; in a typical full-motion video presentation, 30 frames or screens are presented each second. The National Library of Medicine case in Chapter 13 illustrates an excellent implementation of videodisc technology.

In a CAV videodisc, it is possible to store as many as 54,000 frames on each side of the disc. The surface of the videodisc is divided into segments called tracks, and each track holds one frame. In a constant angular velocity videodisc, each frame is at a constant angle to the center of the disc, so its beginning point can be precisely located and the frame displayed. Each side of the disc can store 30 minutes of full-motion video.

Not all equipment for playing CAV videodiscs is created equal. The industry divides the equipment into five levels, from least to most sophisticated. Level-0 devices are for linear-play presentations similar to a videotape; level-I provides the ability to search for a particular frame; level-II discs incorporate limited intelligence for programming instructions, such as continuous replay; and level-III and level-IV devices interface with a microcomputer to provide a flexible platform for video presentations.

A CLV videodisc is able to store up to 60 minutes of video per side, because the inherent features of this implementation permit more than the equivalent of one frame per track the further the distance from the center of the disc. The inability to match a frame with each track among other reasons limits a CLV videodisc to linear applications, such as showing a movie.

Table 3.4 lists storage capacities of selected random access storage media and their relative cost per megabyte.

3.1.3.2. Multimedia Infobases

Coupled with a revolution in the storage media is a new approach to the process of storing data. Known as object-oriented storage, this new approach stores information as objects. These objects are pieces of code ranging in length from short programs to entire graphical applications containing data and information about how to manipulate that data. As an illustration, an object could be the image of a bird with instructions to make it fly or the drawing of a rectangle with instructions to make it revolve. A close cousin of the object-oriented database is a multimedia infobase. There are similarities between the two approaches, but the latter may be considered a refinement of the former. Object-oriented databases consist of objects whose relationships to each other are managed by the software. Multimedia infobases can also define relationships among objects, but they are more general purpose and contain features consistent with the characteristics of the information being stored.

These characteristics include everything from pure text documents to picture-quality images and sound. There are two approaches to managing the contents of an information base containing images and audio. The first stores images in external files on devices, such as CD-ROMs, and refers to them via pointers in a table; this technique is referred to here as Binary Large Objects (BLOB) support. There are shortcomings to managing images outside of the physical database; these shortcomings are related to security issues, data integrity, and recovery procedures. In addition devices holding the contents (*e.g.*, CD-

Table 3.4. Typical Storage Capacities in Megabytes and Their Relative Costs

Disk Type	Capacity in MB	Price/Unit ($)	Cost (MB) ($)	Typical Drive Price ($)	Media Longevity (years)	Access Time (msec.)
Floppy (3.5 in.)	1.44	1	0.69	52	10	79
Hard	200	539	2.98	595	a	16
CD-ROM	660	2	0.00	500	9	500
Floptical	20	20	1.00	599	10–20	80
WORM	6,550	350	0.05	17,000	100	500
Erasable optical	650	250	0.38	6,000	15	100
Videodisc	540(CLV)	10	0.02	2,000	9+	1,000
	270(CAV)	10	0.04	2,000	9+	1,000

aThe mechanical components fail before the magnetic qualities of the storage media deteriorate enough to make a difference.

ROM) have special requirements of their own that may impact on using certain applications development tools.

A second and much newer approach that is just beginning to emerge subsumes the traditional relational model of database theory; however the software has been developed to accommodate data requirements of images and audio. An example of such a product was announced in spring 1992—UniSQL/X from UniSQL, Inc.

Because the BLOB approach is older and more widely used, it is described in detail. It is built on two data types, known as text and byte. Text data types are composed of text characters that form units, such as memos and programming source code, while a byte blob is a set of binary data representing such objects as graphs, pictures, and sound patterns. The existence of the byte data type differentiates a multimedia database from other incarnations of databases.

The need for defining byte data type is due to its ability to assume truly gargantuan proportions. A picture-quality image or a segment of audio from a sound track may be several million bytes large. To accommodate objects of this size, a multimedia infobase defines information in terms of rows and columns. A BLOB-space denotes a logical region of the database containing a column of BLOBs. Although a multimedia infobase can perform many of the functions found in relational and other models of database administration, the way it administers a BLOB-space is unique.

One challenge is to manage these BLOB-spaces and their contents without impairing memory use in the computer. In conventional systems, information is temporarily stored in cache memory. The cache can be thought of as a staging area into which items are moved when it is anticipated they will be called on soon. Once the item is needed, it is copied into a shared memory in RAM. In a multimedia infobase, a byte object of several megabytes would overwhelm cache memory and fill a significant portion of shared memory, thereby affecting other users. In a multimedia infobase, the software places the data in a special user-specific memory that is separate from shared memory; other users can then continue to operate from the shared memory space without being affected.

Storage is also treated differently. Normally database software tries to store the fields of a database in close proximity to another, which facilitates retrieving of the information. A large BLOB would impose constraints if handled in the same way: Its shear size might overwhelm the recording media, such as a floppy disk, and a large number of input/output operations might be necessary to read it. A multimedia infobase puts BLOBs in their own partitions on a disk or places them on an entirely separate device, such as an optical disc. The record to which the BLOB belongs is processed in the usual way; however in the field where BLOB data belong, pointers indicate where the BLOB is actually stored. When the record is accessed, its contents and the pointers are loaded into shared memory; however the BLOB is accessed only when the file containing the pointers is accessed.

Logging and recovery from system crashes are also handled differently in a multimedia infobase. In a conventional system, the integrity of an information base is carefully managed in the following way: When a transaction is completed, the activity is recorded in a memory buffer and periodically written to a log file on disk. The buffer step reduces the number of times the disk has to be accessed and information describing the activity transferred from the buffer to the disk, since the fewer transfers, the better the system performs. In a multimedia environment, a BLOB could easily overflow the buffer used for logging and thus require frequent transfers, which would impair system performance; also the log file containing BLOBs could require a tremendous amount of disk space. Consequently a multimedia infobase does not log changes that have been made to a BLOB; instead changes are made directly to the infobase. A copy of the old and new BLOBs are kept momentarily; if the transaction is successfully completed, the old BLOB is deleted. If the transaction is not successfully completed, the field containing the pointers continues pointing to the old BLOB.

Another aspect of storage is archiving. Making a historical record of a multimedia infobase containing many BLOBs requires a lot of space and time. Rather than automatically store each BLOB every time, an incremental archiving strategy is employed, so that only changes made to BLOBs since the last backup are archived. This can considerably reduce the time and space needed, since BLOBs are often static, *e.g.*, pictures.

These and other features of a multimedia infobase software package support what are known as the ACID properties of data management: atomicity, consistency, isolation, and durability. Atomicity simply requires that once a transaction is started, it must be completed for permanent changes to be made to the infobase. If the transaction is not completed, the contents of the infobase revert to their form prior to the beginning of the transaction. Consistency ensures that transactions are completed before they are viewed by other users; that is, a user sees the information in discrete snapshots and does not view an in-process transaction. Isolation ensures that when one person is making a change to an object in the information base, a second person cannot also be changing that same object. This condition is referred to as contention, and it is managed by locking out one of the users until the other has completed making changes. Durability ensures that once a change is made to the information, it is permanent. If a system failure occurs, that change should be reflected in the infobase when the system recovery features are completed.

Informix Software, Inc., was one of the first companies to introduce software for multimedia infobases. The product is called Informix-ONLine, and it contains the features just described and many more for managing everything from text to photographs and sound.

3.1.4. Communications Environment

Enhanced systems place special demands on the communication environment in which they operate. Usually a large number of terminals must be able to access and either retrieve or update the information base in real time. This requires flexible communication and information exchanges and the ability to interact with different computers in local and often global networks.

Data transmission requirements vary with the type of application to be performed. The greater the bit density per frame, the higher the transmission rate must be. Typically applications involving transactions have a very low density; examples are financial transactions, electronic mail, teleservicing, and order entry. High-density frames have considerable quantities of text, graphics, and photographic-quality imagery.

Interactive video creates a unique need for high transmission rates because of its special characteristics. To present a smooth flow of images devoid of flickers or jerkiness, frames must proceed at a uniform and relatively high rate—30 frames a second usually; in addition the content has a high bit count. Together the frequency and content of interactive video application require a high transmission rate.

Eighteen million bytes of data per second may be needed to produce 30 frames of video per second. Fortunately techniques are available to compress the number of bytes needed to as few as 150,000 bytes per second for transmitting the same 30 frames. Compression is done at the transmitting end, and chips at the receiving terminal decompress the transmission before it is displayed in all of its detail.

The trade-off between the data rate expressed in thousands of bits per second and the time in minutes is shown in Figure 3.2. Acceptable transmission rates for most applications, especially those involving a business activity, begin at 1.2 Kbits/sec and may exceed many millions of bits per second. Transactions taking place over the public telephone network are usually at speeds at or below 9.6 Kbits/ sec; the upper end of this range usually pertains to in-house transmissions over a private branch exchange (PBX) system.

Transmissions between 9.6–56 Kbits/sec use lines maintained for data transport; examples are services provided by value-added networks (VANs), such as Telenet, Tymnet, and the ISDNs being implemented by the regional Bell operating companies. Rates above 56 Kbits/sec rely on data cables and over-the-air communications links; examples of the former are local area networks (LANs) and wide area networks (WANs) installed for in-house communications. Satellite- and microwave-based networks are available through public carriers, and some companies maintain their own networks.

An important feature of many communication services is the ability to bridge to other networks through a gateway. A gateway is simply the hardware and software that form the interface between two networks and enable a user of one system to communicate with a second system. Since networks may support different transmission rates, a gateway provides a change in the data rate. The limiting factor in determining what applications may be supported when several networks are interconnected is the network with the lowest data rate. This point can be illustrated by examining the length of time to transmit a CD-ROM disk containing 660 MB. At 1.2 Kbits/sec it will take 1222.2 hours or over 50.9 days; at 1.544

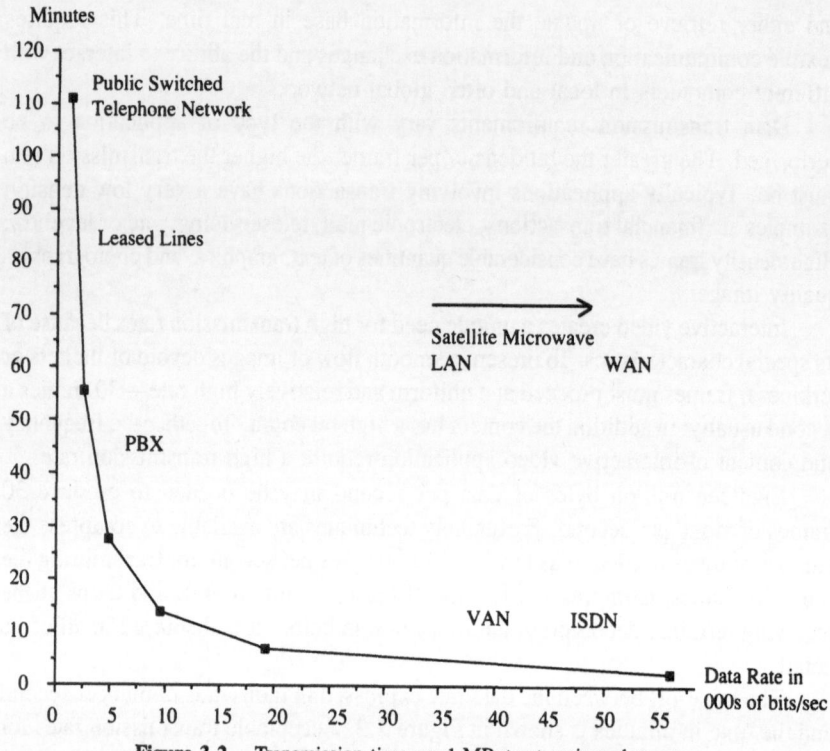

Figure 3.2. Transmission time per 1 MByte at various data rates.

Mbits/sec the time falls to 57 minutes. Obviously the lower speed is unacceptable; especially when overnight express mail guarantees next-day delivery for under $15.

3.1.5. Operational Readiness

The military uses the term operational readiness to indicate its ability to respond in the appropriate way to whatever situation presents itself. The combination of applications and technologies that comprise an enhanced system demand a degree of readiness beyond that found in a typical data-processing department. Operational readiness in such an environment combines system availability and reliability, system consistency, ease of expansion and contraction, and security.

Availability measures the time a system is operable; it can be expressed as the ratio of operating time to the sum of operating time plus downtime. Written in terms of the mean time between failure (MTBF) and the mean time to repair (MTTR), it is

$$\text{Availability} = \frac{\text{MTBF}}{\text{MTBF} + \text{MTTR}}$$

To illustrate, if the MTBF for a microcomputer is 2 years or 17,520 hours and the MTTR is 3 days or 72 hours, availability is 0.996. If the user has sufficient need and a spare microcomputer can be delivered in a few minutes, availability is 100% for all practical purposes. Unfortunately many other components beside the microcomputer are needed; consequently availability of the system is the product of the availability of each component, including the communications link and any special equipment, such as a voice message unit. Depending on which components are needed for a particular application, system availability will change.

Reliability is the probability that the system will continue to function under given conditions for a specified period of time; the formula is

$$R(t) = e^{-bt},$$

where b is the inverse of the MTBF, and t is expressed in hours. To illustrate, if the MTBF is 17,520 hours and the transaction requires 2 minutes, reliability is 0.999998. If there are multiple components, overall reliability is the product of the reliability of each component.

Availability and reliability can be improved by providing backups, good and responsive maintenance, and fault-tolerant hardware and network equipment. A related issue is system consistency: Response time should not vary appreciably at different times of the day. If systems are bridged through gateways; or if a variety of applications are supported, commands should not be dissimilar; for example, a navigation command to advance a frame should not be the key labeled next on one system and the plus symbol on another. Or the function keys on a keyboard on one desk should not be on the side but across the top on the keyboard at the adjoining desk.

Computer systems, especially those supporting a variety of applications, consume varying quantities of resources depending on the type of application and the time of day. It is important for the system to be able to meet these temporal distortions; even more important is the ability to expand or contract the system based on current or anticipated needs without making major changes in the equipment or converting the software.

The system should be secure from unauthorized access to the information base, software, and communications network, and rapid, low-cost content creation must be possible, especially for applications with a heavy graphic or photographic content.

3.2. CONFIGURATIONS

There are almost an infinite number of ways of organizing components of an enhanced system, especially when there is a computer network. An attempt is made here to present three all-encompassing categories: dedicated, shared, and public. A single microcomputer that is not networked is dedicated.

Table 3.5 compares these categories according to the extent to which the

Table 3.5. Relative Comparisons Among Basic System
Configurations When a Computer Network Is Involved

| | Configuration | | |
Factors	Dedicated	Shared	Public
Tailored to applications	Best	Adaptable	Worst
Access	Selectable	Limited	Unique
Cost	High	Low	Medium
Control	Best	Medium	Worst
Speed/response	High	Variable	Low
Security	High	Medium	Low

system can be tailored to meet specific needs, the type and ease of access, capital and operating costs, the degree of control exercised by the service provider, the speed of the communications link, and the level of security attainable.

3.2.1. Dedicated Systems

A system dedicated to supporting enhanced services and some of the possible user interfaces are shown in Figure 3.3. The three features separating it from a typical data-processing installation are the special equipment needed for content creation, the gateway to information providers, and the variety of user interfaces. In a business environment where the system is used only by employees, only the microcomputer, interactive video display, printer, keyboard, and audiovisual interfaces may be present. If the system supports a lobby or convention center information system, a video display with touch screen, and speaker would be appropriate. In an application where information and services are being sold, a money deposit, card reader, and a ticket and coupon printer might be added.

Unlike shared and public systems, a dedicated system can be tailored specifically to the applications it is expected to support. Access to information and services can be selected on the basis of what is needed rather than trying to live with what is already in place; for example, if interactive video is one of the applications, a high-speed LAN may be required; if sound is important, the system may have to be integrated with the organization's telephone network, or terminals with voice synthesizers may have to be provided.

Because a dedicated system is specialized and will not be shared, capital costs are high; however the ability to optimize the system's performance should lower its operational costs relative to a system that tries to be all things to all people. It is much easier to establish operating parameters, such as hours of operation, type of applications that will be supported, the kind of terminals that can be connected, and so on, for a dedicated system as well as modify the system to meet changes in user needs.

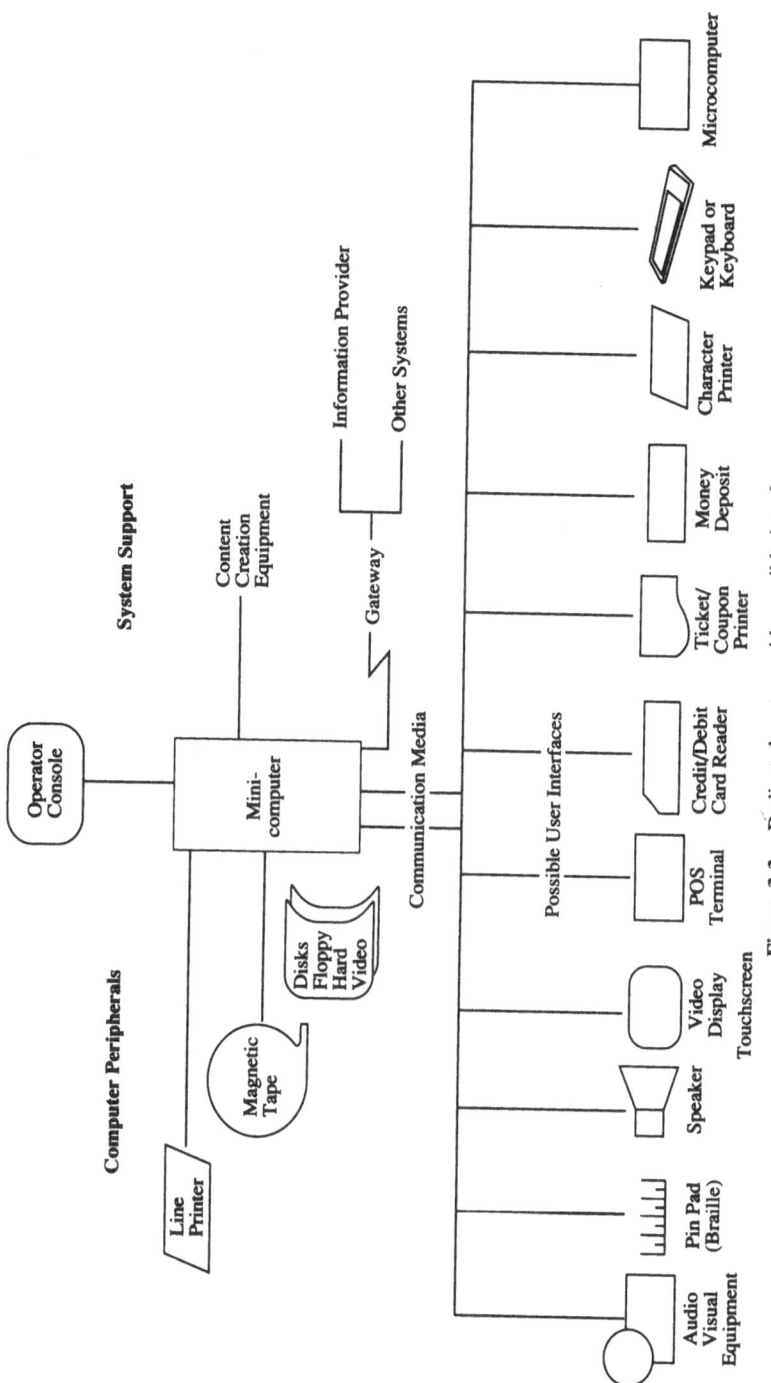

Figure 3.3. Dedicated system with possible interfaces.

Whether a system is accessed frequently by users often depends on the response time. A dedicated system configured to provide a particular level of service, has a more predictable response time than one that may occasionally be bogged down with large data-processing applications involving time-consuming number crunching and file-handling jobs. It is also possible to maintain tighter security because encryption capabilities and special equipment designed to restrict unauthorized access is more likely to be a standard feature of the system. Chapters 14 and 15 contain several cases describing dedicated systems.

3.2.2. Shared Systems

A shared system resides on a mainframe that also supports general data-processing needs. A front-end processor (FEP) is normally included to manage the terminal network and provide such features as error detection, automatic recovery procedures, buffering, code translation, protocol conversion, and other activities that reduce the work load of the mainframe. Terminal response time may even be improved if the FEP provides some preprocessing functions, and the total cost of delivering a frame is reduced if the FEP is capable of adapting to a variety of protocols, networks, and terminals. A shared system with a selection of possible user interfaces is shown in Figure 3.4. Due to the size of the computer, the need for environmental controls, and its use for data-processing applications, many of the interfaces found in dedicated systems, such as pin pads, card readers, and ticket/ coupon printers may not be supported, although technically the system could include them. Chapters 13 and 14 contain cases describing how Clark Equipment Services and E. I. DuPont De Nemours operated services for customers and employees, respectively, in a shared environment.

Access is more constrained in a shared system than in a dedicated system because the communication facilities are usually already in place to support existing applications. Access to external infobases may be especially difficult if it is accomplished through the mainframe, although this does not have to be the case. There are software packages available that eliminate the need to install special protocol conversion equipment or to use special terminals just for external access; one example is ITEX.25, a commercial package from Videodial, Inc., that is described in Chapter 5.

The key to whether a shared system is cost effective is how well enhanced features can be integrated with the existing applications. If the performance of the system is not seriously degraded and only minor additions of hardware and software are needed, a shared system is quite economical. There are few capital costs, especially if the existing terminals and communications facilities can be used, and the operational expense of adding the enhanced applications are negligible.

The availability of a shared system depends on the priority of the applications. In a bank, the support of on-line financial transactions takes precedence

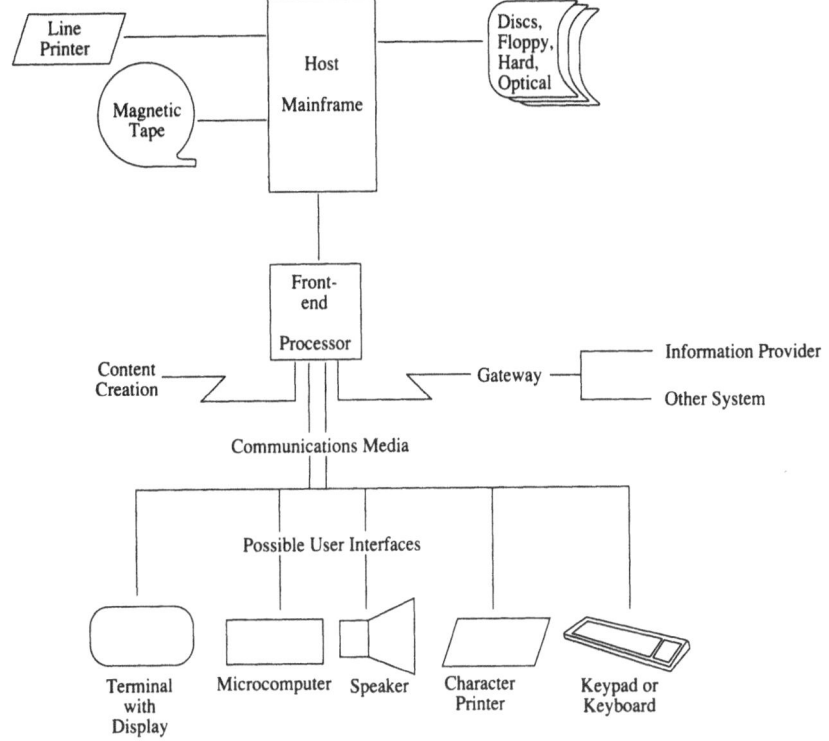

Figure 3.4.   Shared and public systems.

over all else; during class registration, a computer is first and foremost dedicated to meeting the registrar's needs in a university. Speed and response time are directly influenced by the priority established for the enhanced applications relative to the other applications that share the system.

Security is not generally a problem on a shared system, since users for most enhanced applications are employees or authorized customers. However since the organization's database is generally on the same system, the opportunity for mischief is always present; as organizations open their systems for access by the public, the risks will increase. An example of how a company could open its system to the public to promote its public image is explained in Chapters 10 and 11.

3.2.3. Public Systems

A public system is operated by an entity called a service provider, and it is usually made available to organizations and the general public for a fee. There are many services of this type in Europe; examples are Prestel and Teletel in Great Britain and France, respectively; in the United States, CompuServe, Prodigy and

Heartland Free-net fill this role. When only one company is involved, service providers often refer to the application as a private service. Once multiple organizations or people at different locations are involved, it is sometimes known as a closed user group (CUG) or special interest group (SIG). User interfaces that are practical for most applications are shown in Figure 3.4. Since the system is maintained by another entity, the possible applications are limited to what the service provider is willing to support. However the ability and willingness of service providers to customize services for private use is likely to improve as open network architecture (ONA) and basic service elements (BSEs) are fully implemented by the regional Bell operating companies in the mid-1990s. This topic is more fully discussed in Chapter 5.

Access on a shared system is constrained by the public communication services available between the service provider and the customer. In most instances, communication takes place over the public telephone network, public packet data network or data services sold by cable television companies. The former severely limits multimedia applications requiring a wideband, such as video and some graphic applications. The one-way and limited two-way services provided by a majority of the cable systems requires the telephone or some other form of communication to be the uplink to the service provider, which both complicates and slows system performance.

An organization operating under the auspices of a service provider, such as Prodigy, may enjoy several benefits; one of the most helpful is access to a large repository of information. If the organization operates as a CUG on the service provider's system, equipment and communication costs are usually minimal; there is also access to the technical prowess of the service provider for content creation and marketing. Chapter 14 contains cases describing how Ford and American Airlines (AMR) operate within the rubric of a public system.

It is almost always an advantage to have another party saddled with responsibility for dealing with the complexity of operating the system, but this places the ability to dictate what will be offered, when it will be made available, and the terms to be met in someone else's hands.

For many applications the speed of a public service is irritatingly slow. A dial-up procedure takes between 30–60 seconds, and there is always the possibility during periods of peak activity that response times will be unacceptably slow; in some cases, all the lines may be busy. Once access is gained, transmission rates for the public telephone network are painfully slow and as already mentioned, some enhanced applications, such as motion video, cannot be supported. Integrated services digital networks and other communications services described in Chapter 6 reduce the severity of these problems, but most of these services will not be available to the general public until the mid-1990s.

The most feared and least discussed concern associated with using a public system is security. Regardless of the safeguards provided by the information carriers and the service providers operating the system on which it is stored, organizations will always be reluctant to move sensitive information outside the organization where it becomes more vulnerable to scrutiny by a competitor and the public.

3.3. ARCHITECTURE

The term architecture refers to the organizational structure of an entity, such as a computer, a communications system, or an infobase; here the term refers to the basic organizational structure of the computer that makes enhanced services possible. The three architectures discussed here are known as uniprocessor, multiprocessor, and fault tolerant. Fault tolerance may be present in both uni- and multiprocessor systems.

3.3.1. Uniprocessor

The simplest definition of a processor is a device that carries out operations on data. How it accomplishes these operations and what specific hardware and software are involved vary among vendors and lead to lively discussions of just what really constitutes a processor. Here a uniprocessor architecture implies that whatever constitutes the processor, it works by receiving data, manipulating it, and producing a solution, usually based on an internally stored program. An example of a uniprocessor computer is shown in Figure 3.5.

The precedence chart illustrates the tasks involved in a particular job that has been presented to the computer and the order in which they must be done. For example, task AB requires two time units and must be completed before BE can start; BE requires three time units. Task AB uses data element A as input to produce solution B as output, which in turn is the input for task BE.

A sequence of tasks is shown in Figure 3.5a for the uniprocessor. These tasks are to be handled serially, with the processor working on one task at a time. This is referred to as serial processing and represents the predominant architecture since the introduction of electronic computers. Most computer systems work this way, especially smaller micro and mini systems. In this example, 27 time units are needed to complete the job. The design and performance of the hardware and software for such systems are well understood. The systems can be implemented with far less difficulty and at much lower cost than multiprocessor machines, but they lack the computational power, availability, reliability, and ability to support as many simultaneous users as their more complex brethren.

3.3.2. Multiprocessor

A multiprocessor is a computer with two or more processors that may share memory and other system components; multiprocessors are generally divided into

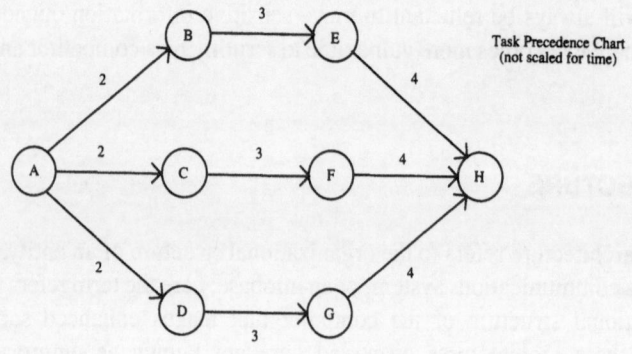

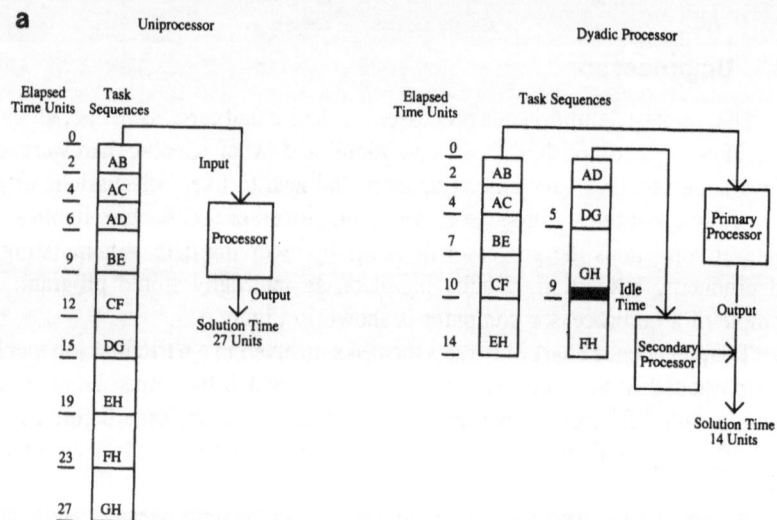

Figure 3.5. Computer architecture: (a) uniprocessor and (b) multiprocessor performance.

two categories: those designed to run a single job across multiple processors and those that run several jobs in parallel on separate processors. The former speeds up single-program processing, and the latter speeds up the total throughput of many jobs. Whichever category is employed the result is greater performance. Figure 3.5b illustrates the impact of two and three processors on the job that was executed on a uniprocessor. A dyadic system reduces solution time by almost half; three processors cut the time to exactly one-third, but the improvement comes at the expense of far greater system complexity. Note that the dyadic processor has forced idle time because task FH cannot start until CF is completed on the primary processor. In the three-processor system, tasks can be divided conveniently into three separate segments of equal lengths, and none of the segments have

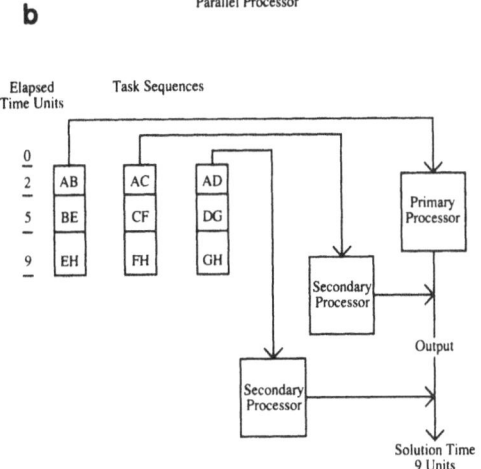

b
Parallel Processor

Figure 3.5. (*Continued*)

interdependencies—an unlikely occurrence in most jobs. A multiprocessor system is referred to as tightly coupled if the processors share memory and other components equally. Generally the more tightly coupled the processors, the greater the degree of system complexity.

Reliability is another major benefit of a multiprocessor system. In the event one unit fails there are sufficient resources to permit the system to continue operating, although at a reduced level of performance.

3.3.3. Fault Tolerance

A system is fault tolerant if it has features to ensure the integrity and continuous availability of data and programs despite hardware and software failures. In systems that support on-line transaction processing, service users who have an immediate need to know, or serve employees or processes that must have constant computer support, fault tolerance is essential.

There are many reasons why a system fails in addition to the obvious causes of disk and processor failures. There may be errors caused by the human operator; various software bugs; new operating system releases; and problems introduced by external sources, such as lightening bolts and computer hackers. It is estimated that downtime causes may be distributed as follows: hardware, 20%; system software, 15%; inability of recovery software to effect automatic recovery, 35%; and operator errors, 30%. Interestingly these figures reveal that between 30% and perhaps as much as 65% of the causes are beyond a designer's control.

What do these figures mean in terms of actual downtime for uniprocessor non-fault-tolerant systems? Many systems boast of reliability around 98.8%. If the computer is operating 24 hours a day, 7 days a week, it will have a mean time between failures approaching one month. If there are two computers in the network, two failures can be expected in a month. In a corporate network of 10 non-fault-tolerant computers, a portion of the network will be inoperative every three days. Examined in this way, 98.8% reliability does not look high enough. It has been calculated that by 1995, non-fault-tolerant systems will cost U.S. businesses a total of 40,000 person-years annually.

Fault-tolerant systems can impact these negative statistics significantly; for example, the AT&T 3B20D is designed so that the expected amount of accumulated processor downtime does not exceed an average of 2 minutes per system per year. Fault-tolerance is necessary for the mission critical systems that are a fundamental part of doing business today. These are applications that automate the essential activities of an organization's operations, as opposed to its clerical, back-office functions; examples are just-in-time inventory control, on-line financial and brokerage services, direct order entry, reservation systems, and automated information and support systems in general. When these systems cease to function business comes to a halt, employee activities are redirected, business is lost, and frequently customers are angered.

When systems were operated in batch mode, that is, all transactions are saved and processed at one time, system failures were more a nuisance than a potential catastrophe. If a computer crashes halfway through a batch job, the program is simply restarted when the system is repaired. But for a system operating in a transactional mode, that is, processing transactions as they occur, making the system operational again takes on the urgency of dealing with a patient who has had a cardiac arrest. Recovery involves restoring the database as well; this means reconstructing the status of all transactions completed, half-completed, and not completely entered at the moment the system crashed. The problems can be severe. If a financial system crashes as a customer is taking money from one account and putting it into another, how does a banker tell the customer the money is missing from the source account but was never received in the target account?

Research on the impact of computer outages was conducted in 1987 among 1444 corporate executives; 160 usable responses were received. More than 85% of the organizations for which data were reported indicated either a heavy or total dependence on computing facilities. Figure 3.6a illustrates the average percentage of daily revenue lost each day a computer was not working. By the sixth day, the typical company experienced a loss of almost 25% of its daily revenue; by the twenty-fifth day, the loss climbed to 40% of average daily revenue. The last figure equals a loss of over $5 million, assuming 250 working days a year and annual revenues of $250 million. Computer outages also increased costs for performing basic business functions and had a very deleterious impact on the ability to perform basic business functions the longer the outage lasted. The severity of this impact is shown in Figure 3.6b. The extent to which the functions are impaired may not be critical or total; Figure 3.6b shows the progressive impairment over time as a percentage of the responding companies.

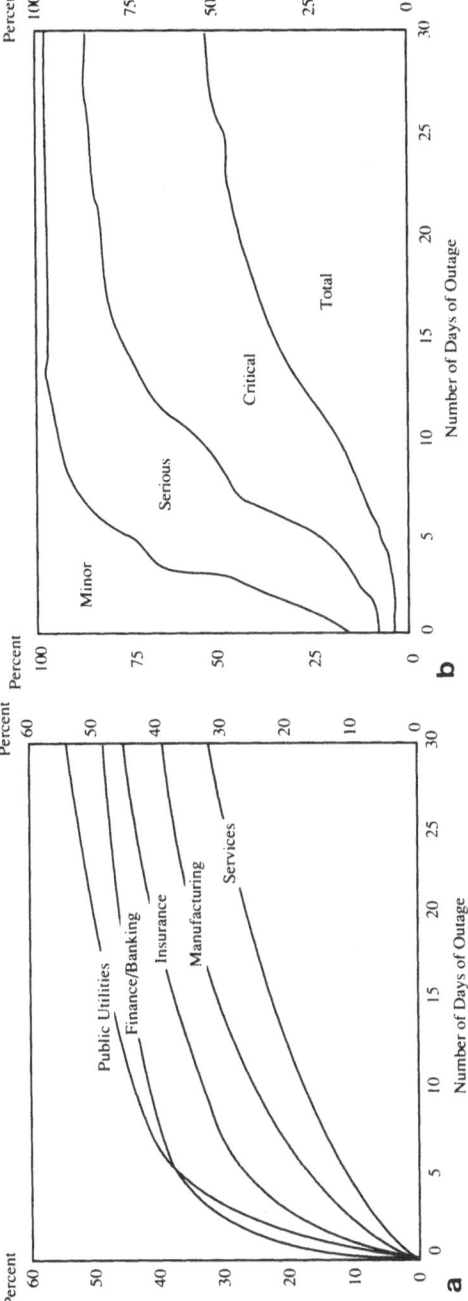

Figure 3.6. Financial and functional impacts of computer usage. (a) Average percent of computer usage. (b) Loss of capability to perform basic business functions. [S. R. Christensen, L. L. Schkade, and A. Smith, *Financial and Functional Impacts of Computer Outages in Business*, Center for Research on Information Systems, University of Texas at Arlington, Arlington, TX (1988).]

Fault-tolerant systems require far fewer service visits because many failures can be handled either by the system itself or by personnel sufficiently trained to exchange equipment modules as directed by the system. In most cases, a fault-tolerant computer continues to function during the diagnosis and repair processes.

3.3.3.1. Requirements

To be considered fault tolerant, a system must be able to detect, isolate, and recover from a fault; these capabilities are implemented differently depending on the vendor. Ideally the computer is designed in such a way that the failure of a single component will not disable the system. If a component fails, normal operation should continue with no loss of data or noticeable interruption. A prime objective of fault-tolerant vendors is to provide this capability with a minimal addition of hardware, without adding significantly to the complexity of the operating system, and by making few or no modifications to the user's application programs.

There are a number of ways of detecting faults. One of the most common uses error-detecting codes as data are being stored and moved from one location to another. These strategies are built around traditional parity-checking techniques, and they may include error-correcting code schemes as well. Hardware duplication and double checking are also popular. Protocol monitors may be present to monitor signals and measure the time between various events; an alarm is generated if deviations from the norm occur.

Faults are isolated as soon as they are detected. Systems with built-in redundancy accomplish this by deactivating the failed component; for example, if a system uses disk mirroring, or shadowing, a procedure where the system maintains a mirror image of disk data by automatically writing it to a second disk, the redundant disk will enable the system to continue operating. A feature called checkpointing or watchdog is present in some systems; in this process, processors running separate programs send messages to each other describing their internal status. If one processor becomes disabled, its partner is notified and picks up the work load.

Once a fault has been detected and isolated, recovery can begin automatically. In some cases, the machine examines its current status, determines what resources remain, and configures the system to make effective use of what remains. Protecting the integrity of the database is especially important, particularly when financial transactions are involved. The concept of atomic transactions originated to deal with this need.

Atomicity was mentioned earlier in this chapter in connection with storing information. It is implemented by defining a sequence of related database actions. These actions are "bracketed," and the system either ensures that it completes all of them or retraces its steps and ensures that none of them are affected. One popular technique is a write-ahead log. The before and after versions of records are written to a mirrored disk log file. If the system is successful in making the change in the database, a success is noted in the log file. If the system is not successful in making the change in the database, perhaps due to a hardware failure, the system examines the log file when restarted. Transactions in the log file that are not flagged as a success are undone by using the before and after images. If the system has multiple processors, transactions can be undone or backed out on-line. Uniprocessor systems usually have to be manually restarted before attempting the back-out.

Another recovery technique is called roll forward. This is used when the on-line database is lost completely. The system periodically dumps the database to a backup tape. When the system is restarted, the tape rebuilds the on-line database, and transactions that were completed after the tape was created are added from the log file.

Providing recovery procedures makes the system more complex and increases the cost to design and build the computer; but the trade-off of downtime, dissatisfied customers, and idled employees make the extra effort and expense worthwhile.

3.3.3.2. FT Architecture

Clever architectures have been designed by manufacturers of fault-tolerant computers. These architectures are significantly different from systems that are not fault tolerant; several factors account for these differences. First fault-tolerant systems normally incorporate two or more processors. This contrasts with most computer systems commonly sold today, which are uniprocessor based. This is especially true of smaller computers. Second the instruction set, input/output procedures, and other operating elements are often different from those found in non-fault-tolerant systems.

Figure 3.7 illustrates the system architecture for the Stratus Continuous Processing System. The system uses a replicated strategy in which the processors, buses, memory, and various controllers are duplicated (see Figure 3.7a). A level of backup is provided by the hardware that performs a comparative analysis of the duplicated components; an example of this appears in Figure 3.7b. During operation all activities, including computation, storage, and input/output, proceed in parallel. Each board checks itself for hardware errors during each cycle, as shown in Figure 3.7b. If the comparison results in different outputs, the faulty board is immediately isolated while the duplicate continues at normal speeds. The faulty board is later repaired during the maintenance cycle.

Disk drives operate with the same information. Whenever an element of data is written on one unit, it is also written on the other. When a read operation occurs, it is normally on the disk whose read-write heads are closest to the data in order to

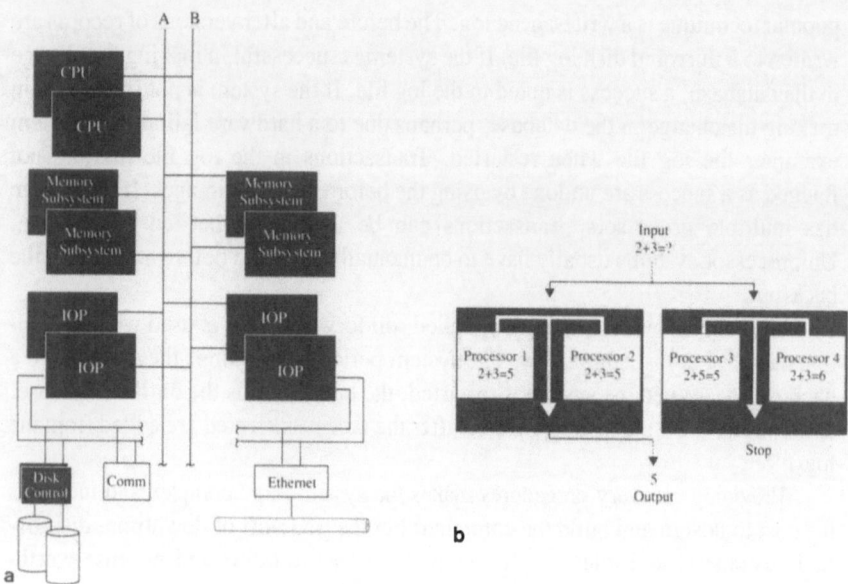

Figure 3.7. Stratus continuous processing. (a) System overview. (b) Hardware checking duplicated components. [Courtesy of Stratus Computer, Marlboro, MA (1991).]

minimize access time. If a disk or channel failure occurs, the operating system uses only the good disk and channel until the problem is repaired. The system then automatically restores the disk content as a background process.

The following process occurs when a failure is detected in a component. The unit immediately removes itself from service, illuminates a red light on the board, and transmits an interrupt command to the operating system. An analysis of the fault is performed automatically to determine if it is transient or permanent. If it is transistory, the board is restarted; if permanent, the board or device is kept out of service, and failure notice is sent to selected terminals and logged in a system file.

A remote customer assistance center is notified automatically of hardware errors via a dial-up modem. If a problem is software related, service technicians from the customer assistance units can often rectify the problem by logging onto the system, studying the problem, and making the necessary corrections. In the event of a hardware fault, the diagnosis is confirmed by analyzing the error log, and a replacement part is shipped by courier. On arrival of the part, the customer installs it. Slots in the chassis are clearly marked, and parts are keyed so that they can be inserted in only one way. After the part has been installed, a self-test is automatically performed to verify the revision level and operation. Synchronization occurs after the test has been completed successfully.

The Stratus fault-tolerant systems may differ in terms of its engineering from systems produced by other manufacturers, but the principles on which it is based are similar. Like its counterparts, it has extensive fault-detection features; it can continue to operate after single- and even multipoint failures; there is a large and well-developed remote customer assistance organization; and the customer is partially responsible for making repairs at the board level. In the words of an old advertisement for wristwatches, fault-tolerant systems are designed to "take a lickin' and keep on tickin'."

3.3.3.3. Trials and Tribulations

On-line transaction processing is expected to play an increasingly larger role in data-processing shops. It is estimated that by the early 1990s as much as 75% of all processing will be on-line. The reasons are apparent: Automated teller machines are becoming the predominant way for the customer and financial institution to interact; manufacturing and monitoring devices are the mainstay of the reindustrialization efforts in the United States; and point-of-sale registers have replaced mechanical registers and cigar boxes of an earlier age. Devices around us are tied directly to computers and depend on their continuous and uninterrupted support.

Most on-line applications are being implemented on uniprocessor mainframes; usually transactions are captured only on-line. Later during the night, they are processed in a batch mode to update various files. This style of processing promotes a tremendous degree of lethargy in the systems area and a resistance to switching from the tried and true hardware and software of major vendors to the smaller and less well-known vendors of fault-tolerant machines designed specifically for on-line transaction processing.

Fault-tolerant computers were introduced to the business world in 1976 by Tandem Computer Systems. Within 4 years, there were more than a half-dozen vendors offering systems. Most of the companies were entrepreneurial start-ups, and they have had a difficult time surviving. Fault-tolerant vendors have experienced two major difficulties. Product delays have been the rule rather than the exception. Although most fault-tolerant systems are based on off-the-shelf microprocessors, complexities introduced by the architecture are considerably greater than in their uniprocessor cousins. Software, as usual in most new computer systems, has been the biggest difficulty. Some vendors have started from scratch and developed totally new operating systems, while others have tried to modify existing software. The goal is to provide software that will make fault-tolerant features transparent to the application software supplied by the user.

Obtaining financial support has been made more difficult because of the long lead times and the inability to demonstrate working models. The lack of a demonstrable product and company track record makes potential customers reluctant to place orders, and a lack of orders compounds the problem of finding ongoing financial support—a true catch-22 situation.

In addition to all of these difficulties, fault-tolerant systems suffer from a three-pronged image problem. First they are viewed as very complex. This is a valid assessment, but complexity should not be apparent to the user in a well-designed system. Second fault tolerance is considered specialized. This may have been a disadvantage before the demand for on-line transaction processing, but a computer's ability to continue functioning under adverse conditions is rapidly becoming a necessity. Third fault-tolerant systems are considered as expensive. Redundant hardware and sophisticated software add to costs; some estimates place the incremental cost between 30–60%, but this may overstate the true picture. A dual-processor, fault-tolerant computer may in fact be less costly than two comparable uniprocessor systems; of course, if only one processor of the dual-processor system is used at a time, then the system is indeed more expensive.

3.3.4. Production System for Content

A tremendous range of production equipment exits for producing content. Since the content may range from simple text to full-motion video, there are many variations in the type of equipment and how it is configured. There are four principles that should guide the selection process: off-line production, automation, modular configuration, and shared operation.

Whenever possible the equipment should be capable of off-line production and down loading the content. This will reduce communication costs if the production facility is separated from the server hardware; it will also free the server for its intended function—delivering the content to the users. The production process should be automated to the greatest extent possible, and equipment should be chosen that contributes to this goal. A few dollars may be saved by picking one type of equipment over another, but labor costs, as explained in Chapter 9, are the principal expense in content creation and updating, and they considerably outweigh equipment costs in the intermediate and long terms.

A modular design is desirable for two reasons. As the infobase evolves, new applications will be added that may require new forms of content; e.g., photographic images will later be complimented by motion video. The ability to add to the equipment as the infobase evolves is less expensive than switching production systems. Second as technology changes, components can be replaced with ones that possess new features. In production systems containing a full complement of peripheral equipment, portions of the system can be shared. Many of the components cannot be used simultaneously by the same operator, consequently they are available to others. This will help spread the cost of what is sometimes a fairly expensive work station among several different people.

Figure 3.8 illustrates a fully configured work station for producing a multimedia infobase. For a variety of reasons related to communication bandwidth limitations and capabilities of the typical user terminal, discussed later in the book, this system would most likely be used for producing content for public

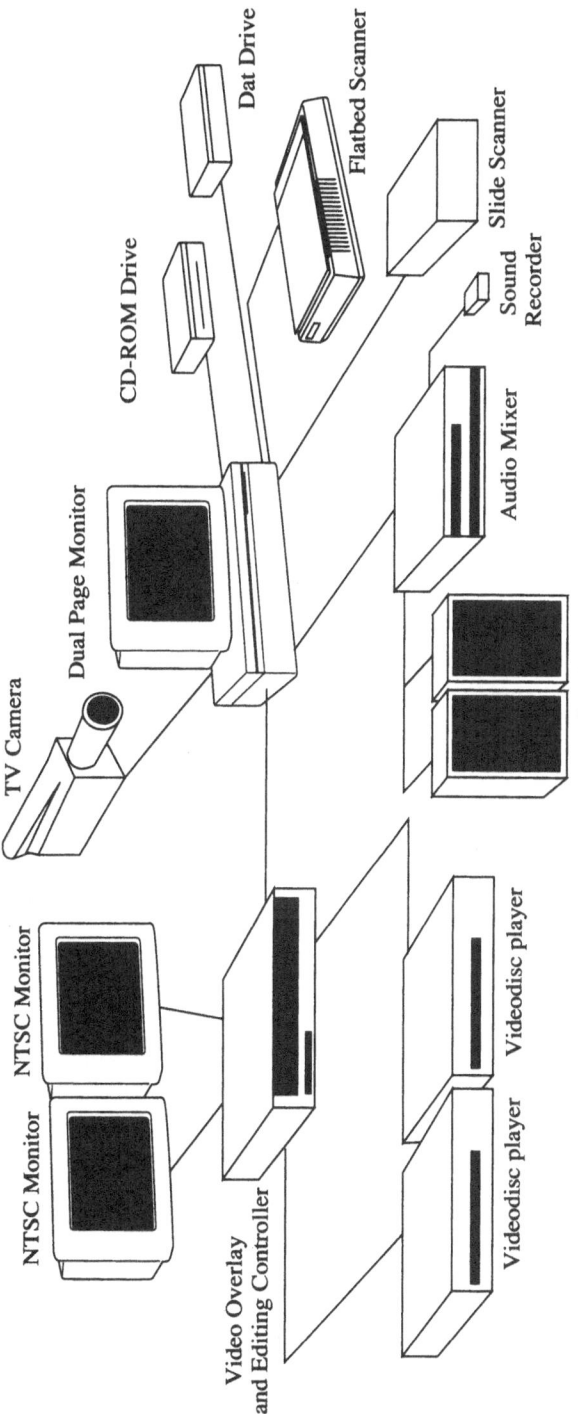

Figure 3.8. Fully configured work station for producing a multimedia infobase.

access terminals. Examples of these systems and their contents are discussed in Chapter 15.

The heart of the production system in Figure 3.8 is a powerful microcomputer with at least 300 MB of storage on the hard drive. The unit will have to have expansion slots to accommodate the peripheral equipment shown, and it will contain cards for digitizing video, processing sound, and for videographic output. Additional storage and material for assisting in the creation of content are contained on an attached videodisc, a CD-ROM drive, and a DAT drive. These devices are essential for integrating clip art, pictures, and motion video into the product being produced.

Scanners for capturing images on paper products and slides are useful attachments. A video camera for capturing both motion video and still frame images will be necessary. A device for overlaying and editing video will be needed if videotapes are used. Similarly an audio recorder, mixer, and speakers will be needed to support sound. If music is added to applications, the unit will require a Musical Instrument Digital Interface (MIDI) and synthesizer. Last a monitor that is similar to what the user will view the information on is necessary to monitor the production process and the quality of the output. In addition to the hardware in Figure 3.8, each of these components will require special software that normally comes with the device. Also a high-capacity back-up system with a fast transfer rate is desirable. The total cost for this system could exceed $75,000. For a more complete description of possible system configurations for multimedia content, see the Rosenthal article in the references at the end of this chapter.

3.4. THE ART OF OBFUSCATION

Measuring the performance of an on-line transaction-processing system is important for several reasons. First it establishes a basis for making comparisons during the acquisition process; it is important for capacity planning and deciding on when to expand the system, and it is important for balancing system components. Comparisons have always been difficult and error prone with conventional systems, and they are even more difficult with on-line systems. Transactions are usually of short duration; transaction-processing programs are generally small and do not require computational horsepower. The performance of a transaction-processing system instead depends on input/output facilities, the efficiency of the system and database software, and the data communications architecture.

Unfortunately when system comparisons are made, it is usually on the basis of processor speed measured in millions of instructions per second (MIPs) or a comparable statistic; examples of measurements used are listed in Table 3.6. The term MIP grew out of a study performed by an IBM employee, J. C. Gibson, in 1960. He performed several dynamic instruction traces on programs written for the IBM 650 and 704 computers. The traces recorded the type of instruction as it was

Table 3.6. Measurements Used
in Comparing Computer Performance

Measurement	Explanation
BIPs	Billion instructions per second
BOPs	Billion operations per second
MIPs	Million instructions per second
MFLOPs	Million floating point operations per second
TPS	Transactions per second

being executed. From these studies, he developed what was called the Gibson Mix, which enabled him to compute the execution time of an average instruction. The reciprocal of this statistic became known as KIPs (thousand instructions per second), the number of average instructions executed per second. By performing dynamic instruction traces on a large and representative body of programs, it was possible to determine how often a particular type of instruction was likely to be used in typical programs. Based on the frequency and execution time for each instruction class, the execution time was computed for an average instruction. A better measure of performance for an on-line transaction-processing system is transactions per second (TPS) or TPS/processor when more than one processor is available. But this measurement as well as other simple indicators of performance are flawed.

It is difficult and perhaps even illogical to reduce the performance of a complex piece of equipment, such as a computer, to just one measurement; it is paramount to using the Environmental Protection Agency's miles per gallon as the sole determinant when buying a car. Processor speed is only one factor that needs to be considered; efficiency of the input/output facilities, software, and communications performance may be more important in some cases. An even greater problem is the fact that there is no uniform standard for calculating these simple indicators. The machine architecture of manufacturers varies, sometimes to a considerable degree. Some machines may have complex instructions hard-wired into a processor chip. In contrast other machines may have simplified instruction sets programmed into the chip that are then linked together to execute data, such as add two bits and skip next instruction or no-op, which does nothing. A microcomputer executing a loop of no-op instructions can produce a MIPS rating that would be the envy of a mainframe.

If simple indicators are not adequate for measuring performance, what is? Over the years, a number of benchmarks have been developed for evaluating various aspects of computer performance. A benchmark is a group of programs run on different machines to establish a basis for comparison. A benchmark for an on-line transaction-processing system consists of representative programs and data that exercise the steps required to process a transaction in a particular environment. For example, in the banking industry, a transaction for a withdrawal or deposit might involve four to six database accesses, some of which may require locking one or more records. The actual computation time is insignificant. As an illustration, assume there is a transaction that has no computation; it simply reads 100 bytes from a teller's terminal, does a one-record read-modify-write to each of three files, and follows up with a write to the history file. The transaction is concluded with a 200-byte message sent and displayed on the teller's terminal.

Two of the best known benchmarks are the Whetstone and the Dhrystone. The former was developed in the early 1970s; it is heavily biased toward numerical computing and floating-point operations, and it is representative of scientific and technical applications. The Dhrystone was first reported in 1984; it emphasizes the type of data and operations encountered in system programming rather than numerical programming, e.g., control statements, procedure calls, and function calls.

A benchmark developed expressly for transaction processing is the debit-credit benchmark. This benchmark assumes the equivalent of 10,000 bank tellers performing an account update transaction once every 100 seconds, resulting in a system load of 100 transactions per second. Furthermore 95% of all the transactions must have a response time of 1 second or less. Response time is defined as the elapsed time between the arrival of the last bit from the communications line and the issuance of the first bit in response. Interestingly most results to date have been obtained at a scale factor of one-fifth normal load, something not revealed by many vendors.

Benchmarking is superior to using simple indicators that measure processor speed, but it is not without problems. The way that input is selected for the benchmark package of programs provides plenty of room for creativity; for example, a UNIX-based system spends approximately half its time on system control functions, whereas many benchmarks focus on user activity. Many benchmarks downplay input/output, but this can be a significant portion of the time consumed by a typical transaction, especially for bit-intensive applications, such as retrieving images from a picture infobase. There is no independent authority to verify performance claims, and manufacturers who so vociferously make the claims are hardly disinterested parties. Another problem is the existence of multiple versions of even standard benchmarks, and when a benchmark is run, it may be under a variety of conditions, e.g., with or without registers, single or double precision, with or without key subroutines. There are many other more subtle ways that benchmark results can be distorted; for example, different releases of the same compiler, the disk transfer rate, using main memory on scaled down versions instead of disk memory, which would be required in actual practice, can all influence the outcome.

An organization called the Transaction Processing Performance Council was created in 1988 to add some objectivity to the process of benchmarking computers. In 1989 the council released the first of what is a family of benchmarks; TPC

Benchmark A. This standard states specific rules about transaction arrival times among other parameters, leaving vendors less room to practice creative manipulation of the input. However despite the Council's efforts, there are no benchmarks of which this author is aware for adequately evaluating computers that manipulate and process the content found in videotex and multimedia applications.

Benchmarks can provide insight about the general capability of a system, but their results should be viewed with a healthy dose of skepticism. The closer the benchmark matches the actual environment in which the computer will be used, the more valid the results will be, but it is always a good idea to ask who ran the benchmark and who paid for it. A clever design can obfuscate the real results.

3.5. SUMMARY

The arcane world of computer hardware is shrouded in technical jargon and complex engineering principles. Such terms as multiprogramming and multiprocessing seldom mean exactly the same thing to different people. On-line transaction processing systems add another level of sophistication to the computer environment.

Enhanced systems are often OLTP systems with additional features, such as integration of heterogeneous technologies, extensive connectivity, special storage and information handling needs, a variety of communication facilities, and a high-level of operational readiness. Enhanced systems can be configured as dedicated, shared, and public systems. Each of these configurations has features that make it especially appropriate for meeting a particular set of objectives.

A computer system can be categorized according to the organizational structure of its components, referred to as its architecture. Enhanced systems are often fault tolerant; that is, they possess an architecture designed to ensure the integrity and continuous availability of data and programs despite hardware and software failures. Comparing systems is an imprecise art due to differences in architecture. A variety of simple indicators, such as MIPS and TPS, exist, but the best approach for comparing the performance of different systems is to use a set of programs and a common database known as a benchmark.

REFERENCES

An Overview of Stratus Computer, Inc., Stratus Computer, Marlboro, MA (1991).

D. Ben-Aaron, One Man's MIPs Is Another Man's Megaflops, *Information Week* (094) (1986), p. 6.

P. Borsook, What Parallel Processing Can Offer Information Networking, *Data Communications* **16**(1) (1987), pp. 121–131.

R. Bowers, CD-ROM Catches on: Optical Technology Hits the Desktop, *Computer Publishing Magazine* **5**(7) (1990), pp. 22–4, 32, 34, 36.

J. Canning, Mastering CD ROM, *Info World* **14**(1) (1992), pp. 55–56.

C. Chen, CD-I and Full-Motion Video, *Microcomputers for Information Management* **8**(1) (1991), pp. 53–57.

S. R. Christensen, L. L. Schkade, and A. Smith, *Financial and Functional Impacts of Computer Outages on Business*, Center for Research on Information Systems, University of Texas at Arlington, Arlington, TX (1988).

Corporate Guide to Optical Publishing, Dataware Technologies, Cambridge, MA (1989).

P. Freibarger, CD Rot, *MPC World* (June/July 1992), pp. 34–37.

H. Fukunaga, Complex Counter Type Videotex Terminal, *Fujitso Science & Technology Journal* **21**(5) (1985), pp. 527–537.

L. Green, How to Rate Benchmarks, *Information Week* (146) (1987), pp. 36, 40.

D. A. Harvey, State of the Media, *Byte* **15** (1990), pp. 275–277, 280–281.

G. Held, Is There a Place for Optical Disks in Data Networks? *Data Communications* **16**(5) (1987), pp. 173–174, 177–178.

S. Holst, Photo/Video Imaging: Into the Database Mainstream, *Advanced Imaging* **7**(3) (1992), pp. 20, 22, 24.

E. Horwitt, Fault Tolerance for the Masses, *Business Computer Systems* **4**(8) (1985), pp. 52, 54, 56, 60–61.

Lies, Damned Lies, and Benchmarks, *Mini-Micro Systems* **19**(15) (1986), p. 50.

R. Lockwood, Multimedia in Business, *Personal Computing* **14**(6) (1990), pp. 116–123, 125–126.

R. Lockwood, Size up CD-ROM, *Personal Computing* **14**(7) (1990), pp. 70, 73, 75–77, 79.

J. Maguire, Movies on Your Screen, *Computer Graphics World* **15**(2) (1992), pp. 51–53, 55–56.

H. A. Maurer and I. Sebestyen, Inhouse versus Public Videotex Systems, *Computer Networks* **7**(5) (1983), pp. 329–342.

W. McClatchy, Will New Technology Bring Interactive Video to Life, *PC Week* **4**(21) (1987), pp. 53, 66.

S. Morse, Multimedia on the Network, *Network Computing* **3**(4) (1992), pp. 58–59, 62–67.

Multimedia PC Hardware Specification, Version 1.0, Multimedia PC Marketing Council, Washington, DC (1991).

G. Newson, Roll Your Own CD-ROMS, *New Media* (May 1992), pp. 9, 11.

The Optical Storage Primer, Hewlett Packard, Palo Alto, CA (1990).

R. Paske, Hypermedia: A Brief History and Progress Report, part 2, *T.H.E. Journal* **18**(1) (1990), pp. 53–56.

R. Paske, Hypermedia: A Progress Report, part 2: Interactive Videodisc, *T.H.E. Journal* **18**(2) (1990), pp. 90–94.

R. Paske, Hypermedia: A Progress Report, part 1: CD-ROM, CD-I, DVI, Etc., *T.H.E. Journal* **18**(3) (1990), pp. 93–97.

D. Rosen, GE/RCA Announces DVI, *CD-I News* **1**(6) (1987), pp. 1, 4–5, 8–9.

S. Rosenthal, Setting up a Multimedia Production System, *MacWeek* **5**(22) (1991), pp. 26–27.

L. J. Seltzer, Floptical Disks Allay Compatibility Fears, *PC Week* **9**(14) (1992), pp. 79–80.

O. Serlin, Fault Tolerant Blues, *Datamation* **31**(6) (1985), pp. 82–84, 86, 88.

O. Serlin, Exploring the OLTP Realm, *Datamation* **31**(15) (1985), pp. 60–62, 64, 66, 68.

O. Serlin, MIPS, Dhrystones, and Other Tales, *Supermicro Newsletter* (April 1986).

O. Serlin, Debit Credit: A Standard? *FT Systems Newsletter* (July 1986).

O. Serlin, Measuring OLTP with a Better Yardstick, *Datamation* **36**(14) (1990), pp. 62–64.

D. Simpson, Multiprocessors Use Radical Architecture, *Mini-Micro Systems* **19**(7) (1986), pp. 77–78, 81–82, 84, 87–88.

G. Smith, and R. White, Multimedia PCs Arrive; Windows Apps to Follow, *PC Computing* **4**(9) (1991), pp. 74, 78, 80.

B. Waring, Snapshots on a Disc, *New Media* (1992), p. 14.

Why Fault-Tolerance Is Imperative, *Business Software Review* (April 1986), p. 6.

XA/R Model 300, Stratus Computer, Marlboro, MA (1991).

Chapter 4

The Workbench

In his keynote speech at the fall 1990 COMDEX meeting, Microsoft Chairman and Cofounder Bill Gates captured the essence of computing and the user interface in the 1990s with his theme of information at your fingertips: "Information at your fingertips exemplifies the concept of making computers more personal, making them indispensable, making them something you reach for naturally when you need information."

In Chapter 4, the term workbench describes the specific place where a job is performed and the equipment and communication services available at that place. The term is reminiscent of an earlier era but accurately reflects the fact that people today work in locations that are often highly structured and organized around a tool set, which, is this instance, is computer based.

Equipment may be an unsophisticated terminal with limited capability connected to a telephone line, or it may be an interactive video terminal with a built-in microcomputer connected to a fiber-optic communications network (see Figure 4.1). The sophistication of the equipment depends on the features it contains. In general the more features integrated within the equipment, the more functions it is capable of performing.

Chapter 4 makes the reader aware of important workbench features beginning with the user's equipment. Quite a few options exist for equipping the workbench ranging from a dumb terminal to a videophone. The task of deciding what equipment to use is made more difficult when there are multiple users, such as in a public location. (See the section on public access terminals for a discussion of the important issues.) The difficult task of tying all the systems together can be made more manageable by developing an initial policy statement governing the acquisition of system components.

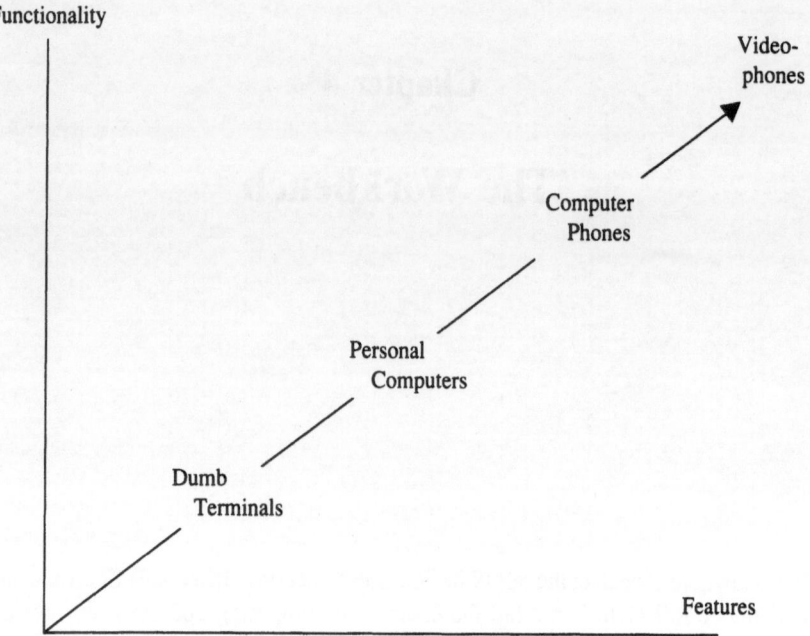

Figure 4.1. Workbench equipment and the relationship among equipment, features, and functionality.

4.1. EQUIPMENT FEATURES

The overriding consideration in selecting equipment is that it adhere to the principles of functional simplicity and technological integration and serve informational and transactional needs. These features are often packaged in an environment that supports images and graphics, points elaborated on in Chapter 2.

Functional simplicity was defined as equipment that is easy-to-use and does not require formal training. Equipment should be able to incorporate different technologies to the degree required to support the activities carried out at a particular workbench. If a manager must be able to access information through the telephone network and in-house computer resources through a local area network, the same equipment and procedures should be followed. The ability to serve informational and transactional needs reflects that fact that people no longer deal with data alone when accessing a computer; they must often retrieve graphics, text, and pictorial displays or engage in financial and messaging transactions.

The general system features just outlined for equipment are implemented in varying degrees. In the business market as in architecture, form should follow function. If the application requires static displays with limited graphic needs, it is neither practical nor economical to install a full video interactive system. The guiding principal is to ensure flexibility and the ability to adapt as needs change. There is no single piece of equipment appropriate for all occasions; instead the system designer must review the user's needs to select the features that are necessary. In an enhanced environment, the features discussed in the following sections would be considered.

4.1.1. Color

Color is essential for conveying information; in fact color itself is informational: It can highlight a point or differentiate among related but separate issues. Color can focus a viewer's attention on a particular area of the screen, and it can suggest priorities, designate responsibilities, and alert viewers to impending problems.

Color monitors like a color television set are not usually sought by the user until he/she has sat in front of one for a few minutes. Color then assumes the same importance in the user's mind as having a window or adequate desk space. Users believe color creates less strain on their eyes; they seem to learn faster, and they definitely find color more aesthetically pleasing than monochrome displays. For graphics-based applications, color is very helpful, and if a color printer is part of the system, the color monitor reduces the number of drafts that must be produced prior to finalizing the document.

There are a variety of color monitors available, but the most popular ones conform to the Enhanced Graphics Adapter (EGA) and Video Graphics Array (VGA) standards. Differences between these standards and some common terminology are shown in Table 4.1. To display more than one standard, some monitors support variable scanning frequencies. The scanning frequency, also called the scanning rate, is the speed at which the electron gun producing the image on the screen scans across the screen to produce one scan line. Scan lines render the text and imagery on the screen. As an illustration, the time to produce one scan line in a television display is 63.5 μsec; therefore, in one second 15,750 scan lines are produced for a frequency of 15.75 KH$_z$. Increasing the frequency and hence the scan lines produces a higher resolution, which is why a VGA display has a higher frequency than a CGA display and thus a higher resolution (see Table 4.1).

There are many variables to be considered when evaluating a color monitor, such as the length of time it will be viewed each day, room lighting, the viewer's distance from the display, type of application, and the importance placed on conveying meaning to the information through color. Often the only way of deciding between a color and a monochrome monitor and of choosing among color monitors is by a side-by-side comparison. Even after such a comparison, the performance of the eye and the psychology of using color to convey intended or unintended biases must be weighed. These issues involving content creation are raised in Chapter 10.

Table 4.1.   Display Terminology

Term	Explanation
CGA	Abbreviation for color graphics adapter, which has a 320-by-200 pixel resolution for color and a 640-by-200 pixel resolution in monochrome and a frequency of 15.75 KH$_Z$. The CGA offers four colors from a palette of 16 but text is fuzzy due to its 8-by-8 dot cell.
Dot pitch	Distance between phosphors of the same color (dots) on the display screen. The smaller the dot pitch, the closer the dots are to each other and the higher the resolution.
EGA	Abbreviation for enhanced graphics adapter, which has a 640-by-350 pixel resolution and a frequency of 21.85 KH$_Z$. The EGA offers 16 colors on screen from a palette of 64, and text is created from an 8-by-14 dot cell.
MDA	Abbreviation for monochrome display adapter, which has a 720-by-348 pixel resolution and a frequency of 18 KH$_Z$. Text is created from a 9-by-14 dot cell.
Multisync monitor	A monitor that can automatically adjust to any scanning frequency within the range of the equipment to which it is attached; e.g., 55–90 H$_Z$.
Pixel	Abbreviation for picture element. A physical pixel is the smallest displayable unit on a screen. A logical pixel is a geometric construct associated with the drawing point. Its stroke width determines the width of a graphic primitive.
Refresh rate	Number of times the scan lines comprising a display are repeated. A television picture in this country is refreshed 60 times a second.
Resolution	Number of distinguishable pixels per linear unit of measure.
Super VGA	An extension of the video graphics array, but there is no industry standard. A common pixel resolution is 800 by 600 for color and 1024 by 768 for monochrome. It may have an almost unlimited palate of colors and a higher frequency than VGA.
VGA	Abbreviation for video graphics array, which has a 640-by-480 pixel resolution for color and 720-by-400 pixel resolution in monochrome and a frequency of 30.5 KH$_Z$. The VGA offers 256 colors from a palette of 262,144.

4.1.2. Graphics

Computer graphics were introduced in the late 1950s. In their original form, these consisted of two-dimensional monochrome displays in which the graphics artist connected points and drew wire diagrams. During the 1960s and 1970s, technological advances coupled with declining costs led to improved displays, new input/output devices, and communication services that provided the throughput needed for the higher byte count required by graphics.

Throughout most of this period, computer graphics has been primarily limited to technical applications usually found in engineering and science. However the proliferation of microcomputers has brought graphics to the desktop to support general business applications often involving the visualization of data. These applications range from illustrating data as tables and charts to the interactive analysis of statistical relationships, such as in a regression analysis. User demand has been the major factor influencing the decision to buy computer graphics, followed by a desire to explore new technology. This demand has been helped along considerably by the ease of using graphics. Until recently producing business graphics required a great deal of special training and creative skill. Pull down menus, well-documented manuals, extensive help utilities, and a telephone number to call for technical assistance have changed the situation, particularly when these features are coupled with other business applications, such as spreadsheets, financial-modeling and statistical packages, database software, and word processing. Standard packages now offer pie charts, histograms, line graphics, and even 3-D displays.

According to a study by Peat Marwick on the competitive benefits of 3-D computing, 3-D displays can be a very cost-effective medium; for example Chrysler Motor Corporation forecast a 20% reduction in the time needed to develop a new car. Other companies in the study eliminated jobs and speeded up design and management decisions. These results have been verified by other research. A study at Xerox Corporation's Palo Alto Research Center on graphical user interfaces found 3-D graphics an aid in helping users visually track information more rapidly than 2-D techniques. In 1991 3-D revenues were expected to account for 25% of the value of work station shipments, and it was estimated that the market would grow at an annual rate of 24% during the next five years.

But the personification of graphics is the ubiquitous graphical user interface (GUI) popularized by Microsoft Corporation's product *Windows* and Apple Computer's Macintosh machine. The GUI, which rhymes with phooey, is popular because it is easy to use, provides a common interface across programs, and it gives users a greater sense of control than an esoteric operating system language like MS-DOS. The basis for these claims rests on many studies. One comprehensive study conducted by Temple, Barker & Sloane, Inc., in 1990 with 120 subjects reported that novice users were 48% more productive and experienced users 58% more productive than their counterparts who were using a CUI (character-based user interface). GUI users also reported less frustration and fatigue.

Standards are usually the hallmark of a mature and relatively stable technology. The limited number of formal standards for graphics suggests that the graphics industry is in the early stage of its development. The Graphical Kernel System (GKS) is a standard for two-dimensional drawing initially developed in West Germany. The American National Standards Institute issues this as Graphical Kernel System (GKS) Functional Description (ANSI X3.124-1985). The Association for Computing Machinery has promulgated the Core System, which supports two- and three-dimensional line drawings as an alternative to GKS. Advocates of GKS and CORE defend their respective graphical system with all the fervor of religious zealots.

The North American Presentation Level Protocol Syntax (NAPLPS; ANSI X3.110-1983) is specifically designed for integrating text and graphical information. It is based on the principal of blind interchange, a desirable feature when a variety of equipment and information services are to be accessed. Blind interchange means that the sender and receiver of the information do not need any prior agreements in order to interchange information. The only requirement is that both parties agree to conform to the standard. This standard is becoming widely used, especially where public information systems or public access terminals are used. A summary of the functions and requirements of NAPLPS is shown in Table 4.2.

Many other graphical systems are in the process of becoming standards, and many companies have systems unique to their company. The absence of a single standard makes selecting hardware and software more difficult because a number of factors must be considered for each graphics application. Several of these factors are

Drawing ability required
Availability of help aids and art work that can be integrated from a disk-based library
Ease of use
Resolution and color
Supported input and output devices
Integration with such popular software packages as Lotus 1-2-3
Data-handling attributes for uploading and downloading to other computers

4.1.3. Voice and Data Integration

Adding voice capability to a device provides a unique way of increasing employee productivity. Such a device is equivalent to a Rolodex, telephone-answering device, note pad, typewriter, telephone, word processor, calculator, slide projector, and computer combined into a single piece of equipment. There are a variety of ways in which this integration can occur. The most common involves physically joining a telephone with an intelligent terminal or a personal computer. This marriage provides telephone management (cost-effective routing, number dialed, length of call); voice communication (voice storage and forwarding); and data communications (automatic dialing, log-on, and session management). At the same time, this integration enables the user to interact with a host computer or perform activities associated with a personal computer, such as spreadsheet calculations and word processing. The integrated voice data market is one of the fastest growing segments of the information industry, and this market provides features that are becoming standard on some equipment.

4.1.4. Audio

People generally consider their eyes and hands to be the human interface to a computer, in many environments and applications, their ears and mouth would be more effective. Sound is an older and often a better method of communicating than sight, and it will become increasingly common in business systems. Sound is

Table 4.2.   Summary List of NAPLPS Functions and Requirements

Functions	Requirements
Character sets	94 primary (ASCII) Characters, 94 supplementary characters
Geometric drawing primitives (PDIs)	Point; line; arc; rectangle; polygon; incremental point, line, or polygon
Domain	Single-value operand length, multivalue operand length, two- and three-dimensional modes
Text	Rotations in four directions; character path movement in four directions; fixed and proportional character spacing; four interrow spacings; variety of cursor styles; column and row specifications; 40×24, 40×20, 40×10, 32×16, 20×10
Texture	Solid, dotted, dashed, and dotted-dashed lines; solid, vertical hatching, horizontal hatching, and crosshatching textures; highlight in all colors
Color	16 simultaneous colors from a palette of 512
Blink	16 blink processes
Wait	Duration of minimum wait interval between 0 and $\frac{1}{10}$ sec, inclusive
Mosaic set	65 block-style graphic characters
Macro set	96 definable macros
Dynamically redefinable character set (DRCS)	96 DRCS characters whose patterns can be downloaded from the host
Physical display	Minimum resolution is 256 pixels horizontal by 200 pixels vertical

not commonly found in most equipment today because the technology is expensive and difficult to implement. As systems evolve, voice mail, voice annotation, and voice-based access and control features will emerge.

Audio has an avaricious appetite for computer memory. The IBM Audio Capture and Playback Adapter card has the ability to digitize audio in three modes, depending on the quality desired; these modes are voice, music, and stereo. The number of bytes of RAM required for each per second are 6K, 11K, and 22K, respectively. To put this into perspective five minutes of stereo quality sound will consume 6.6 MB of hard-disk space.

The three core technologies in sound processing are compression, synthesis, and recognition. Taking voice as an example, compression reduces the amount of information needed to store and transmit natural analog voice. The first step in the process involves sampling the amplitude or height of the analog signal created by the voice; some techniques sample the differences between adjacent points on the signal. The sample is then quantified by converting it into a digital code consisting of ones and zeros. This is followed by applying a formula to reduce redundancies and otherwise reduce the number of ones and zeros needed to maintain the information. The digitized and compressed voice data are then transmitted or stored. A computer cannot process the information contained in the data, but it can reverse the compression and digitization processes to convert the ones and zeros into a sound closely approximating the original voice.

Synthesis artificially recreates the sound by generating waveforms. The human vocal tract accomplishes this with only a few primary elements: The lungs provide the energy that moves air through the trachea; the vocal cords, tongue, velum, and lips assist in forming the sound. In a mechanically produced sound, quality and flexibility are important. The sound must be intelligible immediately, because if the listener requires time to understand the message, information may be lost. Flexibility refers to the ability of the device to synthesize output based on various forms of input—text, voice, analog sensor, system-generated indicator.

Converting text into speech involves several steps prior to the actual synthesis process. In a videotex system in which information has been organized into frames, a page reformatter retrieves the frame requested and reorganizes it into a form that can be read as speech; then a text-to-speech translator converts the reorganized frame into phonetic codes that drive a speech synthesizer. It is possible to listen to a sentence or a line at a time, then to move forward and backward in the frame.

Voice recognition is the least developed of the three core technologies. Voice recognition products are classified as either speaker-independent or speaker-dependent; the latter category permits greater vocabularies, and it is usually more accurate. Factors contributing to the difficulty of designing products based on voice recognition include (1) the changes that may occur to a voice due to illness and degree of alertness, (2) background noise, (3) running together words, (4) the length of time it takes for the system to recognize the word or phrase, and (5) the size of the vocabulary (10,000 words or more) needed for an office environment.

Voice recognition equipment for industrial applications is relatively common and more and more office products are beginning to appear; however for the foreseeable future, it will be overshadowed by voice-messaging systems that digitize sound and then reconstruct it on demand. Voice messaging was introduced in 1980, and it has become a viable alternative to cassette-type answering machines and receptionists.

Audio will make its presence felt as users adopt Multimedia Extensions for Windows. Three different forms of audio are used: CD audio, waveform audio, and Musical Instrument Digital Interface (MIDI). The CD audio is the same as the audio on the CDs that come from a music store; the sound is sampled at 44.1 KH_Z. Waveform is the most readily accessible because it is suitable for recording voice-over and voice mail; it has a lower sampling rate than CD audio. The third form, MIDI, is widely used for interfacing with computers; it consists of special protocols for music synthesis. The CD audio is designed for playback only, while the other two are used for both recording and playback. These three options complement rather than compete with each other.

4.1.5. Imaging

Imaging can consist of text with graphics, photographs, or a sequence of video frames. As the cost of computer resources declines and computers, especially microcomputers, are able to store more and execute instructions more rapidly, image processing will become more common. Capturing, manipulating, and transmitting high-resolution graphics, photographic quality displays, and video are especially important in many business applications. Desktop publishing, graphics presentation, computer conferencing, and computer-aided instruction are a few of the applications that depend on imaging. Images created for desktop publishing using popular software packages merge alphanumeric, graphic, and photographic materials. Graphics presentation might more appropriately be called a picture slide show. It replaces the overhead and 35-mm carousel projectors with a series of frames that may be preselected to fit the need, just as a special mailing list can be created from a much larger database of addresses by selecting a few key parameters.

Computer conferencing may involve the transmission of high-resolution graphics and pictorial quality displays using special terminals or microcomputers equipped with special software and chip sets. Freeze-frame and full-motion videos are also possible with television cameras connected to microcomputer-based terminals that transmit signals through the telephone network or high-speed data transmission services.

Computer-aided instruction supported by photo databases housed on optical discs or hard disks are becoming common. Other applications using photo databases are real estate, security, missing person searches, and similar applications that require photographs to be archived and retrieved on demand and often at scattered locations.

4.2. LIMITED-INTELLIGENCE TERMINALS

A terminal that can not be programmed and is capable of performing only rudimentary levels of a communications protocol is classified as having limited intelligence; such terminals are often referred to as "dumb terminals." Dumb terminals can be categorized according to the human sensory stimulates on which the unit depends for its communication; these categories are tactile, audio, and display. Tactile units rely on physical contact with the user; one example is a Braille pin board for the visually handicapped. Tactile units, infrequently found in business environments, are not discussed here.

Figure 4.1 shows the relationship between equipment features and functionality; obviously the more features supported by a terminal, the more functionality it has and the more enhancements possible. However whatever features are provided, they must be incorporated without compromising the human considerations involved in using the equipment. After all room lighting, physical discomfort, and the quality of the instruction provided can make a terminal just as useless whether it costs $10,000 or $300. Dumb terminals are divided here for purposes of discussion into audio and display based units.

4.2.1. Audio

The telephone is the quintessential device of audio technology, although it got off to a rocky start. In 1876 few people saw a need for Alexander Graham Bell's invention. The *London Times* called it "the latest American humbug." Western Union was offered exclusive patent rights, but its chairman quickly refused the offer saying, "What use would this company have for an electrical toy?"

The toy has emerged as a medium in its own right. It is well established with a 93% penetration in the consumer market and close to 100% in business. The telephone can switch from site to site, and it has addressing capability; next to satellite networks, it has the broadest coverage of any communication service. The telephone permits varying speeds and full duplex two-way communication. But perhaps most importantly, it has tremendous flexibility and the ability to support a variety of enhancements.

Telephone technology accelerated rapidly when the Bell system was broken up in 1984. Immediately following divestiture, customers returned 70 million leased telephones to AT&T, then began buying equipment and services from a host of new companies. A majority of the telephones purchased today have one or more sophisticated features. Some telephones can memorize a hundred numbers so that users dial only one or two digits or type in a name on an attached keyboard to make a call. Other telephones come with voice recognition devices so that the name or number can be spoken to the telephone. If the telephone network is also intelligent, the IQ of the telephone increases dramatically. Call blocking automatically intercepts calls from specifically identified numbers and informs the caller that his her call is not being accepted at this time. Another feature identifies designated callers with a special ring or by displaying the caller's number on a small screen. These are just some of the features that are available. The discussion of open network architecture in Chapter 7 provides a more complete list of many of the features embodied in the increasingly smart telephone network.

Telephone features are being integrated with computer technology in many new and useful ways; often the computer is an inexpensive PC. Some of the products that have been introduced are voice response boards that can answer a phone, playback digitized speech files from a hard disk, respond to signals created by a push-button telephone, and even record incoming speech. This capability is called interactive voice response (IVR), and it includes voice messaging, call directing, and audiotext.

Interactive voice response is not new. In the early 1970s, IVR systems were used by bank tellers to verify customers' account balances before cashing a check. From this humble beginning, the systems have evolved to permit callers with push-button telephones to order goods, retrieve information, and enter readings on utility meters. The newest capability enables callers to request information, enter their own fax number, and have the information faxed to them.

There are more than 50 vendors making IVR systems. A few vendors build their systems around propriety equipment, but the majority incorporate a standard Intel Corporation 80386- or 80486-based PC. This enables them to use off-the-shelf components, such as voice boards, to keep the price down. Organizations adopting IVR systems report considerable cost savings. One estimate is that an IVR system can reduce labor and other operating costs by 20% when the system handles between 10–25% of all incoming calls. Other estimates show that IVR systems have an 8-to-1 cost advantage over operators performing routine data entry and look-up functions because of staff reductions and the greater number of calls that can be handled.

Voice information services, also known as audiotext, are one of the most useful interactive services that may be accomplished with a telephone. Most of the services require a touch-tone telephone for input, but in some cases even a rotary or pulse dial telephone is adequate. Table 4.3 describes the five primary types of voice information services currently available.

To access information from an interactive service, such as Dowphone, subscribers key in their identification numbers on a touch-tone telephone. Important information of general interest is automatically read. If specific information is sought, the subscriber presses the appropriate key at the audio prompt; for example, stock prices are retrieved by obtaining a company identification number on-line or from a print directory issued to each subscriber and then keying it in.

The telephone keypad can also function as a data entry device. In 1992 for the first time, the Internal Revenue Service (IRS) made it possible to file federal tax returns by telephone: A trial conducted in Ohio enabled 23,000 people to file by dialing an 800 number and entering their W-2 information. While the caller waited, the refund or tax owed was calculated, and the caller informed. Filing time

Table 4.3. Audiotext: Public Voice Information Services

Service	Description
976 dial-it services	Information service billed at a flat rate (typical price is 50 cents). The service may be interactive—the caller specifies a birth date for a horoscope reading—or passive—the caller listens to a recorded message.
National 900 services	Nationwide version of the 976 service providing fee-based information and such services as polling by having the caller dial one number to vote yes and another to vote no.
Subscription services	Information services for closed user groups whose members "register." Stock quotes, financial information, and pricing data are typically offered.
Sponsored services	Similar to the three preceding services but free to the caller; access is sometimes provided via an 800 number. A sponsor, such as a governmental unit, medical group, or local newspaper, pays for equipment and information.
Classified advertising	"Talking yellow pages"; advertisers pay a fee to list information about their goods and services. Some services use live operators, but a number of interactive services also exist.

averaged five minutes, and refunds were mailed in 2–3 weeks, half the usual time. Fewer computational errors were reported; there were fewer penalties or fines for underpayment; less data entry time by IRS personnel was required; and tax-processing costs were reduced.

The backbone of an audiotext system is a voice mail processor. These processors incorporate telephone-line coupler cards that connect outside telephone lines with the voice mail computer and line cards that recognize touch-tones from the caller's phone. Line cards also can be used to digitize incoming voice calls for storage. The cards support the reverse flow of information by converting digital signals into analog voice. Line cards are connected to computer cards containing logic for assigning addresses, storing and forwarding messages, and issuing or recognizing commands. Figure 4.2 gives an overview of the hardware required for an audiotext system.

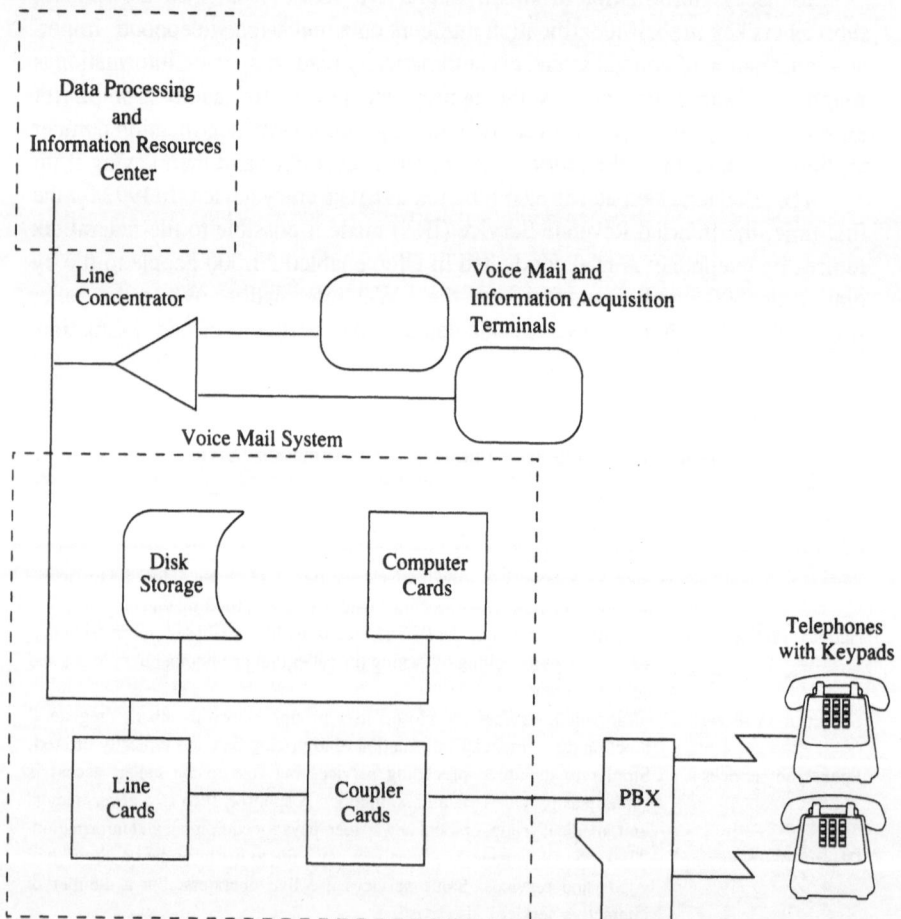

Figure 4.2.   Hardware overview of an audiotext system. (Corporate computer system.)

The Voice-Messaging User Interface Forum (VMUIF) of the Information Industry Association (IIA) developed a specification for voice-messaging user interfaces. The impetus for development of the document is the premise that a consistent interface in the industry benefits users in several ways: It results in faster learning, greater productivity, and greater satisfaction; for the industry, it can lead to wider acceptance and larger markets for products and services. Elements defined by the document are user control and feedback, phrasing statements, menu structure, assigning keys on the telephone pad, and procedures for handling errors.

4.2.2. Display

Equipment that depends on the visual senses can be divided into soft- and hard-copy devices. Interactive hardcopy devices are not so popular as their softcopy counterparts. The most common device is a printer terminal, which costs about the same as a visual display device with an attached dot matrix printer. A printer terminal creates noise, consumes paper, breaks down more often than a softcopy device such as a monitor, cannot support color in many cases, has a very limited graphical capability, is slower than a monitor, and requires considerably more maintenance because of its mechanical parts. Telewriting equipment and sketchphones, which permit the sender to write out a message and the receiver to display and copy it onto paper are even more constrained than a typical softcopy device like a monitor.

Softcopy refers to the presentation of information on a temporal medium, such as a video display terminal. There are many different display technologies— plasma panels, liquid crystals, super twist crystals, and thin-film displays—but the work horse is the cathode ray tube (CRT), which operates on the same principle as a television set (see Figure 4.3).

The five major elements that comprise a typical CRT device and their relative share of the total cost to manufacturer the unit are shown in Table 4.4. The shroud, or housing, accounts for about 22% of the cost of a dumb terminal. This percentage is based on the cost of a plastic composition with injection-molding and tooling costs amortized over a large production run. The power supply is based on a standard transformer. The CRT itself has the shortest life expectancy of all the elements and requires the most power to operate. The CRT also contributes the most to the overall size of the unit. The keyboard is assumed to be a standard typewriter keyboard with standard key sizes, spacings, and travel. The fifth major element is the electronic assembly. In a dumb terminal this may consist of a small number of computer chips; in an intelligent unit, the electronic assembly may involve memory elements, graphics chips, and other components that can increase costs significantly.

Equipment cost is relatively low because television sets operate on the same principles as a CRT device. Consequently the designers and manufacturers of CRTs have been able to benefit from the experiences of the mass market television

There are a variety of display technologies available but the one that is by far the most commonly found is called a raster-scan. It is created by positioning a cathode-ray (electron) gun in one end of an evacuated glass tube which has phosphorous at the other end. When a voltage is applied to the cathode-ray gun, a stream of electrons is produced that excites the phosphorous causing them to glow. The luminance of each spot, called a picture element or pixel, varies directly with the magnitude of the signal.

The stream of electrons moves from the upper left of the screen to the lower-right. As it moves across the screen it is deflected downward slightly so the lines are sloped. Each time the beam reaches the end of a line it is "turned off" or blanked out so it can be repositioned at the left side of the screen without exciting the phosphorous. This process is called the retrace interval. It also occurs when the beam is repositioned from the bottom to the top of the screen.

The number of lines on the screen is dependent on the horizontal and vertical scanning frequencies. If the two frequencies were the same there would be a scanning line running from the upper left corner to the lower right corner of the display. The greater the number of horizontal lines, the greater the number of separate picture elements that can be resolved and then the higher the resolution. In North America the standard (National Television Systems Committee) is 525 lines per frame.

A pixel begins to decay after it has been excited. The human eye does not detect this if the pixel is refreshed at lease 60 times a second. The characteristic of the eye not to detect this decay is called the persistence of vision. In order to transmit 525 lines 60 times a second, a very large bandwidth is required. However, a feature called interlacing cuts the requirement in half.

If 525 lines were transmitted per frame 60 times a second, the horizontal scanning frequency would be 31,500 lines per second. However, instead of transmitting 525 lines, only 262½ are transmitted (called a field) for a frequency of 15,750 lines per second. The next set of lines is positioned so it is displayed between those of the first set. This interlacing of lines coupled with the persistence of vision tricks the eye into seeing more than is really there. The two fields together are refered to as a frame, the interlacing of the two fields can be portrayed in the following way.

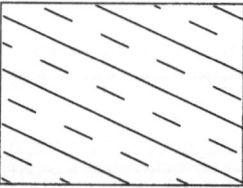

Field 1: Solid Line
Field 2: Dashed Line

Figure 4.3. How a CRT works.

industry. Like a television set, a CRT refreshes the display each time the beam sweeps the surface of the tube. This process requires a refresh store circuit of some kind that contains all the information on the screen. Another circuit, called a random access store, allows changes to be made in the refresh store. Both of these circuits feed a third circuit, called the character generator, which is responsible for translating digital bits into the text appearing on the screen.

Table 4.4. Relative Manufacturing
Costs of a Display Terminal

Components	Terminal Type	
	Dumb (%)	Intelligent (%)
Shroud	22	12
Power Supply	17	12
Display	17	12
Keyboard	22	14
Electronics	22	50
Total	100	100

There are several techniques used for character generation, but the most common is the dot matrix. Characters, such as the letter A, are translated into dots chosen from a matrix. A row of characters is built up on the screen a line at a time. If each character cell is 7 by 9 (seven columns, nine lines), a row of characters begins to take shape as the electron beam is turned on and off. A second line will have lighted and unlighted pixels. This is repeated for all nine lines, and the pixels that have been turned on over the nine lines produce the row of characters.

Graphics capability may be incorporated in three different ways: (1) A hardware device can be attached to a visual display unit; (2) an integrated chip can be built in; or (3) in the case of a microcomputer, a program can serve as the decoder. Each approach has its own set of advantages. A separate hardware decoder enables any acceptable display device, including a consumer television, to act as a full text and graphics unit. Incorporating the chip within the device at time of manufacture is the least costly approach in terms of the incremental cost involved, and it is more likely to yield greater compatibility and improved performance. A software decoder enables a microcomputer to be upgraded selectively and eliminates the need for additional equipment.

4.2.2.1. X Windows

Dumb terminals without processing and storage capability and microcomputers loaded with software are at two widely separated points on the spectrum of end-user equipment. A device called an X Window terminal has emerged on the scene to fill the middle ground. It is similar to a dumb terminal, and it must be connected to a host to function, but it has some of the features normally found in a microcomputer: It can access multiple applications, even on different hosts, and it has the ability to display simultaneously each application in its own window on the screen.

Unlike a dumb terminal, an X Windows terminal incorporates its own server, either in software downloaded from the host or in read-only memory in the terminal, which is the opposite of the normal concept of a server. In a local area network connecting PCs, for example, a server is a device on the network serving (providing data to) the PCs. The X Windows terminal server satisfies requests from the host to create and fill windows on the terminal according to the needs of the application.

Unlike a microcomputer, the X Windows terminal has no operating system; it cannot execute an application. Instead it depends on the host to do this and provide the results in a window. The X Windows terminal's limited computing ability is also highly specialized for creating windows on the display and filling them with the contents supplied by the host. The terminal cannot function like a general purpose microcomputer.

The history of X Windows technology is short. It was developed in 1984 at the M.I.T. Laboratory for Computer Science as part of Project Athena. Like so many organizations, the campus of M.I.T. was home to computers from a number of vendors. A project called Athena, sponsored by IBM and Digital Equipment Corporation, was funded to tie together various work stations and hosts across the campus running under the UNIX operating system. The concept was simple, but the task was difficult. The result of the project was first publicly demonstrated in 1986, and by 1989 a market had developed for the technology.

In 1989 there were 11,000 units sold and the following year 64,000. It is estimated that in 1991, the figure rose to 140,000 X Windows terminals and by 1995 611,000 units will be shipped. To help put these figures into perspective, there were 2.3 million traditional terminals (read ASCII terminals) shipped in 1990. An example of an X Windows environment is shown in Figure 4.4. It demonstrates how an end user can access multiple hosts over the network and display applications from each simultaneously. This approach has several advantages beyond the obvious one of providing a bridge to multiple hosts. The cost of an X Windows terminal is typically in the $1500 range, which is about half the cost of the average PC and considerably less than a workstation, and current prices are becoming competitive with ASCII devices. The X Windows terminals are quieter because they do not require cooling fans or disk drives; they take up less desk space; and since they have fewer components, they require less support. Since data and processing are centralized at a host, it is easier to control the integrity of the data and use of the system; it is also much simpler to upgrade one X-server than dozens of workstations. Some of the extensions being implemented include still-picture imaging and three-dimensional graphics. Software is also available that enables PCs to emulate X Windows technology. As this software evolves, a host of multimedia applications will be possible.

4.3. MICROPROCESSORS

The era of the microcomputer dawned in January 1975 with an article in *Popular Electronics* announcing the Altair 8800. By any standard today, the

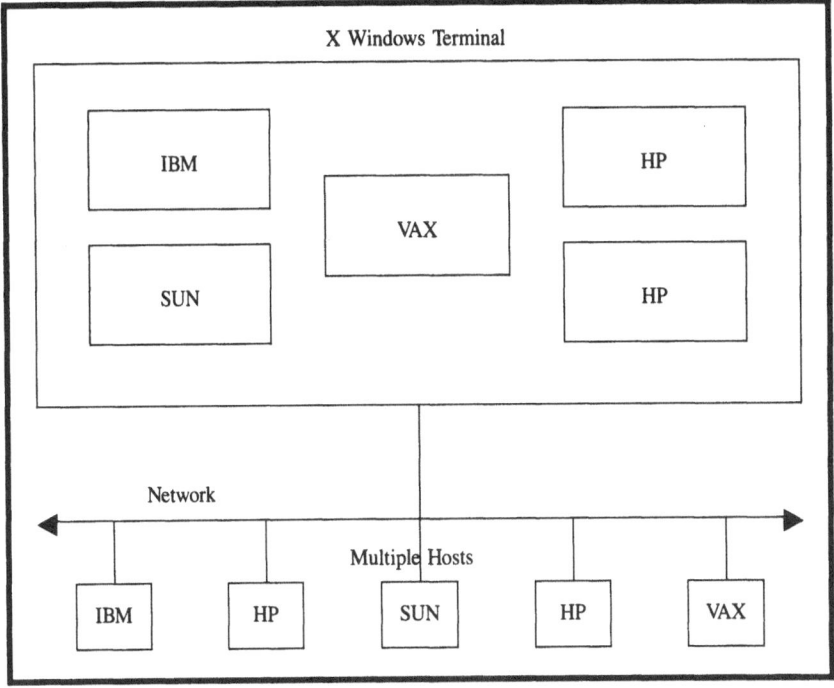

Figure 4.4. An example of an X Window environment with multiple hosts.

machine was a dinosaur: It was built around an 8-bit processor, and it had a maximum memory of 65,000 bytes. Among its list of possible uses were programmable scientific calendar, digital clock, autopilot for boats, and brain for a robot; there was no mention of spreadsheets, word processors or databases.

In the early years, the microcomputer industry focused on home hobbyists and special applications. However by 1981, the microcomputer began to receive plaudits from even its harshest critics as a tool for business. The introduction of the IBM PC on August 12, 1981, changed the landscape forever and legitimized the microcomputer as a worthwhile addition to corporate America. Initially IBM officials are reported to have estimated the company would sell about 250,000 units over a five-year period—a figure that fell far short of the mark. By 1984 IBM's share of the IBM PC-compatibles market had risen to a staggering 63%. However by 1991, it had fallen to 35% due to relentless competition by manufactures of compatible machines. Some of the highlights of IBM's first ten years in the microcomputer business are summarized in Table 4.5.

The price of microcomputers has reached a point where every business, regardless of its size, can afford a computer. In 1956 there were a total of only about 500 computers; in most cases, these machines had far less capacity than today's desktop computer. In 1957 more computers were shipped than existed in

Table 4.5.   Ten Years in the Life of the IBM PC

Year	Event
Aug. 1981	IBM unveils the IBM PC.
Nov. 1982	Lotus announces Lotus 1-2-3.
Mar. 1983	Compaq introduces first PC clone; IBM unveils PC/XT.
Nov. 1983	IBM unveils the ill fated PC Jr.
Jan. 1984	Apple introduces the Macintosh
Feb. 1984	IBM unveils the PC portable.
Aug. 1984	IBM introduces PC-AT.
Oct. 1985	Intel debuts 80386 chip.
Apr. 1986	IBM introduces PC Convertible.
Mar. 1987	Macintosh II debuts.
Apr. 1987	PS/2 debuts; IBM and Microsoft announce joint development agreement for OS\2.
Apr. 1988	Apple sues Microsoft and HP for copyright infringement.
Sept. 1989	Apple and Microsoft announce joint effort on fonts.
Nov. 1989	IBM relaunches OS\2.
May 1990	Microsoft ships Windows 3.0.
Jun. 1990	FTC probes Microsoft.
Jun. 1991	Microsoft ships DOS 5.0.
Jul. 1991	IBM and Apple sign letter of intent to work together.

Source: E. Scannell, The Secret History of the IBM PC "Gamble," *Info World* **13**(32) (1991), pp. 47, 50.

the preceding year. But the real growth came with the introduction of microcomputers for business applications in the late 1970s. In 1987 computer shipments in the United States were 5,745 times greater than they were in 1957, and 97% were microcomputers—nearly four million. In 1991 there were an estimated 60 million microcomputers in the United States. The average yearly decline of the cost of various system elements over the three decades ranged between 11–27%, a remarkable statistic in light of the 400% jump in the cost of living index during the same period.

Microprocessor systems in the form of desktop computers have a number of advantages: They provide functionality needed in different business environments; they contain extensive storage and processing capacity; they can be expanded easily; and they serve as the interface for a variety of equipment and services.

Microcomputers provide a level of functionality and performance not found in a dedicated system. A unit can support general administrative needs, such as word processing, spreadsheet calculations, and database manipulations, but the same device can also support many enhanced applications with the appropriate additions in circuitry and peripheral devices. This incremental functionality reduces manufacturing costs and eliminates the need to replicate equipment on the desk and to learn how to operate different types of equipment.

Business applications depend on storing, manipulating, modifying, and retrieving information to a much greater degree than do consumer applications, and microcomputers are ideally suited for this role. The eight-inch-diameter medium originally used with dedicated word-processing machines in the 1970s gave way to the 5.25-inch medium. It has been supplanted by 3.5-inch diskettes, optical storage discs, and built-in Winchester drives. Each change has been accompanied by increased capacity.

One megabyte of storage is equivalent to about 250 single-spaced typed 8.5-by-11-inch pages, with each page containing 4000 characters. Consequently a standard 5.25-inch floppy can hold 90 pages of text and a 3.5-inch diskette about five times that amount. An optical disc less than 5-inches in diameter can hold more than 250,000 typewritten pages. When erasable optical disc drives were introduced in 1988, they cost approximately $7000, a price that limited their use to commercial applications. But like their magnetic-based predecessors, their cost has fallen rapidly.

Perhaps the most important feature of a microcomputer is the ease with which it can be upgraded to incorporate new enhancements. Its modular nature makes it a simple matter for the user to insert new boards or add software to the base configuration. For example, there are a variety of products that support the implementation of the NAPLPS videotex standard, some of which sell for under $100 per installation; in several instances, information providers have even given away software to generate demand for their services. An additional advantage of many modular approaches is that they allow the user to switch back and forth easily between applications and in some cases, to run two applications simultaneously— one in the foreground and one in the background.

Thanks to IBM, Apple Computer, and Microsoft Corporation there are *de facto* industry standards that enable many vendors to develop hardware and software that will operate with popular microcomputers. This allows common interfaces to be developed and a sufficient base of potential customers to justify developing new products and services. Standards also permit software and information transportability and transferability, since it is considerably easier to transport software than a dedicated terminal, and it is easier to conform to a single set of commands to access centralized services than it is to train users on several sets of commands according to where they are working that day.

4.4. COMPUTER TELEPHONES

Computer telephones combine the two most useful tools on the desk: the personal computer and the telephone. This capability is becoming increasingly possible as the sophistication of the desktop computer permits the addition of relatively inexpensive integrated voice data components. There are three types of computer telephones: PBX propriety computer telephones that function with a specific PBX, standalone devices that may be attached to any telephone line or PBX, and PC add-ons for upgrading a desktop computer.

The propriety version is sold as part of a telephone system. It is protocol and software compatible with the telephone service it is designed to accompany, and as a consequence, the degree to which telephone functions are built in is the highest of the three types of computer telephones. For example, if a call-forwarding function is invoked by pushing a button on the terminal, the computer telephone will interact with the call-forwarding software resident in the PBX by using a specialized digital management protocol.

The greatest advantage of the stand-alone unit is the fact that it can fit into any telecommunications environment regardless of the level of telephone service provided. The stand-alone unit can be attached to any analog telephone line—PBX, Centrex, keyset, or single line. The third type of computer telephone, the upgraded personal computer, is least able to integrate all the telephone features of the PBX, but it compensates for this disadvantage by being easy to implement. In addition corporate workbench standards are not violated, and users do not have to learn a different keyboard or interface.

Physically a computer telephone closely resembles a PC, but a computer telephone often has an attached handset for privacy, a 9-inch or smaller display screen, and a floppy disk slot in addition to the built-in hard disk. The footprint is slightly smaller than a PC, and the unit typically supports high-resolution graphics and simultaneous voice and data transmissions. The latter feature enables a user to send information displayed on the screen and discuss its contents with the party on the other end of the line. The functions commonly found can be grouped into desktop services, telephone management, document integration tools, and external communications (see Table 4.6).

Desktop services are intended to help employees work more intelligently and more rapidly. They consolidate into one device the items commonly found on a desk, such as a clock, calculator, calendar, and Rolodex. Used in conjunction with a spreadsheet, database package, and project manager, these items can save considerable time and increase productivity of the typical worker. The telephone management functions are intended to assist in route selection, billing, looking up numbers and logging calls. Built-in speakerphones, menu-based call forwarding, transferring, and redialing can speed up routine activities and simplify doing infrequently performed ones.

Document integration tools help users package information in formats tailored to specific audiences. This may involve file and general information retrieval and transmission, presentation graphics, word processing, and different forms of input and output media. A host of features are provided to facilitate communication, such as automatic log-on with a single key, an extensive electronic mail capability, and multiple terminal emulations to enable communication to take place with different makes of equipment, facsimile, and even an answering machine.

The major focus of the computer telephone is on access, which distinguishes it from its competition, the applications-oriented personal computer. The computer telephone is especially appropriate for higher level and professional employees who often spend between 60–90% of their time exchanging information and ideas on the telephone; however integrated voice and data terminals (IVDTs),

Table 4.6. Typical Functions Supported by Computer Telephones

Function	Description	
Desktop services	Alarm clock	Notepad
	Calculator	Project/account manager
	Calendar	Rolodex
	Database	Time manager
	File storage	Spreadsheet
Telephone	Accounting (who, when, length)	Last-number retry
management	Call forwarding	On-hook dialing
	Call hold and transfer	Telephone log
	Directory	Speakerphone
Documentation	Multifunction operation to allow switching	
integration	between tasks without changing files	
	File retrieval and transmission	
	Presentation graphics	
	Word processing and text edit	
Communication	Automatic log-on, single-key access	Programmable soft keys for
	Electronic messaging	single-key access
	Integrated modem	Terminal emulations
	Facsimile	Variable-speed communications
	Answering machine	

such as the computer telephone have several shortcomings, the most significant of which is cost. Since a computer telephone often replaces an existing phone, clock, personal computer, and other desktop staples, it is an incremental cost of as much as $5000 per unit. Adding to the cost is the communications link. Since there are few integrated voice and data roadways presently installed to link the desktop with other users, it is usually necessary to install two access lines—one for voice and one for data. This is expensive and cumbersome to administer. Despite problems like these, the computer telephone will continue to be a contender for space on the desktop.

4.5. PICTURE AND VIDEO TERMINALS

Picture and video terminals provide the ultimate in display capability. Their photographic-quality displays fall into two categories—desktop presentation systems and teleconferencing systems.

4.5.1. Desktop Presentation

Desktop systems are at the upper end in the evolution in PC graphics, and they reflect the marriage of computer and video technologies to produce low-cost, high-resolution displays.

One form of desktop presentation system simulates audiovisual presentations. Depending on the size of the audience, the system can be situated on a desk or in an auditorium. There are sound business reasons for using multimedia systems to display information, especially for two or more people. In a 1989 survey of 200 corporate vice presidents by Motivational Systems, 43% of all business meetings were judged boring, and 40% of those surveyed admitted falling asleep during a presentation. In a study entitled "A Profile of Meetings in Corporate America" conducted by the Annenberg School of Communications at the University of Southern California, the most widely used presentation aids were paper handouts (47%), overhead transparencies (13%) and chalk boards (13%). However the study found that the most effective presentation aids in descending order were video, product samples, and tied for last place, computer-based aids and slides. Clearly presentation using audio, video, color, and so forth, is more likely to inform and stimulate the audience than a piece of paper.

There are two basic types of desktop presentation applications. The first is an audiovisual presentation with moving text and data, animation, sound, and television-quality video. Text, data, and animation can be supported on plain vanilla types of PCs with the appropriate software. Sound and video require more sophistication, and they considerably increase the cost of the unit; consequently these features are often found in public access terminals, where their cost is more easily justified, or in conference settings, where there are several viewers.

The second principal type of application for desktop video is supporting picture databases. This is an application where storing and retrieving photographic-quality imagery is important, such as in the real estate industry and fashion design. Picture storage and retrieval are also used for electronic encyclopedias, training systems, and for security clearance to restricted areas.

Presenting picture-quality displays as either one or more still pictures or in a video format requires special equipment to capture and store the images. To use a desktop presentation, format images and data must be captured through a combination scanning device, illustration software, and digitizing tablet. Once captured on the screen of the creation device, the images can be freeze framed, titled, cropped, restyled, colored, and enhanced with fades, dissolves, wipes, and animation. Borders, special fonts, background colors, and audio to accompany the pictures and charts complete the process.

Low-cost image capture is a crucial part of every system. Generically there are five types of scanners: optical character recognition (OCR), image scanners, image scanners with text-conversion software, software that performs image editing and text conversion, and scanners in fully configured desktop publishing systems. Text converters, commonly referred to as OCR scanners, detect an alphanumeric character, identify it, and then convert it into machine readable form. If the item to be scanned contains images and halftones as well as text, an OCR is inappropriate, and an image scanner with text-conversion software is needed.

Image-scanning systems work much like a copy machine. Photosensitive cells in the scanner record the appearance of the page by reflecting light off the paper and measuring the reflection of light back to the cell. White paper reflects the most light and black the least. The photosensitive cell measures the reflected image, and the scanning unit generates digital codes corresponding to the intensity of different portions of the reflected image.

If photographic-quality images are to be viewed at a rapid rate, as in a video display, a considerable amount of storage is needed in the terminal and a high-speed transmission link is required. For example, a studio-quality picture may use as much as one million bytes and require 111 minutes for transmission at 1.2 Kbits/sec. Storage and transmission requirements may be ameliorated in several ways: (1) The portion of the screen occupied by the photograph can be reduced, which is very common; (2) the quality of the resolution can be decreased; (3) the byte count can be compressed; and (4) encoding the image can be simplified. Whichever approach is followed, the fact remains that floppy disks and vintage transmission speeds from the 1980s are often inadequate.

Another approach to solving the storage problem involves the optical disc technology described in Chapter 3. It has been estimated that the entire holdings of the Library of Congress could be stored in 12 file-cabinet-sized drawers of discs. This high storage capacity is achieved by recording data at extremely high densities. A compact disc read-only-memory (CD ROM) can hold 660 M/bytes; in addition, a CD-ROM is removable, practically indestructible, and relatively inexpensive to replicate. Mailing a CD-ROM in an overnight delivery envelope is a very cost-effective way of storing and distributing information. Furthermore the ISO 6990 CD-ROM file format has become the industry-accepted standard, which, with other industry agreements, enables CD-ROM devices and microcomputers from different vendors to work together.

The Commodore Dynamic Total Vision (CDTV) unit combines elements of a CD-ROM and a television set. A CDTV unit resembles an audio compact disc player, and it can be connected to a television set or home stereo system. The CDTV contains the processing power of a small microcomputer and the interfaces for a keyboard, mouse, trackball, and disk drives. The unit runs multimedia software stored on CDTV discs. The discs can store 14 hours of speech-quality sound or up to 250,000 pages of text. The video, which is stored on the disc fills about one-quarter of the screen, in contrast to CD-I, which fills the entire screen. The CDTV was introduced by Commodore in April 1991. Another CD-related product is the Kodak Photo CD introduced in 1992. One Photo CD can hold approximately 100 photographs and future product upgrades are expected to support sound with the photographs enabling a verbal identifier to be stored with the image.

Perhaps the greatest limitation to the growth of full-motion desktop video is the lack of knowledge of what is available, how to bring together all the necessary pieces, and proof that it is worth the cost and effort. The first limitation will be overcome with time, but the other two are not easily countered. A major technical gap exists between computing and video environments because of the difference in their respective signals.

The video-recording signals in the United States and Europe do not match computer video signals. In the United States, the National Television Standards Committee (NTSC) signal is built on 525 lines of resolution and 60 interlaced half-frames per second, as mentioned earlier; a standard that varies slightly is used in Europe. In contrast most computers have 512 lines of resolution and 30 frames per second. Displaying the video on a PC monitor requires matching and mixing the computer video signal with the NTSC signal. Boards added to a PC can resolve the inconsistent signals, thereby enabling a monitor to display full-motion video, graphics, and text. Some of these systems simply transfer the video to the screen by using a device called a genlock and other circuitry. Other units digitize the full-motion video in real time, then merge it with other information to be displayed. One form of the latter is called digital video interactive (DVI).

A DVI can significantly reduce the amount of storage needed per frame on a CD-ROM. A DVI disc is a CD-ROM disc containing data that have been reduced using DVI technology. The degree of compression is especially apparent in a video application, where 30 frames a second must be displayed. Since CD data are available at only 5120 bytes each one-thirtieth of a second, a frame of video must be compressed from approximately 500 KB to less than 5 KB. Audio data must also be compressed to approximately 500 bytes and stored within the 5 KB. This process reduces the storage for each second of video to 150 KB. A DVI will be integrated with PC motherboards by the mid-1990s providing a level of full-screen, full-motion broadcast-quality video only dreamed about today. Compression is accomplished by digitizing the video and running it through a compression algorithm. After the information has been compressed, it is stored on the CD. Later when the compressed data are read, they are run through a decompression algorithm that recreates the original displays at 30 frames per second. The ability of the DVI to reduce the amount of data significantly to represent a frame will be important in bringing full-motion video to the desktop.

Another approach for delivering video to the desktop involves a video-capture board installed in a PC. An image-capture board alleviates the need for storing the digitized video; instead it functions as a gatekeeper importing signals from other sources and displaying them. There are approximately 18 vendors who sell video-capture boards at prices ranging from under $500 to over $4000. Many of the boards operate under Microsoft Windows and allow users to view video from a VCR, laserdisc, cable, or standard television source.

4.5.2. Desktop Conferencing

Conferencing in this text refers to communication between two or more individuals or groups at different locations; conferencing may involve audio, image, text, or video (see Table 4.7). In its most productive form, conferencing is interactive and synchronous, although forms of broadcast or one-way conferencing are a viable part of the conferencing medium. Different types of conferencing can be combined, with audio and video being the most common mix. In 1991 the desktop videoconferencing market in the United States was $9 million; according to the Conference Board, it is forecast to grow to $110 million by 1995.

The type of conferencing system chosen by an organization is often based on the amount of money available. Although this is important, it is not the sole factor in reaching a decision. The decision of which system to choose depends also on the application and the corporate culture: High-resolution imaging may be better in a technical environment where the emphasis is on technical applications, and full-motion video may be better for a planning application where interaction and compromise are more important.

The purpose of building conferencing into the workbench is to create a shared work environment. Within such an environment, participants share a common audio and visual space. Depending on the features supported by the equipment, this sharing enables them to converse while interactively viewing, creating, and modifying text, and various forms of imagery, which means that whatever one user does is communicated to everyone else. This is comparable to a group of people working together at a chalk board—talking, erasing, and writing together.

Ideally a shared work environment for an image-based system has the following attributes so that multiple locations can work together:

Communications in real time over public data and voice networks
Interactive manipulation of images
Remote control writing and pointing mechanisms
Document retrieval and display
Files easily exchanged between locations

Table 4.7.   Conferencing Categories

Category	Description
Audio	Uses amplified telephone speaker devices to permit individuals or groups to converse with remote locations.
Image	Allows people at dispersed sites to write or draw on special tablets or to share pictures and graphs that have been digitized and transmitted.
Text	Employs a personal computer or terminal and consists of shared transmission of text-based messages in a real-time forumlike format.
Video	Freeze frame or slow scan: Transmits photographs and images in a staccato fashion. This form of video conferencing allows a tradeoff between seeing the other participants, albeit in a less than real-time form, and less expensive equipment.
	Full motion: Live transmissions between two or more sites resembling in terms of its quality a commercial television broadcast.

08   Chapter 4

The video portion of a conferencing system is very similar in concept to the Picturephone unveiled by AT&T at the 1964 World Fair. This precursor of today's version of conferencing encountered the twin problems of exorbitantly priced equipment and bandwidth requirements that could not be provided by the voice-grade network. It was only in the late 1980s that advances in technology reduced costs to an acceptable level and succeeded in compressing video information so it can be accommodated by existing communication services. The tortured history of the picture telephone is summarized in Table 4.8.

The AT&T videophone that went on sale in May 1992 for $1495 is a remarkable achievement but a far cry from conference-room-quality equipment. The videophone has 128 pixels/line and 112 lines, which provide a screen of 14,336 pixels; in contrast a super VGA screen of 600 by 800 produces 480,000 pixels. (The ISDN standard for video images is 101,376 pixels.) Also the AT&T video-phone has a fixed focus lens for 1–9 ft and transmits 10 frames a second as compared to 30 frames a second for full-motion video.

The cost of conference room equipment is in the $200,000 range, and equipment that can be transported on a cart is available for under $50,000. But the most significant drop in cost has been in the communication channels for transmitting the signals; in fact since 1984, bandwidth requirements have declined by a factor of two every 2 years. In the 1980s, bandwidth requirements for teleconferencing were measured in megabits, but by 1990 teleconferencing was being done over dual dial-up 56 Kbits/sec lines for a fraction of the former cost. Today a full-motion video conference can be conducted over a pair of 56 Kbits/sec circuits (112 Kbits/sec) for 19 cents a minute.

Three kinds of videophones support desktop conferencing. Those operating over ordinary analog voice circuits and having the same bandwidth requirements as telephones offer conferencing that ranges from single displays of images at the rate of four or five per minute to rates that produce a series of jerky displays in less than full motion. A second category of videophones use ISDN transmissions,

Table 4.8.   Tortured History of the Picture Telephone, a.k.a., the Videophone

Year	Event
1964	AT&T introduces Picturephone at New York Worlds Fair.
1970	Picturephone offered for $160/month.
1973	AT&T drops Picturephone.
1982	Compression Labs introduces $250,000 videoconference system, using $1000/hr lines.
1986	PictureTel unveils $80,000 videoconference system, using $100/hr lines.
1987	Mitsubishi offers $1,500 still-picturephone.
1991	PictureTel unveils $20,000 black and white videoconferencing system using $30/hr lines.
1992	AT&T announces $1495 videophone using regular telephone lines.

Source: From W. M. Buckeley, The Videophone Era May Finally Be Near, Bringing Big Changes, *The Wall Street Journal* 73(103) (1992), pp. A1, A10.

and it is designed to run on the primary rate interface, or 2B+D. These units require one 64 Kbits/sec B-channel for video, a second 64 Kbits/sec B-channel for voice, and a 16 Kbits/sec D-channel for signaling. The third category of video-phones is a PC-embedded device requiring a bandwidth typically greater than that available through a single ISDN channel. Some are designed for LANs, while others are designed for high-speed circuits spanning considerable distances. Three examples of conferencing equipment that provides picture or video capability are shown in Figure 4.5.

Each of the units in Figure 4.5 represents a different level of sophistication. The Photophone (Figure 4.5a) is designed for applications requiring pictures and voice communications, but full-motion video is too expensive or impractical. The unit captures, sends, and receives crisp images of objects and scanned text over a single, standard telephone line in seconds. The basic unit is the size of a small television and has a camera attached on one side. The operator points the camera at the image or text, then pushes a button to freeze and send the image. Transmission time can vary from as little as 10 seconds to as long as 45 seconds, depending on image content, resolution, and whether the image is gray scale or color. The unit is intended for those who want an imaging system that can also run standard PC software. The version shown has an annotation package that allows users to type notes onto the image and transmit the composite display. Images sent or received are stored automatically in the Photophone memory, but they may also be stored on the unit's floppy drive or hard disk. An optional document scanner allows users to integrate standard-sized documents with images.

The HV-100 is the first videophone to contain all the essential components in a single unit: codec, camera, display, and handset (see Figure 4.5b). The unit employs an ISDN basic rate interface (ISDN is discussed in Chapter 6), which allows simultaneous image and voice communications. The device consists of three functional parts: a video codec, a multimedia communication/telephone controller, and capture/display capabilities. The multimedia controller enables voice, video, and data to be transmitted simultaneously. A specially designed television camera is built into the top right corner, and there is a 5-inch color display. Since the maximum transmission frequency is 15 frames a second, the motion is not quite so smooth as a standard television transmission.

Figure 4.5c illustrates the Tel-EYE-Vision System by EyeTel Communications, Inc. The unit is a PC-based still image or motion system. In both configurations, the computer monitor is the display terminal with picture-in-picture capabilities and windows of video images that may be sized and moved. During a visual conference, images can be exchanged, and free hand drawings and notations can be made directly on the display screens of both parties with the mouse. Motion video conferencing is implemented by attaching the device to a high-speed communications link. The system is built around an IBM compatible PC with 4 MB of memory and a super VGA monitor. Since the system operates under Microsoft *Windows*, it permits users to cut and paste documents from any *Windows* application and transmit these with captured screen images.

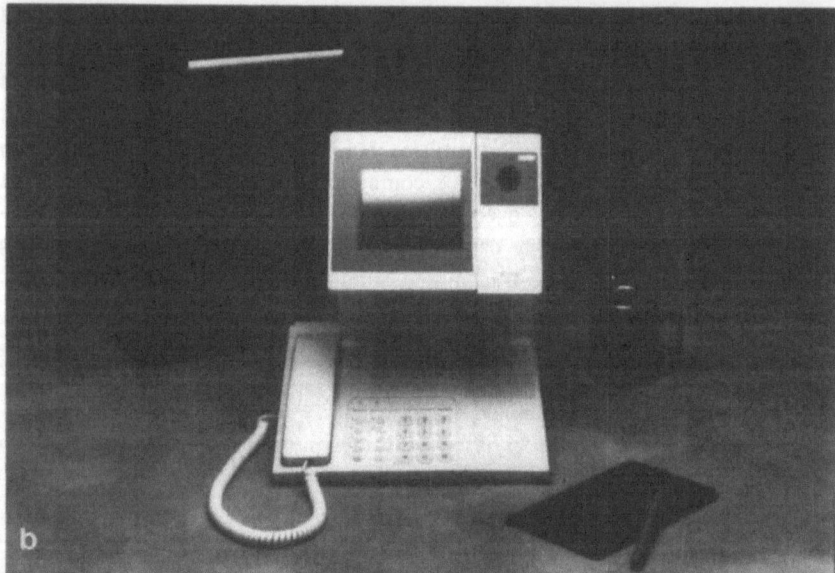

Figure 4.5.   Several examples of videophones. (a) Photophone Workstation: A desktop videophone that can send and receive images and text on a standard telephone line. Courtesy of Image Data Corp. (b) Videophone: ISDN based and full color. Courtesy of Hitachi Ltd. (c) A PC-based motion- or still-image communications system: The Tel-Eye-Vision System. Courtesy of EyeTel Communications, Inc.

Figure 4.5. (*Continued*)

The desktop videophone market is beginning to expand rapidly. At the 1991 Telecom trade show, more than 25 manufacturers were displaying units. Such large manufactures as IBM and AT&T have introduced products, and dozens of small firms are starting to carve out a niche. Compression technology has advanced to the point that limited-motion video over 19.2 Kbits/sec is possible, and the prices of the add-in boards for a PC are in the $300–$500 range, while stand-alone units are available for under $1500. As for standards, a multilayer protocol is under development that will permit transmission of nonvideo data from spreadsheets, word-processing files, and other desktop applications that can be integrated and displayed with the video data.

4.6. PUBLIC ACCESS TERMINALS

A public access terminal (PAT) is an easy-to-use interactive electronic information and transaction unit intended for public use. Such terminals are located in building lobbies, shopping centers, airline terminals, grocery stores, and tourist rest stops; since most systems are operated as a public convenience and are therefore free. The cost of providing and maintaining the system is borne by the corporate sponsor in whose facility the terminal is located or by advertisers who pay some type of fee for their listing. Chapter 15 discusses three cases involving public access systems.

The PATs support a number of applications, usually serving a targeted audience needing location-specific information and transaction services, such as the following examples:

Building/convention center information kiosks: Provides visitors with a list of occupants, calendar of events, facility maps, and general information.

Transportation: Displays transportation schedules, facilities, and services, weather forecast, available lodgings, and sites/events of interest. The system may permit a traveler to book a room or buy a plane ticket.

Retail: This category may be subdivided into the following:

Mall and store directories: Lists stores and departments within each store, maps, list of sale items, and major events. Some facility directories create a route from the viewer's present location to the viewer's desired destination, then print it out.

Coupon dispenser: Dispenses or prints coupons on request based on information displayed by the system.

Focused information: Describes available merchandise; systems supporting this application are often found in the bridal, children's, and gift departments.

Automated sales: Presents information and dispenses specific products; an example of such a system is the free-standing kiosk located in many Florsheim shoe stores.

City and tourist guides: Lists popular city attractions, night life, major events, civic and cultural activities, maps, restaurant guides, and public transportation schedules.

A more detailed list of PAT applications is given in the *Public Access Videotex* reference by the Newspaper Advertising Bureau at the end of the chapter.

4.6.1. Benefits

A public access system has considerable appeal for a corporation, since it can reinforce the corporation's image by first enticing the viewer with attractive visual and audio displays and then command the viewer's undivided attention with interesting and a highly personalized interactive exchange. A public access system can also educate consumers by describing specific product features and the advantages of one product over another.

In-store PATs can reduce the number of personnel needed to serve customers by answering questions or taking credit card orders for merchandise that is mailed. PATs can present complete merchandise selections; especially in stores too small to carry the full line of goods carried at larger stores in the chain. Public access terminals can promote special offers by dispensing coupons, and they can generate impulse sales by permitting the customer to make a purchase after viewing the product or service information.

The PATs enable advertisers to present in-store specials and sale items on a much more frequent basis than is possible with print-based media, since very little lead time is required and sales can be tied to inventory levels on a day-to-day basis. The PATs are also a marketing research tool. Surveys, special needs, complaints, and suggestions can be collected easily and inexpensively. A PAT system can even interact with a customer who requires help in making a purchase by offering suggestions of what is popular and then specific help regarding the final selection.

4.6.2. Equipment

Public access systems integrate a variety of technologies to support the different applications they are called upon to perform. Since such systems serve many users and their costs are often spread among several sponsors, it is much easier to justify video, optical disc storage, sound, and other less-common technologies than when the system is designed to sit on a desk and be accessed by one or a few people at most.

Figure 4.6 shows a typical PAT configuration. Many systems today are in stand-alone kiosks that are updated with a floppy disk that is inserted by hand in each unit. In malls, convention centers, and building lobbies, the system is often networked to an update or host system via a telephone line or local area network. The heart of the system is a microcomputer. Video and audio tape are sometimes integrated with information kept on disks to provide 30-second or longer information clips.

The most common form of PAT input is a touch screen followed closely by ruggedized keyboards and specially configured keypads. Automated sales units often accept credit cards or money. Transaction-oriented PATs sometimes incorporate a telephone for making reservations or contacting someone in the sponsoring organization. Normally a series of menus assists the viewer in making a decision; a particular menu option dials the desired number for a telephone conversation. Some vendors are developing voice recognition features to facilitate the human–machine interface.

The most common form of PAT output is a color display. Printers are often built into the kiosk to provide hard copy of what is on the screen or to print

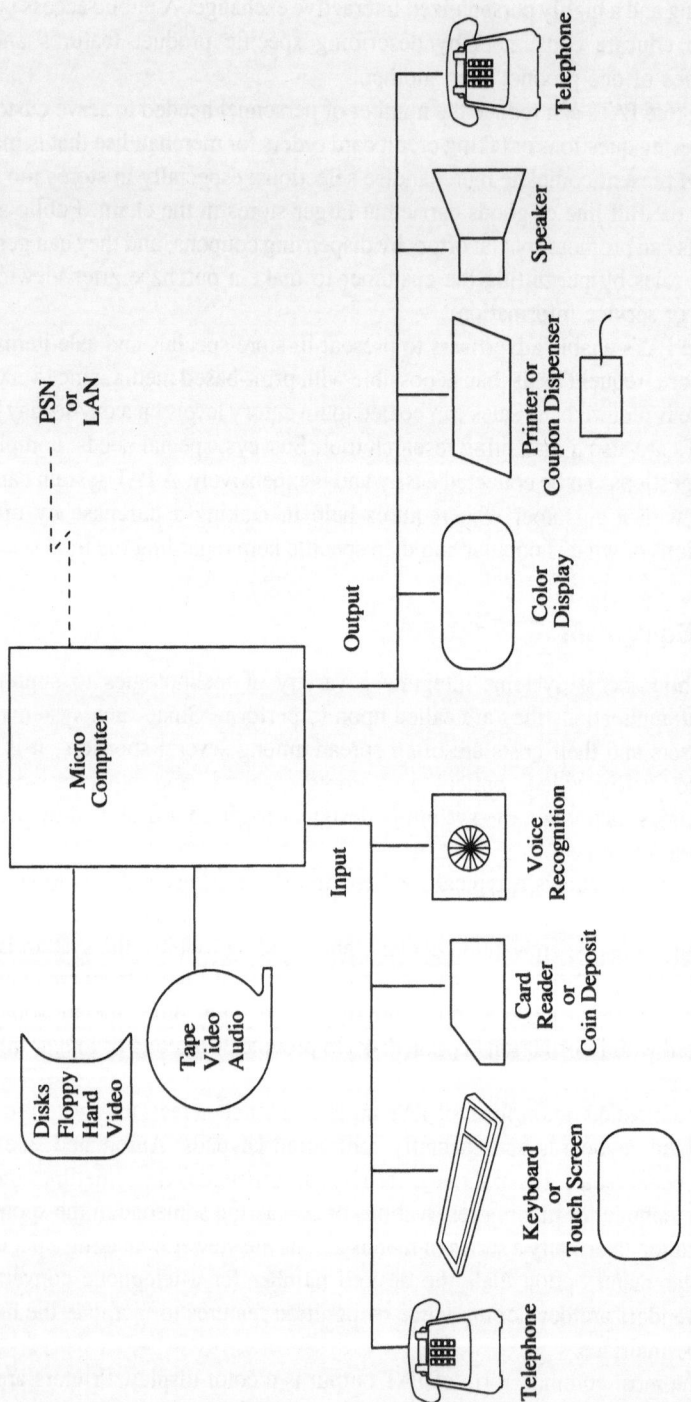

Figure 4.6. A typical public access vidiotex system configuration.

coupons. Speakers are also used, and a telephone may be added for recorded messages from advertisers, perhaps describing restaurant specials or new products. The audio is often synchronized with a video or graphic display on the monitor.

4.7. DEVELOPING A CORPORATE POLICY FOR ENHANCED TERMINALS

The previous discussion of enhanced terminals indicated the variety of technologies and range of capabilities possible. If no effort is made to guide the selection of units, users will not be able to share data; communication among units will not be possible; the cost of maintaining the hodgepodge of systems will be prohibitive; and no one will be able to operate anyone else's equipment. A policy for acquiring and supporting equipment with enhanced capabilities is not only necessary, it is crucial for establishing a functional and efficient work environment. Furthermore the policy should be developed as early as practical, since it is much easier to shape events than correct problems.

Often the systems side of the house is at odds with the user side concerning just what form policy should take. The easiest position for the systems people is simply to limit the acceptable hardware and software. This strategy shows a much greater concern for technology than the application involved. At the expense of user effectiveness, such a policy forces everyone to try and fit into the same pair of shoes, which is often neither desirable nor possible. On the other hand, users tend to lose sight of the larger picture as they focus on their own needs. They want to acquire whatever currently appears to fit their individual needs, then have someone else make sure it works and interfaces with the rest of the system. Clearly compromise is necessary. A policy is needed that is business- and not technology-oriented. Such a policy should prescribe a minimum number of restrictions to ensure the integration of technologies and leave as much room as possible for user needs and innovation.

A collaborative relationship between systems personnel and users must be established, since the low cost of hardware and software and its ready availability at the corner computermart, through the mail, or from a variety of vendors means that users can do what they want. Consequently the best policy states:

> These are the general specifications of the present system and the software, hardware, and communications protocols that are currently supported. If you are considering acquiring equipment and software that you wish to integrate with the present system or have the systems unit support, we will tell you whether it is compatible or how much it will cost to provide what you need.

This approach is a positive expression of the importance of compatibility, and it leaves room for user innovation.

4.7.1. Formatting the Policy

The policy adopted should indicate both a course of action and the reasons for it. Providing the reasoning is important to make the policy appear less arbitrary. It presents the organization's viewpoint and explains why doing something a particular way is important. Employees can accept restraints if there is clear and understandable business logic behind them. Furthermore as time passes and conditions change, a particular way of doing things may no longer be appropriate. People find themselves in a position of following guidelines that no longer make sense and eventually reach the point when the only way to get the job done is to ignore the policy. Including the reasons for a policy decision builds a trigger that signals management when it is time for a change.

Needless to say, the explanation of the reasons and rationale behind the policy should be simple and short. If the reader gets lost in the details the reasoning will neither be understood nor accepted.

4.7.2. Policy Outline

An outline of a written policy is presented in Table 4.9. It begins with a statement of purpose that emphasizes the importance of this type of equipment for employee performance and the organizational goals behind developing a policy that will foster increased levels of performance. A scope statement should be included to provide guidance about the breadth of the policy. Such a statement pertains to equipment, software, and communications that may be located at one person's workbench and the range of costs. Failure to define the scope, will lead to questions about whether the policy encompasses departmental or building computer facilities and perhaps even the mainframe and its peripheral resources.

A comprehensive policy considers technological factors, support services, user obligations, and implementation procedures. It is impossible to list everything that could be included, so the effort should not be made; instead the message should convey the spirit behind major issues that are to be considered. If users wish to violate the spirit of the message, they are on their own and must face up to the consequences of their actions alone. Technical considerations begin with hardware integration—ensuring the ability of various components, especially from different vendors, to work together in the same workbench and among workbenches. Software consistency applies to the ability to use the same software at multiple locations if permitted by the software license. Even more important, however, is the existence of the same software at different locations, so the databases and products produced can be accessed at multiple locations without being rekeyed.

It is just as important to indicate file transport and content standards if the files are to be transmitted to multiple locations. If the devices are to communicate with one another, they should at least be capable of a common terminal emulation

Table 4.9. A Policy Outline for Acquiring
and Supporting User Equipment

Section	Topics
Purpose	Importance for user performance
	Corporate goals in developing a policy
Scope	Equipment limitations
	Budgetary constraints
Technical consideration	Hardware integration
	Software consistency
	File transport
	File content
	User interface
Support services	Identifying appropriate systems
	Evaluating at user's request hardware and software
	Acquisition
	Volume discounts
	Training
	Installation
	General maintenance and spare parts
	Supplies store
	Help desk
User obligations	Physical, data, and communications security
	Adhering to copyright and licensing agreements
	Maintaining integrity of databases
	Removing hardware and software
Implementation	Distribute to current system users
	Give to newly hired employees
	Review annually and solicit input

and of transmitting ASCII files in both directions. Another element in the exchange of information is file content compatibility; that is data and information used by application packages at multiple locations should be widely accessible.

A frequently overlooked technical consideration is the user interface. Similar equipment should provide the same physical and system interfaces. Nothing is more annoying than to find within the same work area different keyboards, main menus, and access instructions. The importance of a consistent interface is not recognized by a user until he or she has to sit at someone else's desk. It is especially troublesome for maintenance personnel who must figure out the system prior to figuring out what is wrong with the system.

The systems area of an organization can control what equipment is purchased by providing an extensive array of support services and advertising these services widely. The service begins by responding to a user need with an appropriate system or systems; additional services include evaluating and acquiring components, volume discounts, installation support, general maintenance, stocking operating supplies, and staffing a help desk.

The user is accountable for the security of the system; for adhering to copyright and licensing agreements, maintaining the integrity of all databases, and conforming to general policies about removing corporate assets. These issues extend beyond matters of employee convenience and performance.

Since nothing is more useless than a written policy no one knows about, the policy should be distributed to current system users and included in the information every new employee receives. To guarantee the policy's continued applicability, it should be reviewed annually, with input solicited from the people it affects most.

4.7.3. Caveats

There are several caveats that the drafter of the policy should follow: (1) Do not require a user to justify computer equipment, especially upper level management, since units costing several thousand dollars are about the same percentage of a manager's salary as a telephone was in the 1950s; (2) Do not tie technology to the policy, technology is constantly changing and tying bit rates, processing speeds, and other technological factors to the policy constrains users and limits innovation. (3) Do not request cost/benefit statements, complex assessments of machine performance or other types of information unless the individual is capable of responding with a valid response. (4) Do not create personnel problems; establish a collaborative rather than an adversarial relationship with the user.

4.8. SUMMARY

Enhanced computer equipment integrates a variety of technologies. The features within the equipment can range from simple color and graphic presentations to voice/data integration, sound synthesis and audio, photographic-quality static displays, and full-motion video. By their very design limited intelligence terminals incorporate very few features—perhaps color, graphics, and voice/data integration. The more extensive memory and logic elements in microcomputers and video terminals enable many more features to be incorporated.

Terminals intended for public use are becoming visible extensions of an organization's information resources. Public access terminals are found in building lobbies, shopping centers, and even grocery stores. These units may provide information about items on sale, directions to the public rest rooms, or they may execute a sale and dispense a receipt or a ticket.

A policy guiding the selection and implementation of equipment is needed to prevent chaos. Users must be able to share data and communicate electronically with each other, and the organization must be able to acquire equipment and maintain it in an economical and efficient way. A collaborative effort between users and systems personnel permits each group to achieve its objectives; thus a written policy is a major step in this direction.

REFERENCES

J. Alston, More Features, Lower Prices for Business Graphics Market, *Software News* **5**(1) (1985), pp. 65–66, 68–70.

AT&T Videophone 2500, *Imaging Magazine* **1**(1) (1992), p. 16.

C. Babcock, X Marks the Terminal, *Digital News* **5**(13) (1990), pp. 3–5, 12.

S. Baker, Custom Applications for PCs and Telephones, *Telecommunications* **25**(7) (1991), pp. 69–74.

L. D. Beran, Hello London! Let's Have a Meeting Now: Teleconferencing, *Telecommunication Products + Technology* **2**(9) (1984), pp. 1, 33–34, 38, 42.

J. Borrell, Voice–Data Integration Enters War for Desktops, *Mini-Micro Systems* **18**(3) (1985), pp. 99–102, 104–113.

L. Brennan, Interactive Videodiscs Gain in Popularity, Especially for Training, *PC WEEK* **4**(30) (1987), p. 8.

W. M. Buckeley, The Videophone Era May Finally Be Near, Bringing Big Changes, *The Wall Street Journal* **73**(103) (1992), pp. A1, A10.

P. W. Bugg, *Microcomputers in the Corporate Environment*, Prentice Hall, Englewood Cliffs, NJ (1986), pp. 115–130.

P. Campbell, All of the News That's Fit to Print, Dial, Listen to, and Interact with, *Infotext* **4**(1) (1991), pp. 38–40, 42–46.

J. Canning, Weighing the Three Sound Resources of Multimedia, *InfoWorld* **13**(42) (1991), p. S104.

K. Chang, Videotex through Microcomputers, *Videotex World* (September 1984), pp. 10–14, 16–20, 43.

W. Conhaim, Talking Newspapers Voice Services Help Newspapers Provide Information on Demand, *Information Today* (June 1990), pp. 42–43, 48.

J. E. Cunningham, *Cable Television*, Sams, Indianapolis, IN (1981).

L. Day-Copeland, Clinical Research Finds PC Users Are More Productive with GUI Interface, *PC Week* **7**(9) (1990), p. 142.

S. Dickey, Diskettes: Smaller Sizes Pack More Punch, *Today's Office* **21**(11) (1987), pp. 27, 31–34.

J. Duffy, AT&T Looks to Build Video Market with Codec Chipset, *Network World* **9**(16) (1992), pp. 13–14.

B. Francis, Workstations Enter the Third Dimension, *Datamation* **37**(17) (1991), pp. 34, 36, 38.

B. Francis, X Terminals Prices Plunge, *Datamation* **38**(1) (1992), pp. 41–42, 45.

H. Fukunaga, Complex Counter Type Videotex Terminal, *Fujitsu Science and Technical Journal* **21**(5) (1985), pp. 527–537.

J. Gantz, Office Automation and DP: A 40-Year Side Trip, *Telecommunications Products + Technology* **5**(11) (1987), pp. 54–55.

J. Gantz, The Dichotomy of the Video Age, *Networking Management* **8**(11) (1990), pp. 46, 48–49.

B. Gates, *Information at Your Fingertips*, Keynote Address at Fall COMDEX Meeting, (November 12, 1990).

E. Gold, Trends in Desktop Video and Videophones, *Networking Management* **10**(6) (1992), pp. 44, 46, 48.

E. Gold, Unified Systems Integrate Voice, Data and Images, *Networking Management* **8**(12) (1990), pp. 29–30, 32, 34.

E. Gold, Videoconferencing Closes in on the Desktop, *Networking Management* **10**(1) (1992), pp. 42, 44–46.

L. Helgerson, Scanners Present Maze of Options, *Mini-Micro Systems* **20**(1) (1987), pp. 71–74, 77–78, 81.

E. Heichler, X Station Sales to Double; NCD Leads the Field at 36%, *Digital News* **6**(9) (1991), pp. 1, 6.

P. Hodges, Three Decades by the Numbers, *Datamation* **33**(18) (1987), pp. 77–78, 82, 86–87.

S. Hopkins, Commanding Productivity in an Information Network, *Telecommunication Products + Technology* **3**(3) (1985), pp. 30, 32, 41–42.

Interactivities **1**(1) (1987).

Introduction to Public Access Videotex, Public Access Committee, Videotex Industry Association, Rosslyn, VA (1986).

S. Janus, CD-ROM Applications Are Expected to Target More General PC Audience, *PC WEEK* **4**(47) (1987), pp. 141–142.

S. Jelcich, Color Capability: Is It Easier to Use, or Is It Just Another Pretty Screen?, *PC WEEK* **4** (38) (1987), p. 105.

E. Kay, Color Outshining, Not Eliminating, Monochrome Monitors, *PC WEEK* **4**(38) (1987), pp. 97–98, 102.

G. Keogh, Stealing a March, *Management Computing* **13**(10) (1990), pp. 88–89.

E. Leinfuss, 3D Graphics Are Moving to MIS, *MIS Week* **10**(18) (1989), pp. 1, 8.

D. Lewallen, AVC Delivers Multimedia Muscle to PCs, *PC Week* **7**(17) (1990), pp. 100, 102.

R. Lockwood, Multimedia in Business, *Personal Computing* **14**(6) (1990), pp. 116–123, 125–126.

J. Majkiewicz, Will Desktop Video Play in Business?, *Datamation* **36**(1) (1990), pp. 53–54, 56.

M. Mann, Video-Capture Boards Deliver Savings in Time, Expense, *PC Week* **8**(26) (1991), pp. 78–79, 81.

J. McMullen, PCs Double As X Terminals, *Datamation* **37**(13) (1991), pp. 41–42, 44.

E. Messmer, New Devices to Boost Desktop Video Mart, *Network World* **9**(2) (1992) pp. 9–10.

N. D. Meyer, How to Design a Nonrestrictive Microcomputer Policy, *Data Communications* **14**(11) (1985), pp. 237–240, 243–244, 247–248.

D. Moeller, Has Videoconferencing Found Its Niche? *Networking Management* **8**(3) (1990), pp. 49–50, 53–55.

O. R. Omotayo, Performance of Videotex-to-Speech Convertor, *IEEE Proceedings* **131**(5) (1984), pp. 328–334.

S. Osmundsen, The Price of X Terminals, *Digital News* **6**(20) (1991), pp. 1, 39–40.

A. W. Paschal, Optical Publishing: Issues Facing Publishers and Subscribers Today and in the Future, *Proceedings of the Eighth National Online Meeting*, Learned Information, Medford, NJ (1987), pp. 387–395.

Photophone (the) PC, *Background Information*, Image Data Corporation, San Antonio, TX (August 1991), p. 5.

Public Access Videotex, Newspaper Advertising Bureau, New York (1987).

Public Access Videotex: A Survey of System Operations, Newspaper Advertising Bureau, New York (1987).

C. W. Reed, Using the Telephone to Deliver Information, *Information Times* **6**(9) (1987), p. 11.

S. Richardson, Videoconferencing: The Bigger (and Better) Picture, *Data Communications* **21**(9) (1992), pp. 103–104, 106, 108, 110–111.

M. Robins, IVR Systems Laden with Advanced Features Fill Mart, *Network World* **9**(17) (1992), pp. 1, 37–40, 52.

J. Rothfeder, Desktop Video Ensures a Good Show, *PC WEEK* **4**(38) (1987), pp. 45, 50.

S. Saunders, For Videophones, Light at the End of a Very Long Tunnel, *Data Communications* **20** (17) (1991), p. 116.

W. Sawchuk, H. G. Bown, C. D. O'Brien, and C. W. Thorgerson, An Interactive Image Communications System Using Narrowband Lines, *Computers and Graphics* **3** (1978), pp. 129–134.

E. Scannell, The Secret History of the IBM PC "Gamble," *InfoWorld* **13**(32) (1991), pp. 47, 50.

Scanners Allow Users to See Text and Graphics in a New Light, *PC WEEK* **4**(47) (1987), p. 58.

J. Takeyama, The ABCs of Audiotext Technology, *Infotext* **3**(10) (1991), pp. 32–33, 36, 38, 40, 42–44, 47–50.

M. Takizawa, H. Goto, and M. Takahashi, Desktop All-in-One, Color Moving-Picture Videophone, *Hitachi Review* **40**(3) (1991), pp. 257–262.

Tax Returns by Phone, *Infotext* **4**(11) (1992), p. 9.

R. Taylor, Videoconferencing—Building Ubiquity, Cutting Costs, and Communicating Early, *Telecommunications* **25**(8) (1991), pp. 46–48.

R. L. Veal, The Integration of Telephone and Computer Technology, *Telecommunications* **25**(6) 1991), pp. 29–32.

Videotex/Teletext Presentation Level Protocol Syntax, North American PLPS, American National Standards Institute, New York (December 1983).

Voice Messaging User Interface Forum: Specification Document, Information Industry Association, Washington, DC (1990).

R. M. Wallach, More Users Seeing the Light of GUI, *Computer Systems News* (478) (1990), pp. 21, 26.

X Windows System: A Brief Guide to Understanding X Technology, Spectragraphics, San Diego, CA (1990).

Chapter 5

Connectivity

Connectivity is the ability to join together equipment in a communications network. The ultimate communications system is a network that supports universal connectivity. In such a network telephone lines, airborne communications, cable systems, and any other type of media are joined together to support universal access to different types of equipment.

Chapter 5 discusses the subject of universal connectivity and touches on a number of points with which the reader should become familiar. Without a basic knowledge of how the pieces fit together, the system can never be understood. Chapter 5 begins with an overview of a business's communication needs and the role of standards in satisfying these needs. A short explanation of storing and transmitting information is provided for the reader who is not familiar with the principles behind these processes. The explanation begins with a brief discussion of the building blocks that make information storage and transmission possible. The concept of communication protocols is introduced, and the two most common protocols are explained. The chapter ends with a discussion of transmission rates and the typical speeds needed for various types of communication services.

5.1. COMMUNICATIONS

Communications is the essence of business; it ties together employees in dispersed locations and it links together a company and its customers; communications is the conduit through which an organization deals with its suppliers and creditors. The nature of the business activity, the organization's size, its financial resources, industrial competitiveness, the sophistication of management, and the progressiveness of the organization are some of the factors that shape the role communications will play. In general, the communication needs of any organization encompass items listed in Table 5.1.

5.1.1. Business and Residential Needs

The communications system must always be available. In an era of voice and electronic mail, electronic document interchange, telecommuting, and direct customer order entry, a system cannot operate only between the hours of 8:00 A.M. and 5:00 P.M. External access from different time zones and by people who work different hours precludes this.

The system must be pliant. As people change locations in an office building or factory, it must be possible to move equipment. Changing addresses should not require new cabling or major network revisions. In addition the network must be able to grow and shrink without incurring major costs as usage patterns change.

Data and voice should be integrated and provision should also be made for incorporating video into some environments. It is no longer economically sound to establish separate systems, nor is it desirable to do so from a technical perspective. Maintaining an analog system for voice, a digital system for data, and a wideband service of some kind for video replicates equipment and needlessly adds to the cost and complexity of managing the network. To make matters worse, some of the duplicated services may emit electromagnetic interference and other forms of disturbance that lead to cross talk and related problems.

Business needs vary tremendously. Not only is the communications system expected to support the information services normally thought of as integral to operating a business, but the system may also be required to support alarm systems, environmental building controls, sprinkler and other fire suppression systems, and even the building's elevators. The variety of demands placed on the system necessitates different levels of support. An alarm system must be sure of access to the network at any time, whereas user's access to an electronic mailbox

Table 5.1. Business and Residential Communication Needs

Entity	Requirements	Entity	Requirements
Business needs	100% availability Pliant Expandable Voice/data integration General applications Variable speed Secure General interface	Residential needs	Voice focused Universal service Reasonably priced Easy to use Specific application Nonblocking

can occasionally be delayed without dire consequences. On the other hand, an alarm system will consume very little bandwidth compared to a video transmission.

The system must be secure, although this is not usually a problem if communication takes place internally. Passwords and such access devices as cards and keys normally provide adequate levels of protection. Problems occur when the system supports off-site users, including customers and the general public, who may intentionally or unintentionally gain access to proprietary information or erase files. Viruses that reach the corporate database via electronic mail are one of the biggest problems today; a virus program can be sent to an unsuspecting employee. Once the program is in the system, it may propagate to other computers, changing addresses and deleting files as it passes through the system. This is one of the risks that must be borne and guarded against as an organization establishes the gateways to external computers and interfaces to the electronic systems within.

Residential communication needs are different. They are now and for the foreseeable future voice focused; the emergence of public videotex services and the appearance of "smart houses" will not change this. In the home, the communication system through which users interact with the outside world is the telephone, and not the cable television nor backyard satellite dish; take away the telephone, and a person feels isolated. Universal service is the basis on which the telephone industry operates in the United States. The operative theory is that anyone can afford a telephone and telephones will be deployed throughout society regardless of socioeconomic standing.

The residential communications system and services obtained through it must be easy to use. Little or no prior training should be needed, and the system must be able to change without a need to reeducate the user. For example, when the telephone industry changed from mechanical crossbar switches to digital switches, the public was unaware of the change and continued to use the system in the same way. Contrast this to the short but necessary training needed when an office using terminals connected to a computer via modems and telephone lines begins using a LAN.

Most residential services are application specific. If a customer wants call forwarding on a data line, something must be done to the system to accommodate the change. In the mid-to-late 1990s, the proliferation of ISDNs will broaden the options, but until then, most residential services will be designed to accommodate specific applications, and if they are used for one particular application, this will preclude their use of the communications link for something else; for example, one person cannot be speaking on the telephone while someone else is trying to access an infobase.

5.1.2. Standards and Standards Issues

Connectivity is made possible through the creation and adoption of standards. In a broad sense, standards may encompass specifications, codes indicating procedures and material, guidelines, recommended practices, grading rules, and so forth. The larger the number of organizations accepting and conforming to a specific set of standards, the greater the level of connectivity that is possible among them.

There are many reasons unrelated to connectivity for promoting standardization: Standardization leads to economies of scale in producing equipment and services; large investments in research, development, tooling, and plant modifications cannot be justified if production quantities are small. Standardization fosters the development of larger and more uniform markets for products; without standardization niche markets would be the rule rather than the exception, and competition among suppliers would be reduced, thereby leading to higher prices for end-users.

Standardization eases the problem of finding an experienced labor force for selling and servicing products, and it often helps simplify the task of training users. Standardization provides a platform of reference for procurement and legislative purposes, and it helps establish an orderly and widely accepted basis for adding enhancements over time.

The standard-setting process is not without its pitfalls: It is frustratingly slow, and the cost of participating in the process can be significant. Some companies try to foist their own product specifications on standards-setting entities to protect or expand market share and raise road blocks in the path of competitors.

There are five stages in the process of setting a standard. The first is the preconceptualization period during which the need for a standard is first recognized. This stage is followed by a conceptualization. At this time, the idea is presented to a standards-developing organization for sponsorship. If adopted, the discussion stage then commences during which interested parties serve on a committee to hammer out a consensus. This is followed by the writing stage during which the committee documents its position. The process ends with implementation of the standard by vendors in hardware or software. In the end, vendors may or may not choose to implement the standard, unless it is for a regulated commodity, such as radio spectrum allocation. In that case, everyone wishing to manufacture or sell products must adhere to stated regulations.

5.1.2.1. Standards Organizations

There are a surprisingly large number of standards-setting entities throughout the world. In the United States, there are over 420 nongovernment organizations that maintain an estimated 32,000 standards. There are an additional 49,000 standards developed by Federal agencies, such as the Department of Defense and the General Services Administration. On the international scene, there are more than 270 standards organizations.

These domestic and international organizations often work together because of enlightened self-interest, government and trade association encouragement, and sometimes government fiat. Table 5.2 lists several important organizations in the standards arena. Some of these organizations develop standards, while others serve as coordinating and approving bodies for their members.

The International Standards Organization (ISO) is the principal standards body in the world, with more than 2000 active technical committees, subcommittees, and work groups. Its membership includes representatives from most major nations. The United States is represented by the ANSI. The International Telecommunications Union (ITU) plays a major role in standardizing international telephone and telegraph networks. One of its most important arms is the International Telegraph and Telephone Consultative Committee (CCITT), which makes recommendations affecting the communications industry. One of the organizations that has played a major role in developing videotex standards is the Conference of European Post and Telecommunications Administrations (CEPT); it is composed of the Postal Telegraph and Telephone Administrations (PTTs) in 26 countries.

The ANSI is the major standards organization in the United States; ANSI does not develop standards; but it coordinates their development and approves them through its membership, which consists of 220 professional and technical societies and trade associations. The ANSI also participates in the ISO and other international standards bodies and interacts with various state and federal governments at all levels. The Computer and Business Equipment Manufacturers Association (CBEMA) is typical of the many nonprofit organizations that represent a particular industry. Its 43 member companies are active in standardization activities, and all of its standards are processed through ANSI.

The Corporation for Open Systems (COS) is a newcomer to the standards arena. It was formed in 1986 to create a way of accelerating the introduction of products and services that support the model for Open System Interconnection

Table 5.2. Organizations Important in the Standards Arena

Acronym	Organization
International	
CEPT	Conference of European Post and Telecommunications Administrations
CCITT	International Telegraph and Telephone Consultative Committee
ISO	International Standards Organization
ITU	International Telecommunications Union
Domestic	
ANSI	American National Standards Institute
CBEMA	Computer and Business Equipment Manufacturers Association
COS	Corporation for Open Systems
ECSA	Exchange Carriers Standards Association
EIA	Electronic Industries Association
IEEE	Institute of Electrical and Electronics Engineers

(OSI). The COS offers a variety of test products as well as testing and information services. It has established alliances with government agencies, industry vendor associations, educational organizations, and research groups worldwide in the furtherance of its objectives.

The Exchange Carriers Standards Association (ECSA), founded in 1983, has become increasingly important in the post divestiture era of the telephone industry. It is concerned with technical standards and related areas regarding the public telephone network; ECSA represents its members' interests on ANSI-accredited committees and standards boards and participates in other organizations.

One of the largest standards developers is the Electronic Industries Association (EIA). It has 3000 member representatives who participate in 750 committees, subcommittees, and task groups. A majority of its standards are processed through ANSI, and its membership is composed of companies involved in manufacturing electronic components, equipment, and systems.

The Institute of Electrical and Electronics Engineers (IEEE) was founded in 1884, and it has over 225,000 members. The IEEE has developed 500 standards, and approximately 800 standards projects are underway, many of which have been or will be processed through ANSI.

5.1.2.2. Pertinent Standards

These and other standards organizations in the United States and abroad are creating the platform that will someday make connectivity among most electronic systems a reality rather than a dream. Table 5.3 gives a partial list of the standards

Table 5.3. A Partial List of Standards That Foster Connectivity and Information Transfer[a,b]

Organization	Standard	Description
CCITT	V.24[a]	List of definitions for interchange circuits between data terminal equipment and data-circuit-terminating equipment
	V.26[a]	2.4 Kbits/sec modem standardized for use on four-wire-leased circuits
	V.26bis[a]	2.4/1.2 Kbits/sec modem standardized for use in the general switched telephone network
	V.29[a]	9.6 Kbits/sec modem for use on leased circuits
	X.21[b]	General purpose interface between data terminal equipment and data-circuit-terminating equipment for synchronous operation on public data networks
	X.25[b]	Interface between data terminal equipment and data-circuit-terminating equipment for terminals operating in the packet mode on public data networks

(Continued)

Table 5.3. (*Continued*)

Organization	Standard	Description
	X.29[b]	Procedures for exchange of control information and user data between a packet mode DTE and a packet assembly/disassembly facility (PAD)
	X.95[b]	Network parameters in public data networks
	X.96[b]	Call progress signals in public data networks
ISO	646-1973	7-bit-coded character set for information-processing interchange
	2022-1973	Code extension techniques for use with ISO 7-bit-coded character sets
EIA	RS-232c	Interface between data terminal equipment and data communication equipment employing serial binary data interchange
	RS-449	General purpose 37-position and 9-position interface for data terminal equipment and data-circuit-terminating equipment employing serial-binary data interchange
ANSI	X3.110-1983[b]	Videotex/teletext presentation level protocol syntax (North American PLPS)—NAPLPS
	X3.92-1981[b]	Data encryption algorithm
	X3.139-1987[b]	Fiber distributed data interface (FDDI)
	X3.140-1986[b]	Information-processing systems—Open Systems Interconnection, Connection-Oriented Transport Layer Protocol Specification
	X12.2-1986[b]	Invoice transaction set
	X12.4—Draft[b]	Payment order/remittance advice transaction set
	X12.1-1986[b]	Purchase order transaction set
IEEE	802.3-1984	Local area networks Carrier Sense Multiple Access with Collision Detecting (CSMA/CD) access method and Physical Layer Specifications
	802.4-1985	Token Passing Bus Access Method and Physical Layer Specifications (broadband, coaxial)
	802.5-1985	Token Ring Access Method (baseband, coaxial, and fiber)
	802.9-1992	Multimedia LAN (expected adoption 1992)
Under development	JPEG	Joint Photographic Experts Group
	JBIG	Joint Bilevel Imaging Group
	MPEG	Motion Photographic Experts Group
	SONET	Synchronous Optical Network
	B-ISDN	Broadband ISDN
	FDDI II	Fiber Distributed Data Inferface II (for sound and video)
	—	Common Interface Commands (for videotex navigational functions)

[a]V designates analog.
[b]X designates digital.

that foster connectivity and information transfer. These are but a small number of the 1000 standards in information technology adopted or under review. When these standards appear alone in the literature, they are usually preceded by the name of the organization maintaining that particular standard, *e.g.*, CCITT X.25. Many of the same standards are endorsed by more than one organization which sometimes causes confusion; for example, the 7-bit coded character set for information-processing interchange is ISO 646-1973. The slightly Americanized version is ANSI X3.4-1977. A whole family of standards may be contained within a common identifier—approximately two dozen standards for electronic business data interchange appear under the designation ANSI X12.XX.

Three emerging standards listed in Table 5.3 are JPEG, JBIG, and MPEG; each will have a major impact on image processing. The first two deal primarily with still pictures and the third with motion video. The JPEG is a general-purpose compression technique for images of all types, while JBIG compresses bilevel images, such as black-and-white pictures or pages of text. The need for these compression standards is obvious when we consider that the average picture contains 2 Mbits. It takes 833 seconds or about 14 minutes to transmit a picture with a 2.4-Kbits/sec modem; transmitting the same picture with JPEG and its 16-to-1 compression rates takes about 52 seconds.

The standards in Table 5.3 must be unified in specific combinations to support connectivity. Those dealing specifically with communications may be brought together through one of at least three network standards: Systems Network Architecture (SNA), Transmission Control Protocol/Internet Protocol (TCP/IP), and OSI.

The SNA, a proprietary product of IBM, has become a *de facto* industry standard through IBM's shear size in the market. The TCP/IP is a set of specifications originally developed to tie together networks operated by the Department of Defense in the 1960s. In 1980 the Department of Defense formalized TCP/IP as a standard, and it has been adopted by a number of computer vendors and large users. Many organizations consider TCP/IP an interim measure for tying together different protocols and equipment; their long-range strategy is to migrate to the OSI standard when it is completed. The TCP and IP are two low-level protocols combined to create the unified standard that bears their names.

The OSI model defines the protocols and interfaces for open communication among electronic systems. The term open indicates the ability of a system from one manufacturer to connect to any other system that also conforms to the model and its protocols. The model is designed to serve as a skeleton to which the appropriate standards can be added. The skeletal structure is composed of seven layers, and each layer has its own control information (see Figure 5.1).

In the OSI model, an information processing task, referred to as an application process (AP), is passed from layer to layer. The application process may be manual, such as a person operating a public access terminal; it may be computerized, such as a computer program automatically accessing a remote database; or it may be physical, such as a robot in a computer-automated manufacturing environment. As the application process passes through each layer beginning with layer seven, additional control information is added. The seven layers and their contents are shown in Figure 5.1. When the data unit reaches layer one it is transmitted; when it arrives at its destination, the process is reversed, and it is

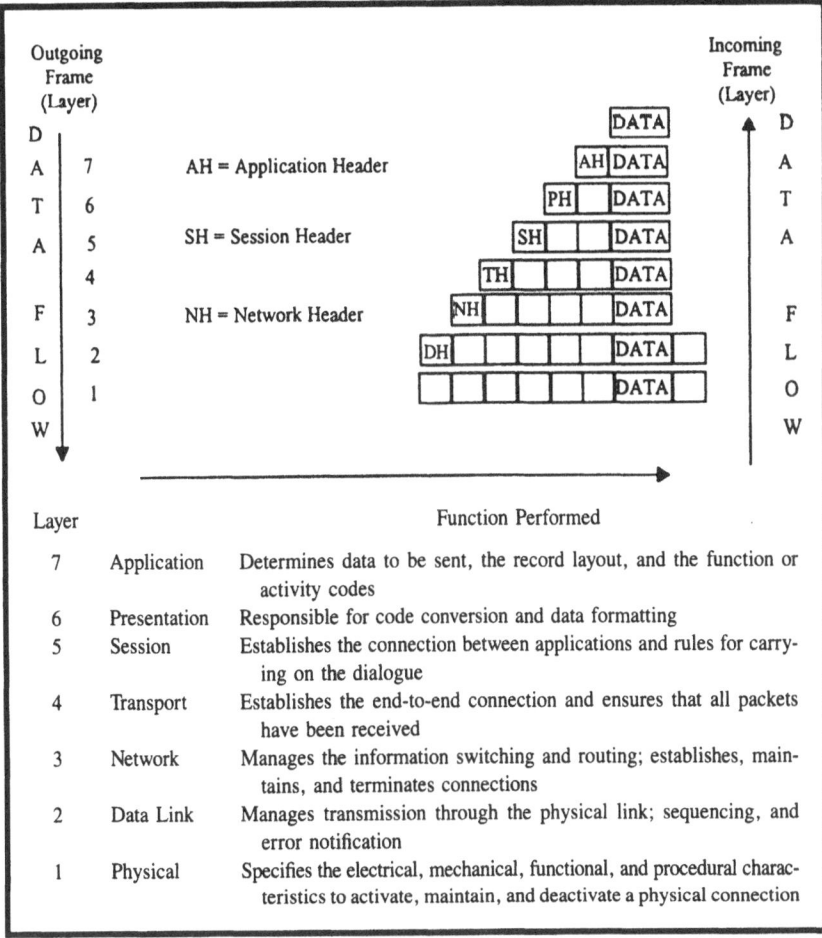

Figure 5.1. The ISO Open System Interconnection Reference Model.

passed upward through each layer. As the data unit passes through each layer, the control information is removed, and the appropriate activity is performed.

Each layer is conceptually an entity in itself: It has defined boundaries that enable the individual functions performed within that layer to change without requiring revisions throughout the software. New network technologies can be added, and revisions can take place as necessary within a particular layer. This modular arrangement allows layers to be exchanged with a minimum amount of disruption.

Examples of popular standards and where they fit into the seven layers are shown in Figure 5.2. Often the demarcation among layers and the assignment of standards to particular layers are not accomplished as neatly as suggested by

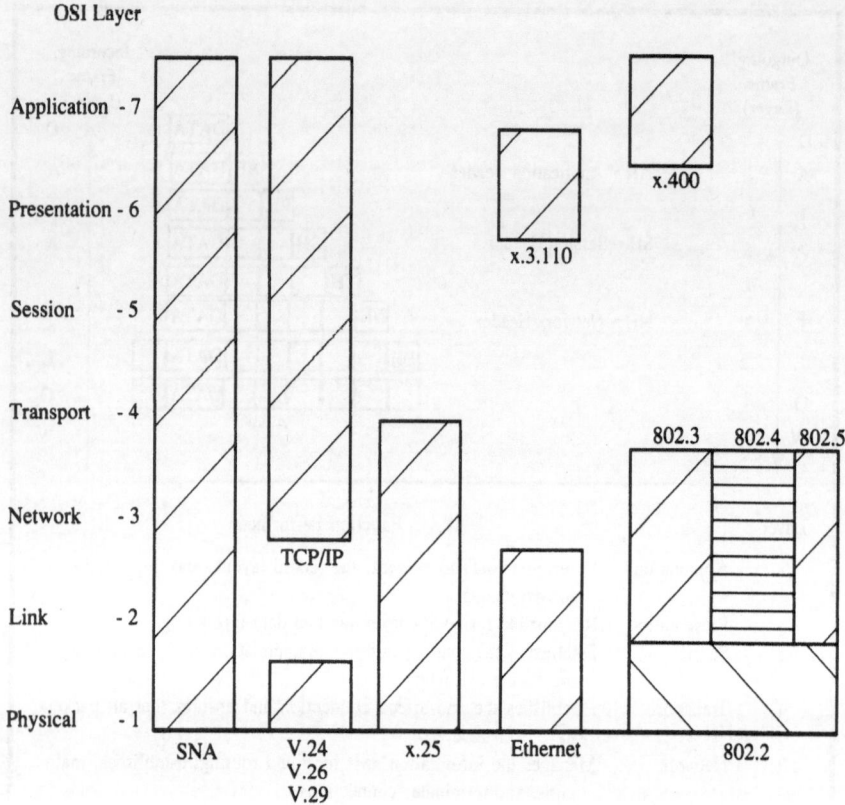

Figure 5.2. Popular standards and here they fit in the OSI reference mode.

Figure 5.2. Clearly the greatest value of the OSI reference model is the organization it provides for a confusing and very complex aspect of the communications industry.

5.1.2.3. Standard Operating Procedures

The standards just discussed relate primarily to communication interfaces among computers and to a lesser extent to transactions, such as ordering and billing, that may take place among companies. Standards for general information transfers among purveyors of information, such as those found in the on-line information industry and the user interfaces do not exist; their absence creates numerous difficulties.

Information providers and system operators spend considerable time and energy writing programs to satisfy the system du jour. Information providers must often alter their frames and the format in which they appear from system to system;

this prevents a consistent look and quality. Worse yet users must often learn a new set of commands and procedures to access various services or even different information bases from the same system operator. This problem is caused by the multitude of technical parameters used by different companies in the information services industry. If a company provides information and services to internal users, there is no problem, but if its users are customers of other services or if it provides information and services to other system operators, a set of industrywide standard operating procedures (SOPs) is sorely needed. These SOPs should deal with the issues shown in Figure 5.3.

Users should always be able to follow the same access functions. Access functions are divided into four types: control, display, input, and navigational. Control functions deal with accessing the system, logging-on, exiting from the system, and so on. The gateways established by the regional Bell operating companies come closest to providing a consistent set of access functions.

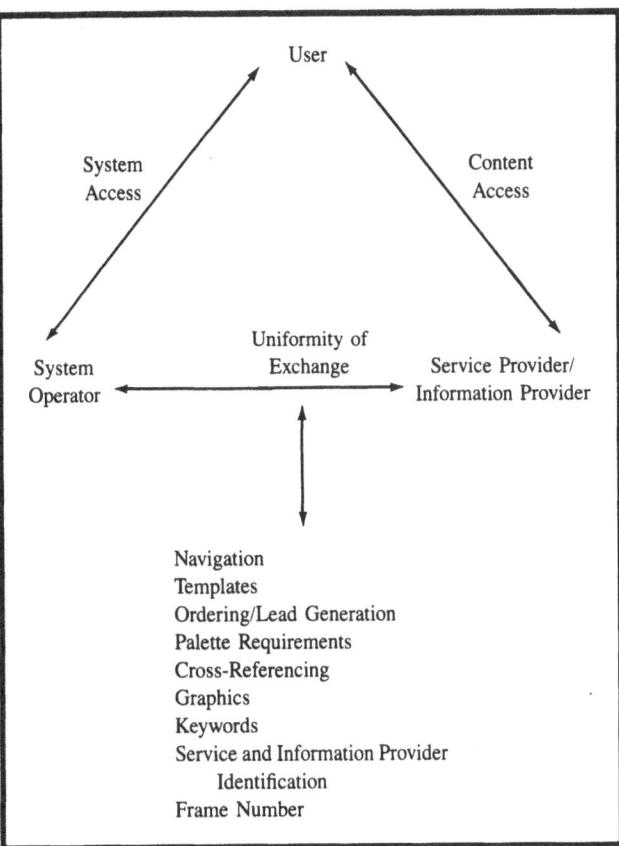

Figure 5.3. The areas where standard operating procedures are needed.

Display and input functions clear and repeat the display, confirm previous entries, and transmit the keystrokes. The set of functions that create the greatest user frustrations are those needed to select and initiate an application and to select and retrieve information. These are referred to as navigational functions, and they enable the user to move through the infobase. Differences in navigational functions affect the symbols used to prompt the user, their placement, the frequency of navigation, and even the structure of the infobase.

Variances in how a user accesses a system and its contents is symptomatic of the lack of uniformity among system operators, service providers, and information providers. This lack of uniformity is evident in many ways.

Some systems use templates or predefined frames on which information is updated periodically. However the templates are seldom identical so the contents have to be translated by each system operator. Ordering procedures differ significantly from service provider to service provider. Some ordering tasks are based on the displayed form, while others lead the user through a series of frames. Palettes differ among systems: One system's red is another's blushing pink. Palette inconsistencies affect the color positions on the frame. An even bigger problem occurs because some service and information providers reserve certain colors to communicate a special meaning. Cross-referencing affects the tree structure, navigation strategies, overall page construction, the theme carried on each frame, and the response time among other things. A variety of graphic standards are used, and the placement of images in the infobase is inconsistent. Some information providers have a heavy graphics content, which slows down a system. Graphics also affect the addition of a status line and a transient indicator; an example of a transient indicator is a blinking cursor telling the user the system is working on the request.

Some system operators restrict the number of key words, which may pose a problem for some information providers because it can affect the availability of some key words and their use in supporting promotions, advertising, and consumer education. System operators normally require service and information providers to identify themselves. The lack of consistency among system operators may require placing an identifier on the first frame or on every frame. This can affect frame design, overlays, the design of the identifier, the number of lines available for content, in addition to causing many other problems.

A variety of type styles are employed by system operators. Service and information providers are often encouraged to conform, which affects design and maintenance of the contents and leads to tremendous costs if multiple system operators are involved. Last even the presence of a lowly frame number can cause a problem. Some system operators can not or do not want to have them on the display and may not support retrieval by frame number.

Standard operating procedures are important for a company that has deployed enhanced information services internally. They are crucial for organizations that support external users, especially public access users; unfortunately for the most part, industrywide SOPs do not exist.

5.2. STORAGE AND TRANSMISSION

Converting information into a form that can be stored in computer memory may be a time consuming process, and it can require extensive amounts of storage device capacity, especially when the content includes such things as audio and picture quality images. Once stored there are a variety of options available for transmitting information to a second location.

How information is stored and transmitted ultimately depends on the user's application. Business users accessing internal corporate information may be assigned dedicated workstations and a dedicated line to the corporate computer. The same user accessing information provided by the government, other businesses, and special information providers, will have different needs. Calls outside the organization might number two or three a day and last several minutes to perhaps an hour. Transaction types of applications, such as verifying credit cards in a retail credit check, may number in the hundreds, but each last for less than a minute. The quantity of information being conveyed will typically be much less than for internal applications, and the response time will be much longer.

Residential users are unlikely to make more than a few calls a day, and the length of a call is 5–20 minutes. The information being conveyed is even smaller than that of a business user, and the response time and number of unsatisfied inquiries that will be tolerated is greater than for a business user. Residential users with sensor-based services have another set of needs. A sensor-based service, also called a teleservice, may monitor for fire or burglar intrusion and perform meter readings. These types of applications may require a transmission every few seconds and involve as few as a dozen characters. Transmission is automatic, and it must be allowed to occur even if the transmission media is already being used.

Businesses that are the repository for the information that others access constitute a third category of needs. These businesses must be able to reach other infobases, locate the desired information, in some cases edit it, and then convert it to the format that is appropriate for its own users. To create, maintain, and update the contents of an information base can require frequent access to other information bases and the transfer of millions of characters of information on a regular basis. To support this level of activity may require an extensive array of special equipment for creation purposes and high speed data transmission facilities. Needs that are considerably different from what it must provide to serve its own users.

5.2.1. Building Blocks

People who work in the computer industry use jargon that to the uninitiated is confusing. While it would be impractical to provide an explanation for all these terms, the following are often encountered.

Computers are in some ways like people, they process symbols. The terms chair, table, car, and so forth conjure up an image or a symbol in the mind of the listener. Just as people work with symbols, so do computers; but at a much lower level of abstraction. A computer operates in a binary world where there are only two choices; for purposes of explanation, the choices could be 1 and 0. These are referred to as binary digits, since there are only two; binary digit is abbreviated as *bit*. Computers work in a binary world for two very important reasons. It is possible to represent a bit (0 or 1) as a physical state, such as two magnetic polarities, two frequencies, the existence or absence of an electrical pulse, and even the presence or absence of a light. A second reason is that bits are an efficient way of representing symbols, such as characters; this latter point can easily be illustrated.

Imagine a light bulb connected to a power source; let 1 represent the light bulb when it is illuminated and 0 when it is not illuminated. One light bulb may then express two states—1 or 0, depending on whether or not it is illuminated. Each of the states could convey a message just as lanterns were used to notify Paul Revere prior to his famous ride, and two light bulbs could express four states or messages. The possible bit representations are 00, 10, 01, and 11. In a similar fashion, three light bulbs could express eight states in the following combinations— 000, 001, 010, 100, 011, 101, 110, 111. To represent the upper and lower characters of the alphabet, the numbers 0 through 9 and special symbols used for punctuation and the dollar sign, for example, 96 different combinations of 1s and 0s are needed. Seven light bulbs would meet this need and provide a few extra combinations.

The letter A may then be defined as 1000001, the letter B as 1000010, the dollar sign as 0100100, and so forth. To detect errors, an eighth bit, called a parity bit, is normally added to the other seven. Combinations of eight bits are often referred to as a *byte*; therefore eight bits, a byte, and a character are the same. A microcomputer with a 640-KB (K= thousand) memory is capable of storing 640,000 letters, digits, and special characters.

To facilitate moving and processing data within a computer, bytes are usually combined to form what is referred to as a computer word. Words may be two, three, four, and even eight or more bytes long. To complete the building process, all the bytes necessary to construct a display on a screen may be combined and treated as a single entity, called a *frame*. In some database applications, larger blocks of data are referred to as *cards*, and the database environment is called a *hypercard* environment.

5.2.2. Transmission Considerations

A bit or any of its combined forms can be transmitted in either parallel or serial fashion. Inside equipment, thirty-two or more bits are often transmitted at one time in parallel; which is considerably faster than transmitting bits one after the other in a serial fashion. However once outside the equipment, transmission is serial. Parallel transmission over long distances is not generally practical because of the expense involved in providing multiple communication channels and the timing problems encountered in having all the bits for the same character arrive at their destination simultaneously. The rules governing how information will be transmitted between two locations is referred to as the protocol; a fundamental issue of protocol is whether transmission will be asynchronous or synchronous.

Asynchronous transmission sends one character at a time. The advantages of asynchronous transmission is that the equipment is relatively simple and inexpensive. However the disadvantages of a low throughput of data per unit of time and its relative inflexibility severely limit its ability to meet today's needs. Consequently synchronous transmission is replacing asynchronous transmission. Synchronous transmission involves sending and receiving blocks of characters instead of one character at a time.

When a character is leaving a terminal or computer and entering the transmission network, the parallel flow of bits in the equipment must be converted into the serial format of the communications medium; the process is illustrated in Figure 5.4. In this figure, an asynchronous protocol is being used. The component that performs parallel-to-serial and serial-to-parallel conversions is called a universal asynchronous receiver/transmitter (UART).

In Figure 5.4, the character G is contained in an envelope that has a start bit and stop bit added to the bits representing the character. When nothing is being transmitted, the line transmits a steady signal designated in Figure 5.4 as a mark; the other state in this two-state scheme is called a space. The UART is constantly sampling the signal to determine whether a mark or space is being transmitted. In this example, the sampling rate is 16 times the bit rate. If the medium is transmitting bits at the rate of 1200 each second, one bit interval is 1/1200 or 0.833 msec. If the interval is sampled 16 times, a sample is being taken every 52 μsec.

When a change from a mark to a space is detected, the sampling process is modified so the signal will be sampled in the middle of the bit interval. After the bits in the character interval have been read, the line resumes its idle state. Regardless of the speed of the transmission, the composition of the material that carries the signal, or the protocol used, a transmission procedure similar to the one described takes place.

5.2.3. Transmission Rates

The bit rate of a communications medium is the frequency at which binary digits, or pulses representing them, pass a given point. Different applications require different bit rates. For example, a security sensor that protects against burglars requires 0.1 Kbits/sec; a terminal or personal computer accessing an infobase can operate effectively at 1.2–4.8 Kbits/sec. Video information is on the upper end of the scale: A full-motion high-quality video signal may require 50–150 Mbits/sec; conference-quality video can be supported at the 50-Mbits/sec level and high-quality entertainment at the upper end of the range. All of these rates are fairly slow compared to the performance of the human visual system. Scientists

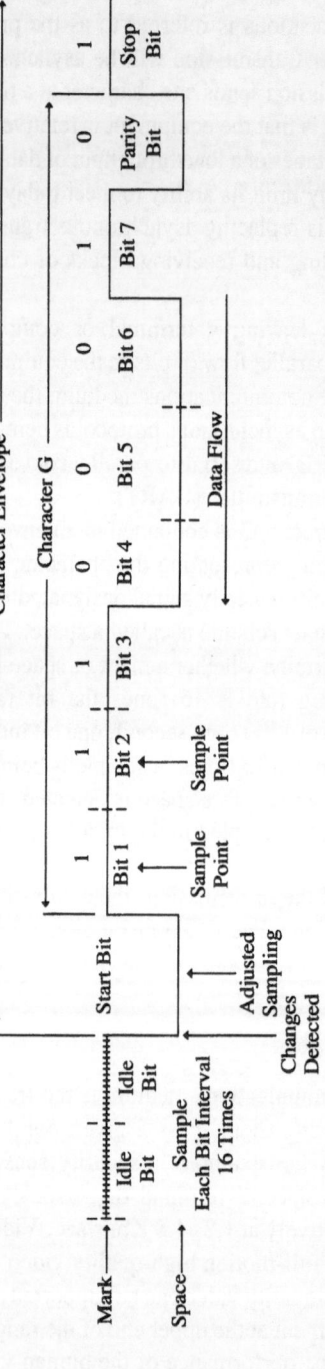

Figure 5.4. How a universal asynchronous receiver/transmitter functions, demonstrating the concept of data transmission, character representation, and sampling.

estimate that the brain's visual input channel processes information at a rate of approximately 2 Gbits/sec. The communication services necessary to support different applications vary tremendously in terms of their capacity. The local telephone network that is the basis for most person-to-person communication in this country has a very limited capacity (see Table 5.4). Side bands on local FM radio stations and unused channels on cable television networks provide additional transport within a community. Integrated services digital networks ISDN support bit rates of 64 Kbits/sec and VANs offer similar rates nationally; optical fiber generally supports the highest bit rates.

The information-carrying capacity of a communication channel is directly proportional to its bandwidth. Bandwidth is the difference between the highest and lowest frequencies that can be carried; the larger the bandwidth, the greater the information carrying capacity. Other factors affecting information flow are disturbances to the signal due to electromechanical interference, the type of modulation used, the capacity of the equipment, and the protocol employed.

As pointed out at the beginning of this section, some applications require high transmission speeds to make the applications possible, as in video conferencing, but generally there are benefits to be realized by providing high transmission speeds regardless of the application. For example, a study of on-line literature searching at 1.2 Kbits/sec and 2.4 Kbits/sec reported a 35% overall savings in time and related costs at the higher speed.

The capacity of the bandwidth can be leveraged by compressing the transmission. Data compression techniques have been around for decades in a variety of forms, but two proposed standards are being developed for compressing images. The Joint Photographic Experts Group (JPEG) is a digital image compression technique for primarily continuous-tone still images. In 1991 a technique was announced for applying JPEG for motion video, but the standard under development for doing this is the Motion Photographic Experts Group (MPEG). The JPEG

Table 5.4.   Representative Transmission Speeds
for Selected Communication Options

Option	Speed	Option	Speed
Telephone	0.3 Kbits/sec	ISDN	64 Kbits/sec
	1.2 Kbits/sec	VAN	64 Kbits/sec
	2.4 Kbits/sec	T-1	1.5 Mbits/sec
	9.6 Kbits/sec	MDS	5.7 Mbits/sec
FM radio	9.6 Kbits/sec	LAN	16 Mbits/sec
CATV		Satellite	50 Mbits/sec
VBI-2 lines	26.88 Kbits/sec	Fiber optics	275 Mbits/sec
Full channel	3.44 Mbits/sec		

is closer to adoption as a standard and may be approved in 1993; MPEG is at least a year behind. In the meantime, proprietary versions of both techniques are being used to enable the presentation of images and video.

Both JPEG and MPEG are symmetrical compression/decompression algorithms that can be handled with relatively inexpensive hardware; this contrasts with other compression techniques, such as digital video interactive (DVI) and CD-I, which are asymmetrical. The algorithms (asymmetrical) required for compression are much more computer intensive than those required for decompression. Consequently DVI and CD-I are limited primarily to interactive playback applications, such as training and storing reference materials. Asymmetrical compression techniques are not suitable for desktop authoring, conferencing, or other forms of interaction involving the creation or transmission of images with a device having only the computational capacity of a desktop computer.

5.2.4. Optical Fiber

Of all the communication options listed in Table 5.4, the one with the greatest potential for providing the capacity needed by systems with enhanced features is fiber optics. Telephone lines and packet networks are sufficient for transporting data, graphics, and low-resolution images, but picture-quality images and motion video have a voracious appetite for bandwidth. A broadcast-quality motion video application requires at least 45-Mbits/sec bandwidth for digital transmission unless it is compressed. Fiber can easily accommodate this need.

5.2.4.1. Advantages and Disadvantages

In addition to bandwidth, there are several other advantages optical fiber has when compared to other forms of transmission, especially copper wire facilities. The signal attenuation, which refers to the reduction in the strength of the signal as it travels through the fiber, is considerably less than the loss incurred traveling through copper, and the error rate is superior. The typical bit error rate for fiber is 10^{-11} compared to 10^{-9} for copper. Optical fibers are immune from electromagnetic and radio interferences, which are a major source of signal corruption in copper and microwave transmissions. Optical fiber transmissions do not radiate signals as do other forms of transmission, so they are much more secure. Optical fiber is tough; it has the same tensile strength as steel wire of the same diameter, which is very important when pulling and handling during installation. Even splicing is no longer a problem due to new techniques employed.

If fiber has a disadvantage, it is the cost, and this is rapidly changing: By the mid-1990s the cost of installing fiber to the subscriber will be competitive with installing copper. Until this point is reached, fiber will continue to be installed primarily in buildings, within businesses, for high-speed lines in the telephone and data communication industries, and in a few fiber-to-the-home trials. However when cost parity with copper is attained, fiber will begin appearing in that last mile to the customer's telephone called the local loop. Once replacing copper in the local loop begins in earnest, it will take decades to complete and cost billions of dollars. For example, in the Wenner article referenced at the end of this chapter, it is estimated that the cost of installing a nationwide residential broadband network in half of all U.S. residences could cost as much as $100 billion. This translates into an average charge per telephone subscriber of $5 each month for 33 years to fund the investment.

The conversion from copper to fiber is occurring in stages. The Cadogan reference cites estimates that by 1999 the trunk segment of the telephone network in the United States will be 100% fiber; the loop segment will be nearly completely converted by 2010. According to a study by Kessler Marketing Intelligence, the private network market was ahead of the public sector in the early 1990s: About 15% of the structured private networks have fiber extended to the desktop, whereas less than 1% of the public network has fiber into the home.

5.2.4.2. Video Dial Tone

Video dial tone has its corollary in the dial tone heard when the telephone receiver is picked up. The presence of the dial tone signals the caller that the system is ready to receive the subscriber's request. For a video dial tone, this request may range from initiating a voice call to requesting a movie that will be displayed on a television set or transmitting a full-motion image of the caller, commonly referred to as video conferencing. All of these features and many others are incorporated into various trials being conducted by telephone companies in the United States.

There are three principal standards that are shaping developments in the industry; these are: Fiber Distributed Data Interface (FDDI), Synchronous Optical Network (Sonet), and Broadband ISDN (B-ISDN). The FDDI defines a physical and data link layer with a data rate of 100 Mbits/sec; a newer version, called FDDI II, designed for interactive multimedia communications that will support sound and video is under development. The Sonet is being developed by the telephone industry to assure consistency of industry transmission; it will support transmissions close to 150 Mbits/sec. The channel structure of B-ISDN is not yet defined, but 150 Mbits/sec and 600 Mbits/sec are channel speeds commonly mentioned.

The challenge facing the communications industry, especially the telephone segment, is how to position itself to implement fiber-based services that will support multimedia, variable bandwidth, and broadband applications to which all users have immediate global access. The task ahead is formidable, particularly for telephone companies that for over a century have catered to the low-bandwidth requirements of plain ordinary telephone service (POTS). Shifting to a strategy where diversity rather than homogeneity is the goal will be slow, risky, costly, and difficult.

5.3. SUMMARY

Connectivity is the ability to interconnect equipment through a communications network. The cornerstone on which the concept of connectivity is built is a broad-based set of widely accepted standards; without standards, connectivity is impossible. There are literally thousands of organizations in the world promulgating standards. Internationally the ISO and the ITU are the most prominent; in the United States, the ANSI is best known.

The basis of information exchange is data storage and transmission. The fundamental building blocks of information exchange are the bit and its extensions, *e.g.*, the byte and computer word. Certain protocols and transmission rates are appropriate for particular applications. Although it may not be true in every case, generally the more efficient the protocol and the faster the transmission rate, the greater the number of different applications that can be accommodated by the system.

REFERENCES

A. F. Alber, *Videotex/Teletext: Principles and Practices*, McGraw-Hill, New York (1985).

J. Altson, Videotex Standard Interface, Preliminary Committee Report, Videotex Industry Association (1984).

S. Banerjee, J. Harrison, and J. Held, Satisfying Bandwidth Hunger, *Network World* 7(9) (1990), pp. 37–40, 42.

J. K. Brown, and P. M. Lew, Critical User Issues for Fiber Backbones, *Telecommunications* 24(5) (1990), pp. 33–34, 36, 38.

J. Brule, and I. Ebert, FiberWorld: An Overview, *Telesis* (Bell Northern Research) 17(1/2) (1990), pp. 5–17.

W. J. Cadogan, Fiber: The Tie That Binds, *Lightwave* 8(13) (1991), pp. 20, 22–24.

Directory of International and Regional Organizations Conducting Standards—Related Activities, National Bureau of Standards Special Publication 649, Gaithersburg, MD (1983).

M. Fahey, FDDI II to Include Video, *Lightwave* 7(10) (1990), pp. 1, 20.

Fiber-to-Home Cost Expected to Drop 40%, *Lightwave* 7(3) (1990), p. 27.

J. Gantz, Standards: What They Are, What They Aren't, *Networking Management* 7(5) (1989), pp. 23–24, 26–28, 30, 32–34.

B. Glass, When It Comes to Protocols, TCP/IP Is Universal, *Info World* 13(31) (1991), p. S63.

L. Haber, New Image Buzzwords: JPEG and JBIG, *Network World* 8(7) (1991), pp. 41, 55.

G. J. Handler, Local Packet Transport Planning: Videotex and Beyond, *IEEE Communications* 22(4) (1984), pp. 12–17.

ISO Reference Model for Open Systems Interconnection (OSI), *Data Communications*, Data Pro Research Corporation, Delran, NJ (1986), pp. cs93.107.101–cs93.107.105.

I. Knight, Telecommunications Standards Development, *Telecommunications* 25(1) (1991), pp. 38–40, 42.

A. Leger, T. Omachi, and G. K. Wallace, JPEG Still Picture Compression Algorithm, *Optical Engineering* 30(7) (1991), pp. 947–954.

S. Levine, Motion JPEG Compression for Desktop Nonlinear Video Editing, *Advanced Imaging* 7(1) (1992), pp. 32, 34, 67.

D. Livingston, Software Links Multivendor Networks, *Mini-Micro Systems* **21**(3) (1988), pp. 43–46, 49–50, 53–54.

E. St. J. Loker, *Fiber Optics—How Soon?*, American Newspaper Publishers Association, Washington, DC (1990).

N. J. Muller, R. P. Davidson, and M. I. McLoughlin, *A Management Briefing on the Emerging Synchronous Optical Network*, General DataComm, Middlebury, CT (1990).

S. Osmundsen, TCP/IP: New Growth, Old Limits, *Digital News* **6**(15) (1991), pp. 1, 29–30, 36.

I. F. Rockman, and B. Williams, Cost Effectiveness of Literature Searching at 2400 Baud, *Proceedings of the Eighth National On-Line Meeting*, Learned Information, Medford, NJ (1987), pp. 413–416.

S. Salamone, Loop Fiber Costs Down, Copper Systems Up, *Lightwave* **6**(10) (1989), pp. 1, 30–37.

D. A. Stamper, *Business Data Communications*, Benjamin/Cummings, Menlo Park, CA (1987).

Standards Activities of Organizations in the United States, National Bureau of Standards Special Publication 681, Gaithersburg, MD (1984).

User Interface Requirements Subcommittee Recommendations, Videotex Industry Association, Silver Spring, MD (March 1, 1991).

T. Valovic, Will TCP/IP Pave the Way for OSI? or TCP/IP? *Global Inte Net* (April 1992), p. 3.

D. L. Wenner, Are You Ready for Residential Broadband? *Telephony* **216**(21) (1989), pp. 84, 86, 88, 97, 99–103.

Chapter 6

Communications Services

A bridge must exist for a user to access an enhanced system. In the public arena, the bridge may be the ubiquitous telephone network and its progeny—T-1 services, the emerging ISDN, or various bypass options; on a broader scale, it may consist of interlinked computer networks. The largest of these is Internet, which is composed of over 5000 networks. Internet operates in over 40 countries and ties together more than 300,000 computers in several thousand organizations. Within an organization or a limited geographical area, a private branch exchange or area network may be the thread that joins the user to the system.

There are many ways of connecting the user and the information. The reader should be familiar with the existing options and their strengths and weaknesses. This is especially important given the inability of many of the options adequately to support the transfer of multimedia content. Chapter 6 examines the major public and private communications services and discusses the role each can play in the area of enhanced information systems.

6.1. SOME TERMINOLOGY

In business as well as in private life, the word communication usually means an oral exchange between two people; this is especially true when the parties involved are separated by great distances, but voice communication is not where the major growth is. Corporate voice needs are increasing by only about 5% annually compared to data needs, which are increasing 35% a year, and video traffic, which is more than doubling every year.

Communication services that operate in only one direction, such as a radio broadcast are referred to as simplex. Communications flowing in both directions but only one direction at a time are called half-duplex. The simultaneous exchange of communications in both directions is full duplex (see Figure 6.1).

Although enhanced services can operate over narrowband and voiceband channels, the need for rapid response time and the amount of data that must be transmitted generally dictate using a wideband service. Fiber optics have revolutionized wideband services in the United States. The heart of a fiber-optic system is a tiny laser that pulses light 565 million times per second or faster; the absence of light represents a 0 and the presence of light, a 1. The glass filaments that form a fiber-optic carrier are grouped as pairs for full duplex transmission, reinforced for strength, and bundled into large cables. An optical fiber network carrying 565 Mbits/sec can support 8000 phone conversations or transport 28,000 pages of text

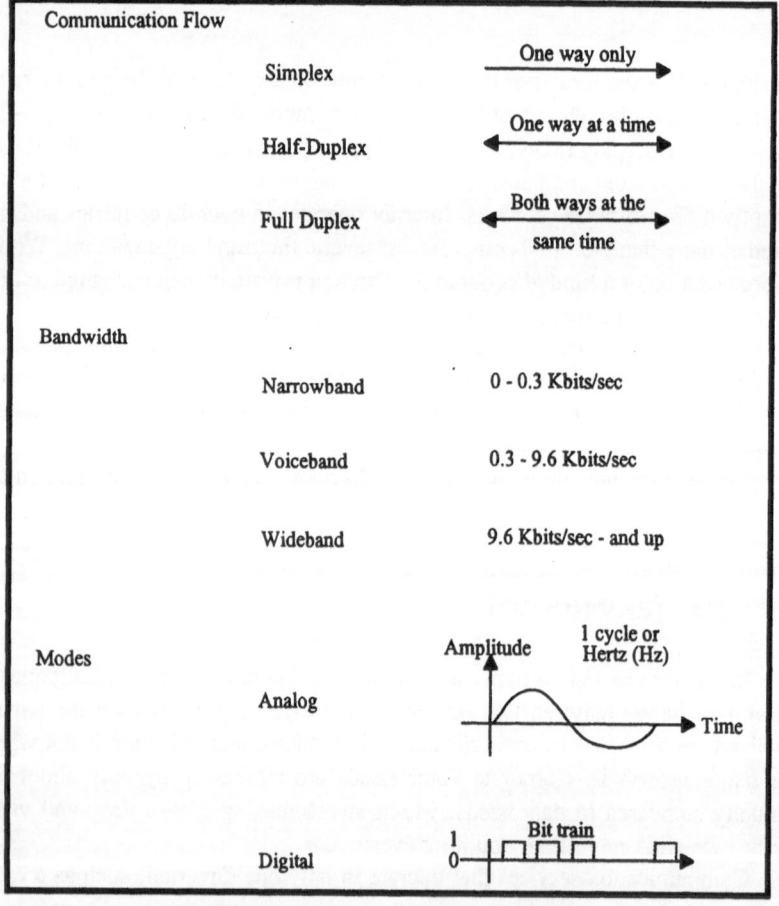

Figure 6.1. Communications terminology.

each second. Prior to the advent of optical fiber, most communications was analog. Although analog transmission is still the most common form of communication, the shift to digital is occurring at a rapid pace.

An analog transmission represents data as a continuous signal (see Figure 6.1). Analog is based on the ability to measure physical quantities, such as voltage levels and variations in three properties of a wave—amplitude, frequency, or phase. Analog transmission has several advantages over digital; the most important is its availability. Dial-up analog voiceband service is available anywhere in the world, and equipment is generally cheaper and less complex. Analog transmission generally does not have the access-line distance limitations of digital; however these three advantages are temporal, and digital is clearly the preferred method.

In a digital transmission, data are represented as a discrete signal, which may take the form of the presence and absence of an electrical pulse or light. The advantages of digital transmission are lower error rates; higher transmission rates; no need to convert from the digital format used by equipment to analog and back to digital at the receiving end; and the ability to integrate voice, data, and video, which is important in enhanced systems environments.

Integration within a communications medium can occur in two ways: (1) Separate channels within the same medium may be reserved for voice, data, and video; this is the approach used in broadband LAN but it is not true integration. (2) The transmission from each source can be intermingled on a single channel; this form of integration is more complex and expensive than simply creating multiple channels. Integrating voice, data, and video conserves bandwidth and more fully employs the existing medium. It reduces the number of wires that must be installed, it facilitates adoption of integrated voice/data terminals, and it supports the concept of a single and unified approach to network management.

The characteristics of voice, data, and video transmissions vary tremendously and make the process of integration difficult (see Table 6.1). Bandwidth can vary from a few bits per second for fire, security, and building environmental controls to millions of bits per second for video. The response requirements for voice are measured in fractions of a second. A delay of more than a half-second makes voice conversation nearly impossible. On the other hand, data exchanges can be interrupted for several hours for some applications. Employees and consumers are intolerant of interruptions to telephone service, but they are much more tolerant of disruptions to video transmissions.

Since voice communication has presumably two intelligent beings as part of the process, the system does not have to provide the same level of error detection and correction as is needed for data exchanges. Human beings are equally adept at dealing with distortions and snowy images appearing in a videocast. The usage level for voice, data, and video is relatively constant and predictable. One of the major contrasts is in the area of protocols. Voice networks demonstrate full connectivity from anywhere to anywhere in the world. There are only three major

Table 6.1.   Voice, Data, and Video Characteristics for Interactive Communication

Characteristic	Voice	Data	Video
Bandwidth	3000 Hertz	1 to several Mbits/sec	112 Kbits/sec plus
Response requirements	Immediate (0.5 sec)	Vary widely based on applications	Full-motion video varies from immediate to 3 sec for a snapshot video
Availability requirements	Highest	Medium	Lowest
Reliability requirements	Highest	Medium	Lowest
Error detection and correction	Participant controlled	Programmable	Programmable
Channel occupancy	Relatively constant	Relatively constant	Relatively constant
Protocols	Worldwide standard	Multitudes	National standards

standards for broadcast video; however a babble of data protocols exist, and more are being developed despite the efforts of standards-making organizations to promote universal connectivity.

6.2. PUBLIC SYSTEMS

The 1990s will be the golden age of communications services: A large selection of public services will be competing for customers, which will be reflected in the variety of alternatives and pricing; Table 6.2 lists many of the options that will be available. The services offered through the public switched network (PSN), better known as the telephone system, will predominate; several of the more important ones are discussed in the following sections.

Table 6.2.   Public Communications Services

Service	Acronym
Community Antennae Television (commonly referred to as cable)	CATV
Direct Broadcast Satellite	DBS
Multipoint Distribution Service	MDS
Public Switched Network	PSN
Local Exchange Carrier	LEC
Interexchange Carrier—long-distance services	IC
Subsidiary Communications Authorization—data radio	SCA
Subscriber TV—may include data channels	STV
Value Added Network	VAN

6.2.1. Telephone Service

Telephone service in the United States in ubiquitous: Virtually all businesses and 93% of the homes have at least one telephone. As is true of most innovations, the telephone was not initially greeted by the world in 1876 with much enthusiasm. The *London Times* referred to it as "the latest American humbug." Western Union, the telecommunications giant of its era, was offered exclusive patent rights to the telephone, but its chairmen declined the offer, saying, "What use would this company have for an electrical toy?"

This toy has emerged as a sophisticated electronic device; today more than 60% of all telephones purchased in the United States have the ability to do more than simply place and receive calls. Programmable telephones can memorize up to a hundred numbers; unattended telephones can dial a list of numbers and play a recorded sales pitch. Some telephones contain voice recognition devices that enable them to dial the person whose name is spoken. Telephones are in cars, on the walls of elevators, clipped to shirt pockets, on commercial planes, in the backyard, and on commuter trains.

The proliferation of the telephone and changes in its appearance have been evolutionary rather than revolutionary. Perhaps the most dramatic changes have appeared in the last 30 years as a result of the development and resolution of the AT&T antitrust suit. Chapter 7 contains a short history of recent deregulation in the U.S. telephone industry. The breakup of AT&T has had a major impact on communication services. The steps leading to the 1984 divestiture had its roots in the 1956 Consent Decree and the subsequent regulatory actions called the computer inquiries. Every 3 years following the breakup, the court reexamines its decision to determine whether additional deregulation is in order. By the mid-1990s, it is likely that the communications industry will be almost totally deregulated.

The telephone has several important advantages as a device for supporting data communications. It is well established, it is found on virtually every desk and in every home; and it has addressing capability. In some parts of the country, it is also possible to acquire a calling number identification feature to indicate who or what is calling. The telephone can support full duplex communications, and the telephone company can serve as the gateway to many services providing data transmission, address translation, protocol conversion, billing management, and introductory information for accessing services.

The U.S. telephone network has significant excess capacity. A 1987 study revealed that only 10% of network capacity was used and that the average household telephone was used for only six minutes a day. This remaining capacity has the potential for generating billions of dollars for the telephone industry; in fact each additional minute the network is used represents an additional $20 billion in revenues for telephone companies. Consequently the telephone industry is anxious to leverage its network resources to begin reaping the potential bonanza.

The major liability of the telephone network is a low data rate, although this is beginning to change with the introduction of ISDN capability and other services mentioned in the following sections. The telephone network was designed for voice communication. A typical voice call lasts three minutes, whereas data communication is very bursty and may last for hours; this can be costly, and it may tie up the telephone, thereby preventing its use by others.

6.2.1.1. How the Telephone Works

A basic understanding of how the telephone set and the telephone network functions is important for understanding the role the telephone plays in implementing enhanced services. In this section and the next, the PSN at the local level is described; private or in-house telephone systems found in many businesses operate in a similar fashion.

At the local level, the telephone system is set up in a hub and spoke configuration. Each telephone set is connected to a central office (CO), which contains switching equipment for a particular geographic area. The connection is made with two wires called the wire pair. One of the wires is referred to as the tip or T wire and the other as the ring or R wire; a diagram of the telephone phone set is shown in Figure 6.2. A more complete explanation of the components is contained in the Fike reference at the end of the chapter.

When the handset is resting in the cradle, the switchhook is forced open, which opens the circuit between the telephone phone set and the CO; this is called the on-hook condition. When the handset is lifted from its cradle, the switchhook closes, which completes the circuit to the CO; this is called the off-hook position. One of the events that occurs in the off-hook condition is the transmission of a dial tone from the central office.

There are two ways of sending the telephone number being called to the central office. Dial pulsing is the older method, which often uses a rotary dial that opens and closes the loop's circuit at a timed rate. The number of dial pulses created is determined by how far (what number) the rotary dial is turned prior to release. The second and newer method is called dual tone multi-frequency (DTMF). Instead of a rotary dial, the telephone has a 12-button keyset for the digits 0–9 and the two special characters, * (asterisk) and # (octothorpe). When a key is pressed, an electric circuit generates two tones in the 300–3400 H_Z bandwidth of the telephone network. For example, pressing the digit 7 produces tones of 852 H_Z and 1209 H_Z. Frequencies for the rows and columns and the layout of the extended keypad are shown in Figure 6.3.

When the person who is called removes the handset from its cradle, the loop is closed on that end also. The CO removes the ringing signal and the ring-back tone, and a continuous loop exists between the two telephones. The sound created by the person speaking is converted by the transmitter in the telephone's transmitter into electrical current. On the receiving end, the electrical current is converted by the receiver into acoustical energy (sound) that the listener hears. When either party hangs up the handset, the loop is broken, and the CO disconnects the lines.

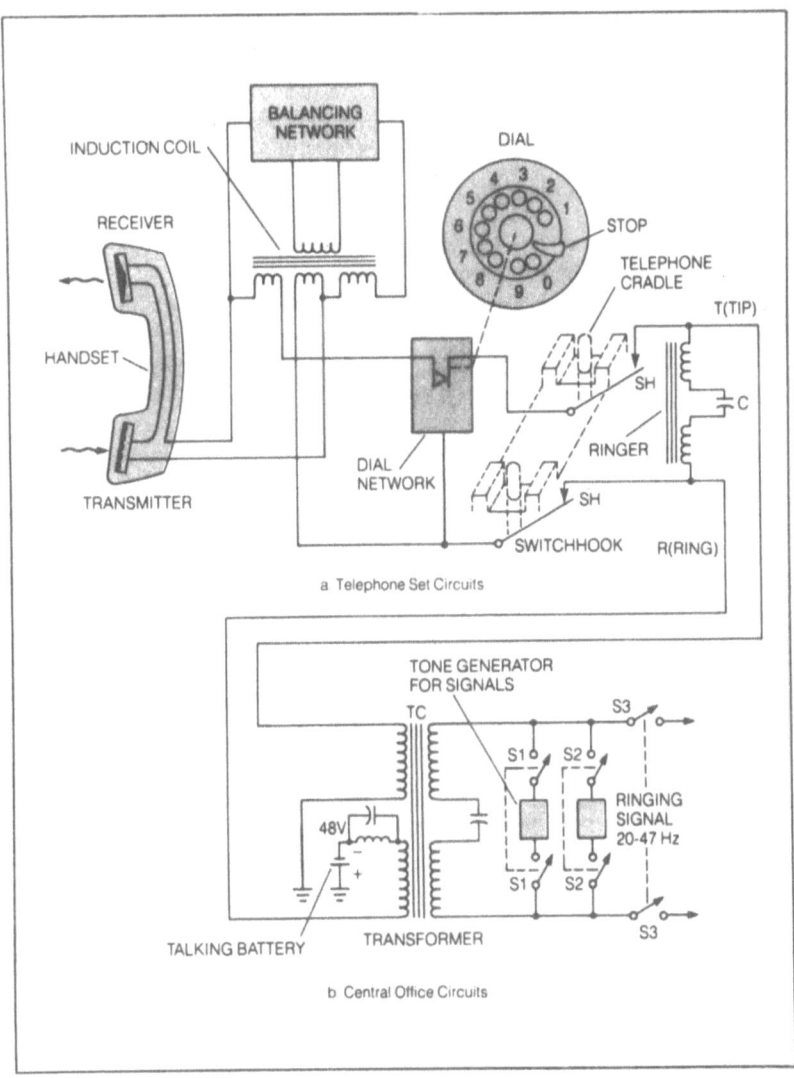

Figure 6.2. The telephone set. [From *Understanding Telephone Electronics*, by J. L. Fike and G. E. Friend, Sams, Indianapolis (1984). © 1984 Texas Instruments Incorporated.]

The DTMF has several advantages when compared to dial pulsing. For an enhanced service, one of the most important is that once the loop to the other telephone has been established, the digit receiver is out of the circuit, and DTMF tones can be transmitted the same as speech. This enables the DTMF unit to function as a terminal for electronic messaging, information retrieval, and transaction processing. Other advantages of DTMF when compared to dial pulsing are

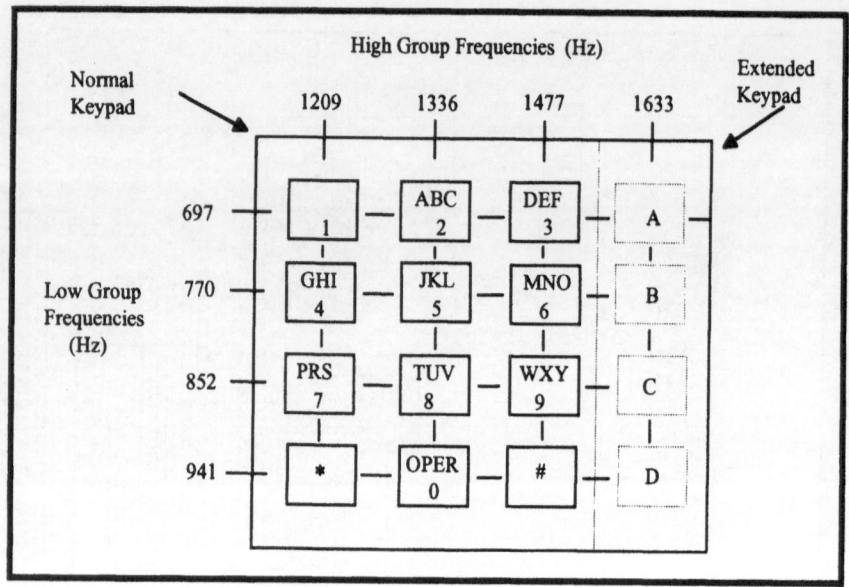

Figure 6.3. DTMF-extended keypad.

faster dialing (approximately 10 to 1), reduced equipment requirements at the CO, and compatibility with electronically controlled exchanges.

6.2.1.2. The Local Loop

The area served by the local telephone company is divided into geographic regions called local access and transport areas (LATAs); at last count, there were 165 LATAs in the United States, and the number is slowly increasing as population changes occur. Interexchange carriers (ICs), such as AT&T and MCI, provide the bridge for communication between LATAs even when the two LATAs are adjacent and within the operating area of the same telephone company (*e.g.*, Ameritech). Within each LATA, there are often several COs containing the switching equipment to help route the call.

The basic layout of the area served by a CO is shown in Figure 6.4; a more detailed explanation of the network structure is given in the Fike reference at the end of the chapter. Telephone lines or an outside plant cable pair connecting a subscriber to a telephone company's central switching office is called the local loop. Wire pairs that form the local loop fan out from a location called a wire center to the serving area; in an urban environment, an average wire center will serve 41,000 phones and 5000 trunks.

The transmission network consists of the physical lines, switches, support cables and poles, transformers, amplifiers, and so forth. Transmission facilities fall into three categories: metallic, analog carrier, and digital carrier. The metallic

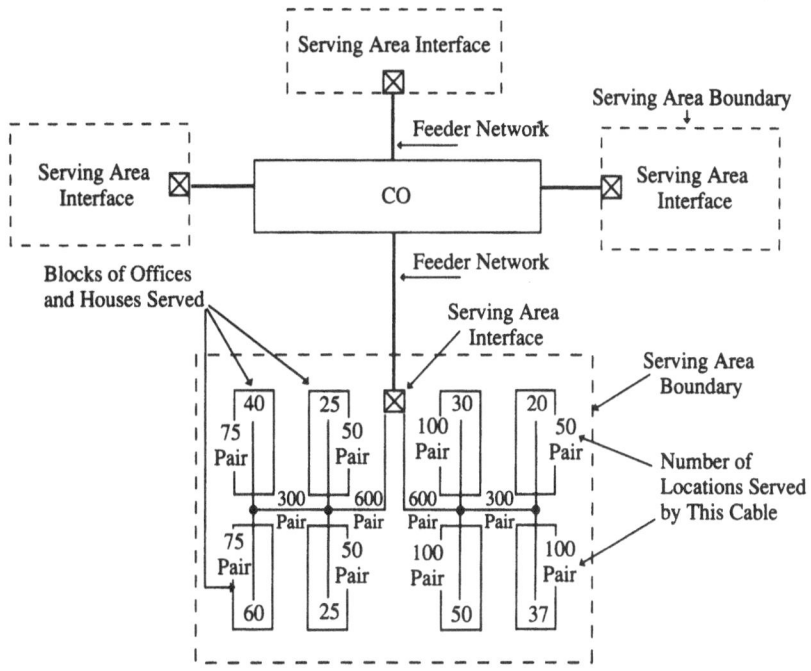

Figure 6.4. The local loop with an expanded view of a serving area.

facility is the simplest of the three; it is composed of the two wires connecting each phone to the CO. Analog carriers consist of wire pairs carrying several channels; analog carriers may also encompass multiwire cable, coaxial cable, microwave, and satellite transmissions. Digital transmission facilities are similar to the analog, but they use a digital rather than analog format.

The necessity of a CO and its switching capability is clear from examining Figure 6.4. In this expanded view of a typical serving area, 287 homes and offices are being served. Without a central switching office, 41,041 ($n[n - 1]/2$) links are required to connect every subscriber, but with the CO, the number of links is 287. The national telephone switching network resembles a pyramid with local loops and their COs forming the base. Central offices are connected together directly and through intermediate and toll points; these are in turn combined into primary, sectional, and then regional centers at the top of the pyramid.

6.2.1.3. Intelligent Network and Universal Access to Information Services

Until quite recently, most of the intelligence found in communication networks was vested in the telephone and telegraph operators who regulated the flow of information. This gradually changed as networks were automated, and it changed significantly with the introduction of electronic switches controlled by software in the 1960s. In the past 30 years, digital switching, satellite communications, integrated-switching digital networks, Signaling System 7 (a control protocol), and network databases have accelerated the distribution of intelligence to the network. The pace of evolution will quicken as optical fiber is introduced.

A 1991 study by Kessler Marketing Intelligence (KMI) predicts that by the year 2001, over 40 million houses will be connected to a nationwide fiber-optic network. Fiber's declining cost will make this economically possible. In 1992 fiber-to-the-curb systems are less expensive than copper wire for voice telephone in some regions. The KMI forecasts that fiber costs will be lower in nearly every case by 1996.

The speed at which optical fiber transmits signals is approaching the transfer rate of data in a computer. This enables computer components to be distributed in the network without a deleterious effect on system performance. Eventually the network will assume some of the features of a giant computer whose far flung components are tied together just as today a computer's internal components are connected by its circuits. In essence the network becomes the computer. This will result in increased control and the proliferation of customized services. It will make available to business and residential subscribers applications that do not presently exist, and it will lead to the development of an information network in place of today's voice network. As the network becomes more intelligent, it will place pressure on many organizations to make information available electronically to an ever-widening group of constituents. Such organizations as the Alliance for Public Technology have already begun lobbying to ensure universal access to information services. This activity is based on the belief that the ability to locate, manipulate, and disseminate information is essential to effective participation in the society of the 1990s and beyond; and electronic information services are viewed as an increasingly critical component for providing access to that information.

The voice-based telephone network is ubiquitous and transparent. It provides simple and effective end-to-end communications. For most of its life this network has been considered a lifeline for business and residential users, which has fostered a policy of universal access that has determined how the industry is regulated and how money for its development should be invested. As the network's IQ rises, the cries to ensure universal access to information services will also increase. To connect terminal equipment to a telephone network, equipment must be registered with the Federal Communications Commission under Part 68, "Connection of Terminal Equipment to the Telephone Network" of its "Rules and Regulations." These rules are intended to protect the network from potential harm caused by the device; They do not guarantee that the equipment will function as advertised.

6.2.2. T-1 Service

Enhanced services can be provided through plain old telephone service or POTs, but not very well. In the next decade, ISDN will become the backbone of the enhanced services industry; T-1 is the intermediate stage in this evolution. The T-1 is a high-speed, digital transmission service carried over telephone lines. Although it is not commonly found in the local loop, the capability it provides will be extended to the local loop in the next generation of communications services through ISDN. The T-1 may be thought of as a huge pipe carrying digitized voice, data, and video.

The T-1 evolved from transmitting voice: To transmit voice in a digital format, the analog voice signal is sampled 8000 times a second. Each sample is then represented by an 8-bit byte. There are 24 channels multiplexed together to form a frame, with one bit added for control and synchronization at the front. This forms a frame of 193 ($24 \times 8 + 1$) bits. Multiplying the 193 bit/frame by the sampling rate of 8000 times/sec yields the T-1 rate of 1.544 Mbits/sec. Because so much T-1 equipment is used in the communications industry, especially within the telephone industry, the T-1 rate has become the *de facto* standard for transmission by satellite and other media.

The principal benefits of T-1 carriers are transmission speed and economy. High-quality video teleconferencing can be transmitted over a single T-1 facility when the signal is compressed. The T-1 is also suitable for high-speed facsimile systems in addition to fulfilling its traditional role in voice and data communications. On average it is cost effective when replacing seven voice grade lines and three to five 56-Kbits/sec circuits. In the next few years, T-1 and its derivative fractional T-1 will provide the bridge to ISDN, which will form the communications backbone of many enhanced systems. It will be possible to migrate from T-1 to ISDN using some of the same hardware. The T-1 and ISDN both operate at the same rate. To make the one compatible with the other, the control information is removed from 23 of the T-1 channels and combined in the twenty-fourth. This creates 23 channels capable of transmitting at 64 Kbits/sec, and one 64-Kbits/sec signaling channel. As explained in the ISDN section that follows, this will bring the integrated voice, data, and video world of the future much closer.

6.2.3. Fractional T-1 Service

A popular derivative of T-1 service, called Fractional T-1 or FT1, was introduced in the mid-1980s. The FT1 enables companies to purchase portions of a T-1 line in 64-Kbits/sec increments. The communications company requires special equipment, and the user must have a multiplexer to accommodate the FT1 format. The service is especially amenable to voice, video, facsimile, and data communication requirements of an organization because it enables the organization to choose the bandwidth needed for a particular application. In addition to the economics associated with being able to pay for just the bandwidth needed rather than a whole T-1 circuit, there are other benefits. Since the channel used is a subset of the full T-1 link, the high quality of voice transmissions is unaffected. The much lower cost for an increment of a T-1 line reduces the time to justify T-1 service. Secondary and backup channels can be reserved in the event of a disruption in communications far less expensively than dedicating a separate link for this purpose. The FT1 also offers another migration path to ISDN: Organizations can incrementally add bandwidth and when their need reaches the capacity of an ISDN link, the organization can upgrade to ISDN.

6.2.4. ISDN

The ISDN is a digital network that carries voice, data, telemetry, slow-motion video, and other forms of communication separately or simultaneously over a transmission medium accessed by a limited set of standard multipurpose user network interfaces. The concept of a limited set of interfaces is important because it means that the multitude of different access points of yesterday's communications environment is replaced with one interface. Another important characteristic is that the information content (speech, data, or video) is separated from the signaling and management functions during transmission.

6.2.4.1. Architecture

The ISDN introduces a new nomenclature created to describe components that make up the system. The most important of these are contained in Table 6.3 with a brief definition.

ISDN may appear in two forms: The basic and the primary rate access. The basic rate access consists of two 64-Kbits/sec bearer (B) channels and one 16-Kbits/sec signaling (D) channel (2B+D). Bearer channels carry voice, data, and video information that is to be transmitted. The B-channels are clear in that they do not carry signaling or control information; instead signaling and control information are carried out-of-band on the D-channel, which is largely reserved for this purpose. There are two benefits from having a separate D-channel: First all of the bandwidth in the B-channels may be used for carrying information; second the network can be controlled more efficiently.

Primary rate access consists of twenty-three 64-Kbits/sec bearer (B) channels and one 64-Kbits/sec signaling (D) channel (23B+D). The basic rate access is designed to satisfy information needs of a relatively small capacity, such as terminals. Primary rate access will meet the needs of large-capacity devices, such as PBXs, LANs, and computer hosts. In both types of access, the bandwidth needed to transmit signaling and control information may be less than the 16 or 64 Kbits/sec that is available. When this is the case, the D-channel can also carry packet data. It should also be noted that B-channels can be combined in multiples of 64 Kbits/sec to form channels with greater capacity. Future generations of ISDN will permit higher data rates than the present 1.544-Mbits/sec maximum achieved by combining all 24 channels. Data rates of 560 Mbits/sec are easily feasible in the future with optical fiber technology.

The general architecture for the user's side of the ISDN network is shown in

Table 6.3. ISDN Terminology

Term	Explanation
ISDN	A digital network that carries voice, data, telemetry, slow-motion video, and other forms of communication separately or simultaneously over a transmission medium accessed by a limited set of multipurpose user network interfaces
Basic rate access	Two 64-Kbits/sec bearer (B) channels and one 16-Kbits/sec signaling (D) channel
B-channel	A circuit-switched bearer channel
D-channel	The packet-switched signaling channel
NT1	A network termination (type 1) device that converts a 2-wire transmission (U-interface) into a 4-wire customer distribution wiring (T-interface)
NT2	A network termination (type 2) device that can perform switching and concentration such as a digital PBX
Out-of-band signaling	Transmitting signals on a channel dedicated for that purpose
Primary rate access	Twenty-three 64-Kbits/sec bearer (B) channels and one 64-Kbits/sec signaling (D) channel
R-interface	The interface between a terminal adapter and a non-ISDN terminal
S-interface	The interface between a piece of network-terminating equipment and an ISDN terminal or terminal adapter
T-interface	The 4-wire, physical interface between terminal gear
U-interface	The interface between the ISDN network and network-terminating equipment
TA	A terminal adapter that permits non-ISDN terminals to operate on ISDN lines

Figure 6.5. The network is designed to accommodate both new equipment that is compatible with ISDN technology as well as older equipment. Compatible terminals are designated as TE1 and noncompatibles as TE2; noncompatibles require a terminal adapter to operate on ISDN lines.

Four reference point interfaces (R, S, T, U) are defined; separate signaling is supported for all but the R interface. The meaning of the four reference points is explained in Table 6.3 and their location is indicated in Figure 6.5. Terminals are connected to the network with network termination equipment designated as NT1 and NT2. The NT1 is the carrier side of the network connection, and it is responsible for terminating the transmission, managing data flow, and sending signals from the CO to test and monitor the system. The NT2 is the user side, and it takes care of such things as multiplexing and demultiplexing the B- and D-channels.

6.2.4.2. Implementation at the Local Loop

Implementing ISDN at the local level requires overcoming technical and economic hurdles. Implementation will not require replacing wires that are

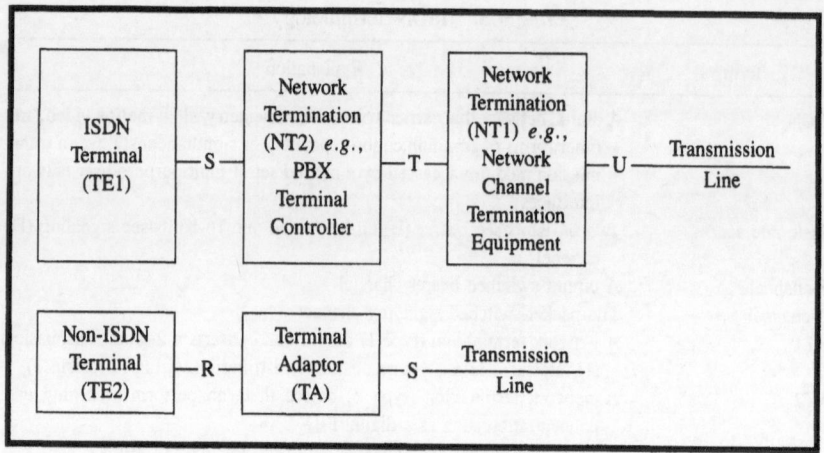

Figure 6.5. ISDN user and network interfaces.

presently strung, since there is nothing inherently digital or analog about wiring; however there are loading coils built into the local loop. These are placed approximately every three miles to amplify voice frequencies, and their existence curtails the transmission of frequencies above 4000 H_Z. A 1983 survey revealed that the average length of the local loops in the United States is 10,400 ft. The survey also revealed that 83% of the local loops are nonloaded; that is, they do not have loading coils. Since the local loop is unloaded, digital signals can be sent only at low speeds and for short distances. It will be necessary to remove all loading coils and install regenerative repeaters for digital transmission. Unfortunately the lines so modified will then be unsuited for analog voice signals.

The solution around this problem is either to install separate lines for ISDN or to digitize the voice signals before they enter the local loop; the latter solution is more practical at this time. A related problem is line noise; for example, 88% of the loops tested in New York were found *not* to be quiet; studies have also found major cross-talk contamination from other high-speed loops. The local loop can be converted to a digital facility, but the cost will be tremendous. It has been estimated that 40% of the telephone companies' total plant investment is in the local loop; in the United States, this represents $80 billion. Business and residential subscribers have also made substantial investments in equipment for both voice and data. Consequently ISDN will appear gradually over a lengthy period as depreciated equipment is replaced and new facilities are built. During the transition period ISDN terminal adapters will be a popular way of implementing service economically. An ISDN terminal adapter connects non-ISDN data terminal equipment to ISDN lines.

Initially the cost-benefit trade-offs of hardware, software, maintenance, and training will attract large businesses. An ISDN access mode will be installed at the CO, and business networks will be connected to offer various services. Within each business a standard interface will be installed to which different types of terminals can gain access. As the ISDN service permeates the community, small businesses that use PCs to access information, perform bank-related transactions, and are involved in electronic document interchange will adopt it. Eventually ISDN will reach the residential market to serve individuals using computers and terminals for videotex-related activities and telecommuting.

During this transition period, there will be quasi-ISDN services requiring a smaller investment that will help develop the market for true ISDN; short-haul technologies and derived channel strategies are two likely candidates. As the name implies, short-haul technology is aimed at providing high-speed data communication over short distances. A popular piece of equipment for doing this is the line driver. This device drives a digital signal directly across a transmission path.

Two examples of derived channel approaches are local area data transport (LADT) and the technology used in Project Victoria. The LADT began operating in Florida on July 1, 1983, as part of the short-lived Viewtron videotex service; it has been reincarnated in various forms since then. There are three types of access with the LADT network: dial, direct, and host. The dial-up version is a 1.2-Kbits/sec service for residential and small business users; direct access (4.8 Kbits/sec) is designed for applications where continuous availability is needed, $e.g.$, security services and credit verifications; the host access (9.6 Kbits/sec and higher) is for high-volume uses, such as business-to-business communications. The direct access permits voice and data communications to occur simultaneously by transmitting the data above 4 KH_Z and the voice signal in its standard frequency range of 0–4000 KH_Z. Special equipment is needed at the user's location and in the CO.

Project Victoria was a trial communications service started in August 1986, by Pacific Bell; approximately 200 homes in Northern California participated in the project. Service enabled a single twisted copper pair in the local loop simultaneously to transmit on seven channels: two voice, one 9.6 Kbits/sec, and four 1.2 Kbits/sec. The trial was a successful demonstration of the technology, but it was beset with regulatory problems during the period it was conducted and afterward. In May 1988, Pacific Bell announced its intention to license the technology to others; shortly afterward the company revealed it would not deploy the service on a trial basis itself due to the regulatory uncertainty confronting the telephone industry at the time.

6.2.4.3. Will It Ever Get to Market?

Introducing ISDN has been a long and difficult process, and there are a variety of reasons why it is taking so long. A major problem has been incompatible products from different vendors, which makes it extremely unlikely that equipment from two different vendors will work together. The basis for this problem has been the failure to have a completed standard. In late 1991, the Bellcore standard to govern implementation among the regional Bell operating companies (RBOCs) was still not 100% defined.

Another factor has been an apparent lethargy among the RBOCs. All of them will have ISDN primary services available by the end of 1992, but only two had services in mid-1991. Even where services exist, they are often islands preventing interconnection with other portions of the network upgraded for ISDN. The power and flexibility of ISDN have at times been its own worst enemy: There are so many options that choosing among them can be time consuming and difficult; for example, there are 120 packet parameters for setting up a basic rate line.

Price is also an issue, but it is becoming less so as ISDN spreads. Ameritech was the first RBOC to begin an ISDN trial in 1986, and it was the first to file a tariff for its non-Centrex business customers. The tariff was filed in June 1990, and enabled a business customer in Chicago to order one line that could simultaneously handle voice and data communications for $22.66/month and a one-time installation fee of $94.50; two business touch-tone lines providing the same service would cost $18.78 and a one-time installation fee of $148.50. But the largest barrier to ISDN deployment is the lack of users clamoring for the service. With so many alternative forms of communication available and the special and relatively expensive end-user devices necessary to adopt ISDN, subscribers are not beating a path to the RBOCs. Those who have adopted ISDN report a variety of benefits; these include reduced communication costs, greater flexibility in allocating bandwidth, practically error free transmission, much higher transmission rates, increased productivity, and a host of new telephone features.

6.2.4.4. Broadband ISDN

Before initial ISDN services have even become common, the next step in the evolutionary cycle has started. Broadband ISDN or BISDN is based on ISDN functionality and the transmission speed made possible with fiber optics. It is one of several next generation network services that will accommodate the bursty nature and bandwidth requirements of multimedia data transmissions (see Table 6.4).

A digital-based television signal can require as much as 90 Mbits/sec, if uncompressed, and high-definition television could even double that requirement.

Table 6.4. Next Generation Network Services

Service	Abbreviation	Data Rate in Mbits/sec
Asynchronous transfer mode	ATM	1000
Broadband ISDN	BISDN	600
Synchronous optical network	SONET	51.84
Switched multimegabit data service	SMDS	45

Fortunately full-motion video is possible at 384 Kbits/sec using compression algorithms. But even when requirements for throughput are reduced that much, the ISDN basic rate access service is inadequate. Combining the two B and the D-channels creates a 144-Kbits/sec pathway, but even that is insufficient. Suitable capacity can be created by combining some or all of the 23 B- and D-channels on a primary rate access ISDN service. The BISDN is designed to satisfy the need for full-motion video and other transmission-intensive applications, such as LAN interconnection. With BISDN the bandwidth threshold is raised to 600 Mbits/sec and beyond. Trials employing this next step in the evolution of ISDN are planned for the mid-1990s, with implementation occurring near the end of the decade.

There are three fundamental differences between ISDN and BISDN other than speed. ISDN makes use of the telephone infrastructure and its copper wire pairs, while BISDN uses optical fiber cable. ISDN employs circuit switching and performs packet switching only on the signaling channel, D, while BISDN uses only packet switching. ISDN rates are prespecified, while BISDN has virtual channels allowing multiple data rates of very high speeds.

6.2.5. Bypass Services

As discussed earlier, a telephone company's services are divided into service areas known as LATAs. The LATAs were created from areas covered by RBOCs following the breakup of AT&T. Consequently the LATAs tend to follow state boundaries, and they are organized around major population centers.

The LATAs establish the boundary between local and long-distance services. An RBOC can provide local service within each of its LATAs, but communication that crosses the boundary of a LATA, even a LATA within the same RBOC's region, must be handled by an interexchange carrier. A long-distance carrier such as AT&T or Telenet establishes a point of presence within a LATA and provides from that location inter-LATA service.

Some common media for communicating within and between LATAs is shown in Table 6.5. Twisted pair is by far the predominant media for inner-LATA

Table 6.5.   Selected Media Found
in Inner-LATA and Inter-LATA Communications

Inner-LATA Media	Inter-LATA Media
Twisted pair	Microwave
Microwave	Optical fiber
Coaxial cable	Satellite
Optical fiber	
Data radio	

communications; however microwave, optic fiber, and coaxial cable are also used. Data radio is gaining in popularity for one-way transmission.

Most residential and business customers are connected to the outside world through the telephone company's local loop. It is estimated there are at the present time 200 million local loops in this country; most of which are analog. When a customer's communications do not use the local loop but instead some non–telephone-company-supplied form of service, it is referred to as a bypass. There are at least three types of bypass: service bypass, facility bypass, and total bypass. A service bypass may be provided by the telephone company for an organization that requires a high-speed communication service, such as T-1. Such a bypass enables the company to avoid the CO within the same LATA. A facility bypass is provided by someone other than the local telephone company. It avoids the local loop altogether by constructing a facility bypass, which enables the customer to transmit to the point of presence and into a long-distance carrier's network. The third form of bypass is referred to as a total bypass, and it occurs when communications are between users, thereby avoiding both local and long-distance carriers.

In 1988 the RBOCs and GTE estimated that user bypasses cost them approximately $3.8 billion a year. Bell Atlantic Corporation claimed the largest loss—an estimated $888 million per year. California has the highest incidence of bypass with an estimated $467 million per year in lost revenue for Pacific Telesis. As an addendum to the topic of bypass, it is important to realize this is not a recent phenomenon; in fact in the early 1900s, farmers in remote regions not serviced by telephone companies discovered they could use bard wire to transmit telephone signals. Railroads, electric companies, and all levels of government have also practiced bypass for decades. However as John Gantz points out, the telephone companies may have only themselves to blame; it is often "their own structural incompetencies which are causing the user infidelity." Some bypass alternatives are discussed in the following sections.

6.2.5.1. Data Radio

Data radio or, as it is also known, subsidiary communications authorization (SCA) is an inexpensive and relatively unsophisticated one-way communication service. Data radio uses a portion of the signal produced by an FM radio station. The portion of the broadcasting cycle from 88—108 MH_Z has been set aside for FM radio stations, and each station is allocated a 100 KH_Z bandwidth. A segment of the bandwidth must be set aside as a guard band, leaving 99 KH_Z for transmission of content. An FM stereo uses ± 53 KH_Z of that envelope, which leaves ± 46 KH_Z for other purposes. A data channel can be created by establishing an audio subcarrier; e.g., at 67 KH_Z and modulating it $\pm$ 6 KH_Z. A subchannel bandwidth of this size is adequate for supporting a transmission rate of 9.6 Kbits/sec.

Data radio has been used to broadcast news, financial reports, price updates, check alerts, inventory changes, and other applications for which a one-way information flow is adequate. It is usually cheaper than transmission services provided over the local loop, and it has the added advantage of communicating with every location simultaneously within the listening area.

6.2.5.2. CATV Data Carriage

The most pervasive bypass technology is already installed and awaiting data customers; it is the nation's cable television system. By 1988 it had been installed in 52% of the television households; by 1992 this figure is forecast to be 57%. Cable television passes or is in within close proximity of most suburban and inner-city businesses, and most importantly it has considerable unused capacity. Community antennae television (CATV), or cable as it is popularly called, is being supplemented with dedicated data-carrying systems that are being built into the business districts of most large cities today.

The first commercial cable television systems were constructed in the late 1940s to serve areas where natural barriers interfered with television reception. In the intervening years, justification for subscribing to cable has changed from better reception to a desire for a greater selection of programming. The next step in cable's evolution is incorporating data services, such as home security, energy management, information retrieval, shopping, messaging, and business-to-business communications.

6.2.5.2.1. Interactive Cable Architecture. Business- and transaction-based consumer services require two-way capability. In 1972 the Federal Communications Commission began requiring major cable systems to include a technical capacity for nonvoice communications from the user to the system operator. However most system operators have incorporated the ability, but not the equipment needed to sustain two-way communications. Consequently the number of functioning two-way systems is small, and only 11% of the 800,000 miles of installed cable was interactive in 1989. The basic architecture of an interactive system is shown in Figure 6.6. The heart of the network is the head end. This is where equipment is located for receiving the television signal, and it may include a satellite earth station, microwave, and feeds from other cable systems. A gateway and perhaps computer storage facilities support communications with users not on the cable system.

Integral to the operation of a system are band pass and band stop filters, processors, converters, modulators, and mixers. Filters allow desired signals to traverse the network and block out noise. Processors compensate for fluctuations in the input signal and introduce a high-level signal to the network. Converters move signals to another channel, and modulators incorporate signals from different media and impress them onto the channel carrier.

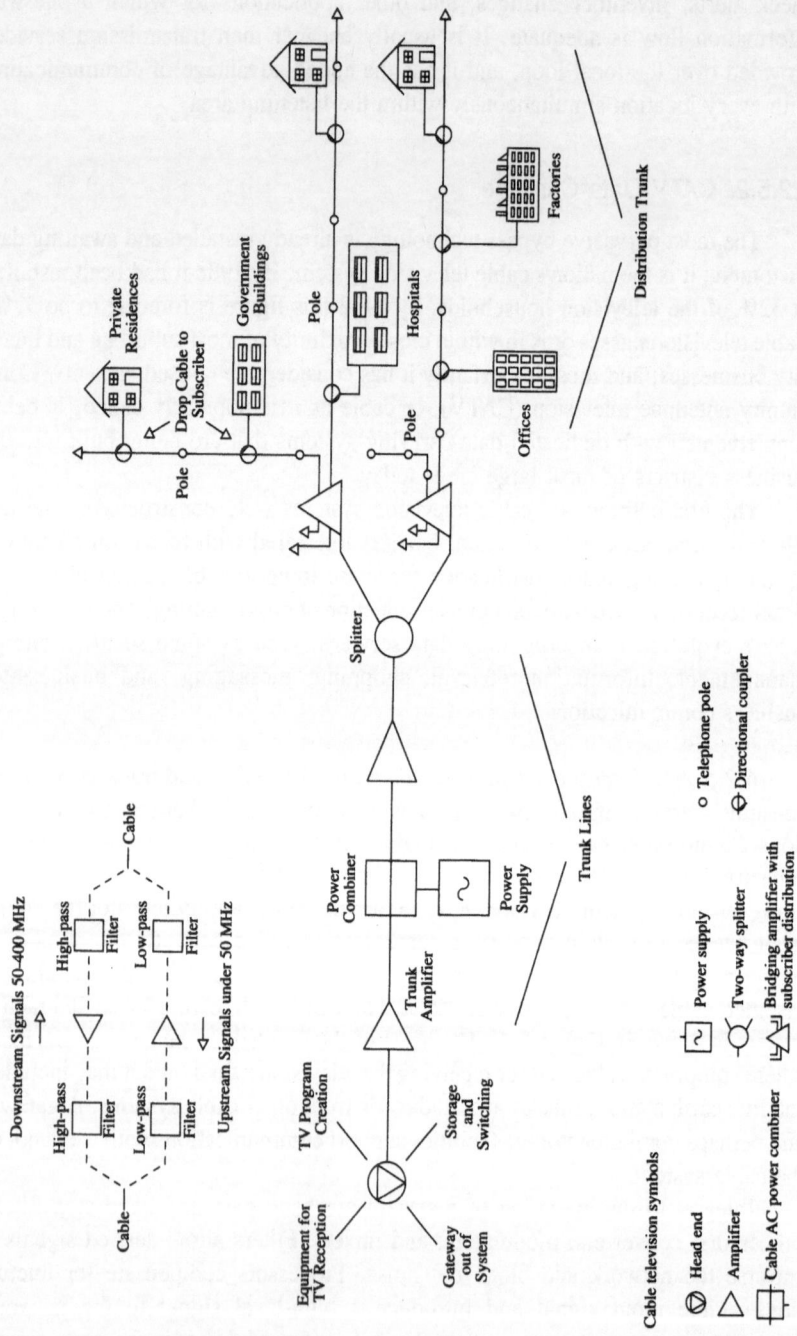

Figure 6.6. Interactive cable architecture.

The network is composed of trunk, distribution, and drop lines. The two-way flow of signals between the head end and each user may take place over a bidirectional cable with downstream and upstream videos and data movement, or two cables may be installed to move data in each direction. Some hybrid systems exist that support a limited form of bypass by using cable for delivery and the telephone network for requests. Trunk, bridging, and distribution amplifiers are located throughout the network. In a bidirectional cable, separate amplifiers must be provided for the downstream and the upstream signals. In a two-cable system, each cable has its own amplifiers, and the trunk cable can be subdivided with splitters. Bridging amplifiers provide the interface between the trunk and distribution lines. An additional element called a line extender raises the signal to a satisfactory level if the distribution line is long. Directional couplers form the final bridge and connect the distribution lines and the customer's drop cable.

In a bidirectional cable, downstream and upstream communications occur on different frequencies. The capacity of a cable is most primarily determined by the amplifiers and their spacing in the network. Current technology usually provides a bandwidth of 400 MH_z and above. The bandwidth for each channel is approximately 6 MH_z, which allows a throughput of between 5–10 Mbits/sec and a bit error rate of 10^{-8}; the telephone network has a bit error rate of 10^{-6}. The downstream amplifiers handle regular television programming and information content within the range of 50–400 MH_z (see Figure 6.6).

The upstream path is typically divided into three subpaths. The range from 5–12 MH_z is for such teleservices as monitoring systems and meter reading and for data transmission; the range from 12–20 MH_z is for packet or block transmission of data; and the range from 20 MH_z to approximately 30 MH_z is allocated for video signal transmission. The range from approximately 32 MH_z to 50 MH_z is not used, because the amplitude and phase characteristics of the filters are worst in this range. Schemes other than these are possible, and a variety of protocols, the set of rules and procedures governing the exchange of information within the system, are possible. Modifications to the carrier sense multiple access collision detection (CSMA/CD) and polling protocols, discussed later in this chapter, are the most common.

6.2.5.2.2. Advantages and Disadvantages. Coaxial cable and optic fiber provide bandwidths tens and even hundreds of times greater than such services as data radio and the telephone network. The CATV in general is an option when particular locations can not be served by over-the-air communication services due to traffic congestion or the existence of natural barriers. In systems with excess capacity, CATV may be the most economical form of point-to-point communications available. Perhaps most important from a strategic standpoint, CATV gives a company an alternative to using the local telephone company's network and thus provides leverage for negotiating and backing up existing services.

Unfortunately public systems supporting interactive forms of data exchange are not widely available. The average cost of construction is approximately $20,000 per mile, and franchises granted by large cities are usually competitively awarded and expensive to obtain. Once installed even those cable systems that do not carry television programming are susceptible to interference and noise if they are not well shielded.

Cable networks can act as antennae and absorb undesirable electromagnetic interference (EMI), created by industrial machinery and other sources of electrical energy. A cable placed too near an alternating-current power line or even a powerful citizen's band broadcast may affect the signal that carries data.

To compete with the telephone network, a high level of availability is required (99.99% is suggested). This can be achieved only with standby power supplies, redundant amplifiers, sophisticated monitoring equipment, and well-trained personnel. Carrying data traffic requires a much higher level of system support than does carrying television programming. A final point is security. In a telephone network, a private link is established between the two locations involved, and the specific link carrying the communications must be intentionally accessed to compromise security. In a cable system, channels reserved for data communication are usually shared with many other users. Messages contain an address, and they are broadcast throughout the system for reception by the designated addressee. Since anyone accessing a particular channel can intercept messages, data encryption is a necessity when financial or proprietary information is being transmitted.

6.2.5.2.3. Upgrading the CATV System with Fiber Optics. In 1989 the cable television industry spent almost $1 billion on rebuilding or upgrading. A significant portion of this money was for fiber-optic cable to replace or complement coaxial cable. Fiber optics is being installed in several locations in the network.

The system design or architecture shown in Figure 6.6 is called a tree and branch design. This is the most efficient way of distributing a package of channels of programming from a headend to subscribers. Fiber optics is being installed for linking separate headends together, which is known as supertrunking; it is also being used as a backbone link to connect the headend with some distant point in the network from which coaxial feeder lines pick up the signal and distribute it to the subscriber's drop cable. Fiber optics is being used to carry the signal from the headend all the way to the curb—the point where each subscriber taps into the system.

It will be some time before the final segment, the connection from the tap to the subscriber's unit, is fiber optic. A study by Rogers Cablesystems estimates the crossover point for the television industry is some years off (see Table 6.6) because fiber optic is not cost effective for short links. The more costly elements of fiber technology are the connections between fiber and the conventional electronic systems that deliver the signal to the receiving device. Since each subscriber needs this connection and thus the connection cannot be spread among many customers, fiber cannot economically make that last leap from the curb to the

Table 6.6. Deployment Costs per Subscriber for Television Signals

Fiber Penetration	1992	1999
Fiber to the bridger	$ 315	$223
Fiber to the node	$ 675	$276
Fiber to the home	$1835	$377
Coax to the home	$ 344	$344

Source: G. Kim, Studies Advance Fiber Penetration, *Lightwave* 9(2) (1992), pp. 1, 20.

set; thus coaxial cable will continue to be used. It should be noted that the telephone industry faces a similar hurdle in installing fiber optics.

6.3. PRIVATE SYSTEMS

Private branch exchanges, LANs, and WANs are often the information bridges within an organization. More and more firms are acquiring their own systems for internal communications rather than relying on outside suppliers, such as the telephone company, because it is much less expensive. In addition it is generally easier to make the information exchange secure, and the company can configure the system according to its specific needs; new state-of-the-art systems also permit greater flexibility, *e.g.*, the integration of voice and data.

The PBXs and area networks are not interchangeable, although each is absorbing more of the others capabilities (see Table 6.7). The single most distinguishing feature from the standpoint of design is that a PBX uses a hub and spoke arrangement with the switch located at the hub, while area networks normally share the same communications channel. A PBX has higher levels of availability and reliability and costs less to acquire and to install per workstation. In contrast area networks have considerably higher bandwidths and transmission speeds; for example, some optical fiber networks have speeds of 800 Mbits/sec and higher.

6.3.1. Private Branch Exchanges

The first PBXs that dealt with voice and data on an equal basis appeared in the early 1980s. These PBXs employed switching techniques that were not designed solely for the fixed bandwidth allocation schemes appropriate for voice-only transmissions.

A PBX is composed of four basic units: common control, switching matrix, lines and trunks, and the user or station terminal (see Figure 6.7). The heart of the system is the common control, which consists of one or more microprocessors and

Table 6.7. The Relative Strengths of PBXs and Area Networks[a]

Features	PBX	Area Network
Availability	X	
Distance	Same	Same
Least cost to install	X	
Least cost per station	X	
Reliability	X	
Transmission speed		X
Voice/data integration	X	
Computer interface	RS232c port	Computer bus specific
Network interface	Telephone jack	Tap
Topology	Star	Bus, ring, tree
Medium	Twisted-pair wire	Coax, fiber optic, twisted-pair wire

[a]Includes local and wide area networks.

associated software. The processor performs the functions contained in the software, such as monitoring extensions and trunks, processing instructions dialed into the system by the user, and routine diagnostic self-checks.

The switching matrix joins the calling party's line with the called party's line. There are three types of matrices used: space division, frequency division, and time division. Space division makes a physical connection that can be traced from beginning to end along a discrete path in the matrix. The frequency division technique sets the frequencies of the calling extension and the called extension to the same channel frequency. The most common form of switching is time division, which supports both analog and digital transmissions. In a time division transmission, a channel is divided into increments that may be thought of as time slots. A voice is sampled, digitized, and transmitted as a series of bytes assigned to a series of time slots. At the receiving end, the voice is reconstructed or data are reassembled into records or frames.

Lines and trunks are the transmission facilities used to communicate with station terminals, computers and other systems, and to the telephone company's CO. A wide variety of station terminals exist, including the standard telephone handset, computer phones, facsimile devices, and visual display units with, and without, built-in intelligence. The capability of the PBX determines which units can be incorporated within the system. For example, to realize the full potential of a PBX with ISDN features, special telephone sets with built-in ISDN features must be used.

6.3.1.1. Advantages

Twisted-pair wiring is installed in most work sites for providing telephone services. When a PBX is installed, at least two pairs of wiring and sometimes

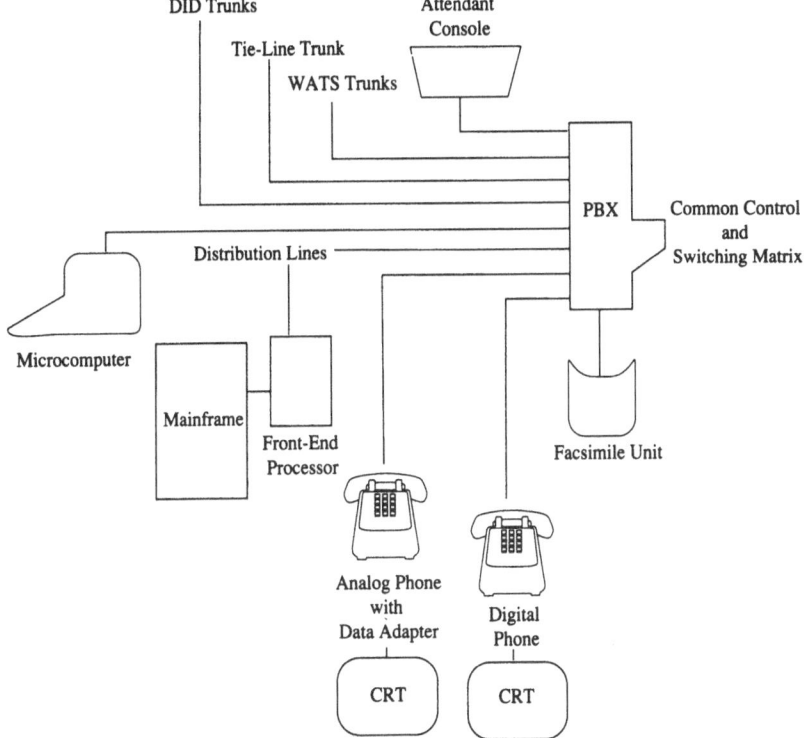

Figure 6.7. Basic PBX configuration with several examples of station terminals.

three are installed as backups. This redundancy creates a ready network for nonvoice needs. Once data has been incorporated, the PBX can be connected directly to high-speed longhaul networks through T-1 interfaces using a variety of media, such as coax, optical fiber, satellite, and microwave. Another seldom considered advantage is the sophisticated administration and maintenance features that are required to manage a complex communications system and come as a part of a PBX.

6.3.1.2. The Major Problem

A high bit rate is required to carry images, picture-quality displays, and large data files. Although most PBXs do not operate at their full potential, it is possible to transmit data routinely at 9.6 Kbits/sec; digital systems may operate at 64 Kbits/sec. In comparison to the local loop, this seems rapid; but it is far below the speed required for all but the most basic data transmission needs.

Compounding the problem are the inherent differences between voice and data traffic. The volume of telephone traffic on a system is more predictable than the amount of data traffic. On voice-only systems, the number of circuits is far fewer than the number of telephone sets; the ratio is determined by an estimate of peak period activity. As soon as the number of callers equals the number of circuits, additional callers are blocked and they hear a busy signal. Incorporating data traffic on a system designed for voice traffic increases the likelihood of blocking. Third-generation PBXs designed to deal with the problem of blocking are still impacted by data traffic. Data calls may last for hours; and data calls can also involve short high-speed bursts of transmission followed by long periods of inactivity. When this occurs, the switching capacity dedicated to data is relatively underemployed.

6.3.1.3. Centrex

Centrex is a communication service provided by the local telephone company. It is an alternative to a PBX or in large companies, a complement to the installed PBX equipment. The service normally originates in the CO and hence the origin of the name central exchange. In some instances, the switching equipment may reside at the customer's site.

Centrex has two basic features: Direct Inward Dialing (DID) and Automatic Identified Outward Dialing (AIOD). The DID enables callers to bypass a company switchboard and dial directly to the desired telephone; the AIOD generates a detailed readout of long-distance charges. When AT&T was broken up, Centrex was one of the few services retained by the operating companies; as a consequence, the operating companies have invested heavily in developing and marketing Centrex, and many new features have been added since 1984.

Incoming calls placed to a specific site (DID) enter the CO and are switched to the desired Centrex extension. Incoming calls to the organization's directory number also enter the CO, but these are then routed over an attendant trunk to the organization's operator or automatic attendant system. Centrex is one of the largest, if not the largest, business communications system. It works well for voice and low-speed data communications, but because of its design and its dependence on technology used in the CO; it poses major challenges for transmitting high-density graphics and supporting other enhanced system applications. Perhaps the best reason for considering Centrex is that it offers a relatively seamless transition to ISDN because it avoids the need to invest in equipment that will become obsolete as ISDN is introduced.

6.3.2. Area Networks

Local and wide area networks have proliferated at an amazing rate since the early 1980s. There are dozens of companies engaged in offering commercial products, and the cost per interface has fallen from the $1200–$1500 range to less than $100 in six years. Area networks are frequently the backbone of an enhanced system within an organization. The reader who wants a more thorough treatment of the subject will find many books and several magazines devoted to the subject.

A multimedia LAN standard is nearing ratification: IEEE 802.9 provides for a 4- or 20-Mbits/sec network that supports three types of channels—an ISDN basic rate channel containing two 64-Kbits/sec channels and a 16- or 64-Kbits/sec signaling channel, an 802 LAN packet service, and one or more wideband isosynchronous channels. The ISDN service carries voice and data; the packet service transports data only; and the wideband service supports digital video.

One form of WAN that has appeared in the last few years is a metropolitan area network or MAN as the telephone industry has begun moving to provide higher speed metropolitan networks to serve its business clients. The backbone of the MAN is a connectionless switched mulitmegabit data service (SMDS). Connectionless means that there is no dedicated end-to-end connect; data transmissions are handled similarly to how voice calls are handled by the telephone network. One proposed application for this form of WAN involves tying together multiple LANs.

Area networks are used for conveying information in a videotex format in shopping centers, building lobbies, convention centers, airports, and college campuses. The basic architectures, two of the forms the networks can assume, and three common protocols are discussed here. Chapter 12 considers how some organizations employ networks to support computer environments containing enhanced features.

6.3.2.1. Architectures

There are four commonly found architectures: star, bus, ring, and tree (see Figure 6.8). The star arrangement connects each terminal to a central, traffic-routing hub. Such a configuration simplifies network management, can be easily expanded to the limit of the hub computer's capacity; permits a relatively high level of security, since switching is through a central hub, and provides through the hub a natural gateway out of the system. A star is the best configuration when the host contains the infobase or when the information is accessed frequently and must be closely controlled.

The major disadvantage of a star arrangement is its extreme vulnerability: If the host breaks down, the individual units cannot communicate with each other. A star configuration is also more costly because more communication links must be installed and maintained, and it is relatively inefficient if there is not much traffic and the circuits are underemployed.

A bus topology connects each unit to a single cable running the length of the network. It is usually simple to install and does not require a central controller. A bus configuration can handle short transmissions among many stations, and it continues functioning if one or more units along the bus fail. Because of the

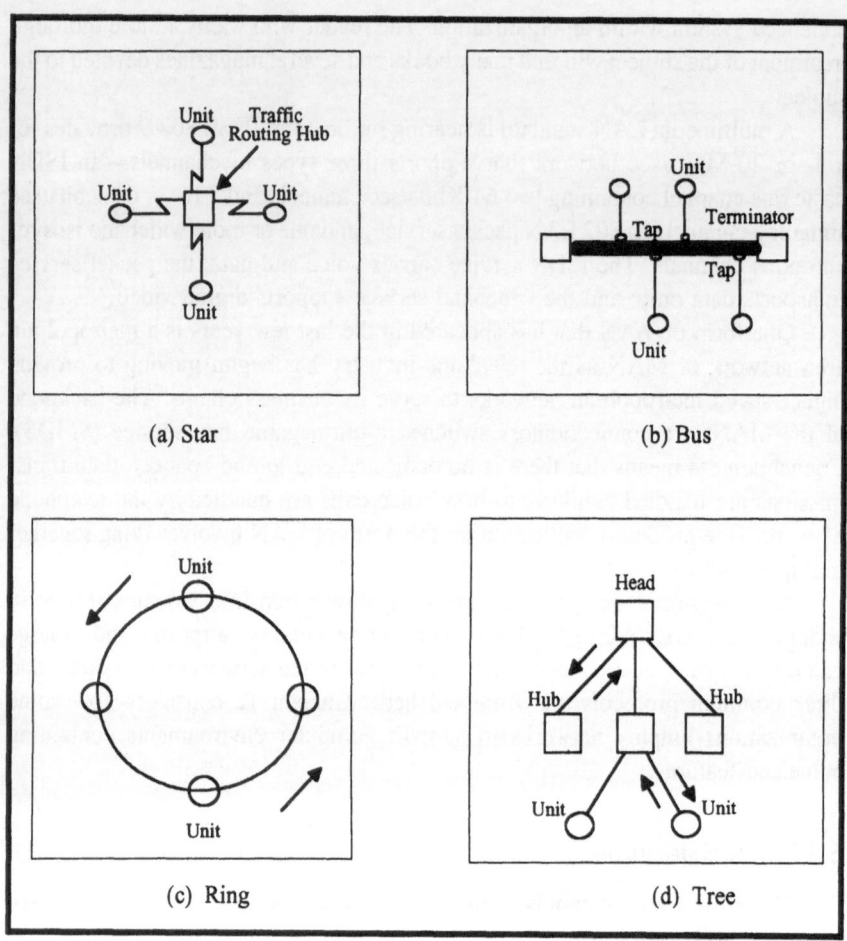

Figure 6.8. Several examples of area network architecture.

bandwidth and the manner in which units tap into the network, a bus system can readily support video, data, and voice on the same system. Its disadvantages are the impact on the whole system if the bus is damaged and unable to pass on transmissions. A bus configuration is also vulnerable to breeches of security.

A ring interconnects the units in a continuous loop. In some configurations, one station may be the master unit controlling the transmissions of other units, or individual units may have the ability to initiate transmission when the network is perceived to be idle. The ring has higher equipment reliability if it does not support a master-slave environment; however, rings may be very vulnerable to failure and security breaches.

The fourth configuration is a tree topology; which has units connected to multiple central hubs that are in turn connected to each other. Like the bus arrangement, a central control is not needed, and a tree topology can economically handle short messages. Portions of the tree will continue operating even though connections among hubs may be broken. Expanding a tree topology is relatively easy, and bridges to other networks may occur at various points.

Regardless of the architecture adopted, multimedia applications will gridlock most area networks. Current capacities and network protocols will often be unable to keep up with the transfer requirements of BLOBs (Chapter 3 discusses BLOBs), even when such compression techniques as JPEG and MPEG are employed. For example, full-motion video requires 30 frames/sec. Assuming each frame consists of 10 Kbytes, a network to support this application will have to handle over 2.4 Mbits/sec. If there are several users engaged in similar applications, very few networks can sustain the transmission requirements.

Typical LAN speeds are in the range of 10–16 Mbits/sec. High-speed networks, such as those listed in Table 6.4 can fulfill the bandwidth requirements, but they are not plentiful. Until networks with suitable carrying capacity are in place, many capacity hungry multimedia applications will be limited to single user environments. The Orfali–Harkey reference explains some of the trade-offs and performance implications of different protocols.

CO-LANS. The telephone company has aggressively marketed Centrex as a less capital-intensive and more flexible alternative to a customer-owned PBX. To continue encouraging customer use and loyalty, telephone companies have upgraded Centrex to provide data and network features. The product is called a central office local area network or CO-LAN (pronounced see oh lan); these are sometimes referred to as Centrex-based LAN service, since they may be offered as an add-on service to Centrex.

A CO-LAN provides data-switching capabilities; in this respect, it is similar to a LAN-based network located on the user's premises. However in a CO-LAN, equipment is located at the telephone company's CO, and existing telephone lines are used. At least two approaches exist for combining voice and data signals over a Centrex access line: data over voice (DOV) and simultaneous digital voice and data systems (SDVD).

The DOV and SDVD pass the voice transmission to the voice switch via Centrex. The DOV technology carves out a data channel from an unused portion of the analog bandwidth in a voice channel by means of an inexpensive data/voice multiplexer. This approach has been used for many years. The SDVD has devices located at the terminal that digitize voice and data and transmit the digital signal to the CO. This technique fully integrates the data and voice. The DOV can support fully synchronous 64-Kbits/sec and SDVD 19.2-Kbits/sec transmission speeds.

For organizations that choose Centrex, a CO-LAN can provide several benefits. One of the most fundamental is that the telephone company takes care of installation, support, and maintenance. No extra wiring or complicated cable needs to be installed, since twisted wire is adequate. A CO-LAN is more flexible than a PBX, since the number of PBX ports is predetermined. With CO-LAN, the customer simply leases the number of ports needed on a Centrex system, which permits the size of the system to grow according to need. A CO-LAN can also be a viable alternative to Ethernet and token ring LANs if the network is not used for such bit-intensive applications as transmitting picture-quality images. Lastly, a CO-LAN provides an interim step on the migration to ISDN. It is in the twilight zone between an analog present and a digital future and thus serves as a transition mechanism to ISDN for the telephone industry.

A CO-LAN has several limitations too: Most CO-LAN services are restricted to asynchronous devices, which limit transmitting and receiving equipment to relatively low-speed applications. Another limitation is the transmission speed: High-frequency signals can be transmitted only over unloaded, unfiltered local loops that are less than approximately 18,000 ft long; this restricts transmission to 19.2 Kbits/sec which is probably suitable for most applications not involving time-consuming file transfers or graphical and pictorial transmissions. Some vendors do not offer a file server, which prevents the CO-LAN from functioning as a traditional LAN.

6.3.2.2. Broadband and Baseband Transmissions

The two basic transmission techniques employed in area networks are referred to as broadband and baseband. In a broadband transmission, bits are represented by either the amplitude, frequency, or phase of radio frequency (RF) carrier. Several channels may exist on a single cable separated by guardbands that prevent the signals on one channel from interfering with those of another channel.

A broadband network has two main parts: a headend and a distribution system. The headend is a multiplexer, and the distribution system is twisted pair, coaxial cable, or optic fiber. If device A wants to communicate with device B, the digital signal goes through device A's interface unit where it is converted into an analog signal and placed on the proper channel. The signal travels to the headend where the message is retransmitted through the distribution system on a different frequency to device B. On receipt the signal is converted from analog into digital in device B's interface unit and passed to device B.

A broadband network may be a single cable or a dual-cable system. In the former, signals move in two directions separated by the guardband previously described. In a dual-cable system, no guardband is needed. In a single-cable system, there are three ways of separating the forward and return paths: subsplit, midsplit, and high-split. Placement of the split determines the bandwidth of the upstream and downstream communication path and depends on the application. If an equal amount of information is being sent and received, a midsplit is appropriate; if a menu selection is being sent and a still video image is being received a high-split is appropriate.

Broadband is best for medium-to-high-speed applications and for integrating voice, data, imaging, and video applications on the same network. Since the technology is based on CATV technology, it is tried and proven, and the equipment is on the market and very reliable. The transmission media usually has a high bandwidth (500 MH_Z and up) and can span 50 miles or more. Expansion is relatively easy, and the network is relatively fault tolerant, which allows portions to be taken out of service for maintenance and repair without shutting down the entire network; its only real disadvantages are cost and complexity.

In a baseband network, the signal is transmitted by means of voltage fluctuations, and the entire bandwidth of the medium is used. The network is less expensive to install and maintain than a broadband system because line drivers and simple electronic receivers are used. However its overall information-carrying capacity is more limited, and it does not allow the wide integration of applications that are possible with a broadband network.

6.3.2.3. Protocols

A protocol was defined earlier as the set of formats, rules, and procedures governing the exchange of information between devices in a communications system. There are three widely used protocols: CSMA/CD, token passing, and polling. Since these protocols are referred to in later chapters, a brief explanation of each follows. The CSMA/CD is sometimes referred to as the listen before you talk protocol, and it works in the following way. A device preparing to transmit follows the same rules of etiquette as two people engaged in a conversation. The one wishing to speak first listens to make sure the other person is silent. If no dialogue is sensed, the speaker begins; if the other person is speaking, the person waits to begin. If both speakers begin at the same time, both sense the other's dialogue and stop for a random length of time before resuming. Eventually one or the other speakers gains the floor, and the exchange resumes. The CSMA/CD operates in the same manner with one minor exception: If two devices are contending for the network at the same time and their signals interfere with one another, all devices sensing the signal interference momentarily add to the cacophony by also briefly broadcasting, further jamming the circuit; however even this aspect of the protocol has its analogy if one recalls ever watching the Muppets.

The token-passing protocols for broadband and baseband are shown in Table 5.3; these are two of the many variations which exist. Token passing is often implemented in a ring configuration. For any unit to initiate a transmission, it must first have possession of the token, which consists of a special character (*e.g.*, 11100111). The token is passed from device to device until it reaches one with a message to transmit. The device transmits its message once it has possession of the token. When the intended target receives the message, it notifies the sender who then releases the token, permitting it to continue its journey from device to device.

Polling is probably the most common of the communications protocols. In a polling environment, one device is designated as the primary or supervisor, and all other units are referred to as secondary devices. The primary device is in charge of the communications flow, and secondary stations may transmit only after receiving permission from the primary.

In the most common type of polling, called roll call, the primary is given the list of addresses for each secondary. The primary picks one of the addresses and transmits a brief message asking the secondary device if it has anything to send. If the secondary has data, it either responds with the data or a positive acknowledgment, followed by the data. If it has no data it responds with a negative acknowledgment and the primary proceeds to the next address on its list. If the primary has data to send to a secondary, a similar process is followed, but the message inquires whether the secondary can receive data. If the secondary responds with a positive acknowledgment, data are transmitted; if the acknowledgment is negative, the primary waits for a period of time, then tries again.

Polling eliminates line contention, but it makes every device that has something to send wait until it is polled. If there are a large number of terminals and some have data to transmit, the wait can be quite lengthy. There are many variations to roll call polling, but in general, as the number of secondary devices rises, response time worsens.

The efficiency of a polling strategy depends to a great extent on the location of the polling activity. For example, in an interactive cable system, such as the one shown in Figure 6.6, several possibilities exist. In one configuration, the bridging amplifier in the subscriber network can be equipped with a special code-operated switch that serves 500 subscribers, so that the central computer would scan the network, activating one switch at a time. Each switch is the repository of transmissions for its set of subscribers, and when activated, it passes transmissions forward to the central computer. The computer then moves to the next switch and in this manner indirectly polls every subscriber in the system within a few seconds. A more brute force type of approach, and one that would take considerably longer, is to have the central computer poll each subscriber's terminal either at the distribution system and drop cable interface or at the terminal itself.

6.4. SUMMARY

Communication services form the web that units components of an enhanced system. These services have a unique terminology that must be learned to understand how the services function; for example, words such as duplex, wideband, analog, bandwidth, protocol, and so on. The vocabulary is so extensive that dictionaries of many hundreds of pages exist so that people from different branches of the communications industry can communicate with each other. A popular 402-page dictionary used by people in the telephone industry contains 16,000 words and terms.

Regardless of where you are when you read this page, you can probably see the instrument that is the termination point for the most ubiquitous of all communication services, the telephone. The growth in the number and variety of activities the telephone can perform has been evolutionary rather than revolutionary. This evolution is most evident in several of the high-speed services that have been introduced in the past three decades and are being introduced today; two examples are the aging but widely used T-1 service and ISDN.

The telephone is facing an onslaught of competitors; especially since the monolithic telephone industry was broken up in 1984. These competitors for data and even voice enable customers to bypass the telephone network partially and occasionally totally. Examples of bypass technologies are data radio, cable television, microwave systems, private optical fiber networks, and satellite systems.

An organization has several ways of supporting internal electronic communications. The popular PBX, the venerable but newly rejuvenated Centrex, and the growing use of area networks are the principal contenders today. Each has its own set of advantages and disadvantages that must be considered when making a decision.

REFERENCES

American Express Advance NAPLPS Service Providers Production Standards Proposal. Version 1, American Express (1984).

H. Armbruster, Retrieval Services with Broadband ISDN: Access to Information and Information Processing, *Information Journal* **27**(4) (1989), pp. 408–417.

I. Brodsky, Tapping into ISDN, *Data Communications* **19**(5) (1990), pp. 117–122, 124.

D. Bushaus and P. Travis, Users Find ISDN Pays off, *Telephony's ISDN Special* **218**(6A) (1990), pp. 12–13, 15.

W. J. Cadogan, Fiber: The Tie That Binds, *Lightwave* **8**(13) (1991), pp. 20, 22–24.

J. Castro, Telephones Get Smart, *Time* (March 30, 1987), pp. 50–51.

CATV, Part 1, *GTE Linkurt Demodulator* **28**(7 & 8) (1979).

V. G. Cerf, Network, *Scientific American* **265**(3) (1991), pp. 72–81.

Communications in the Workplace, *Fortune* **116**(9) (1987), pp. 143–144, 148, 150–152, *passim*.

Z. Cvijan *et al.*, ISDN Computer-Aided Telephony, *IEEE Network Magazine* **5**(1) (1991), pp. 46–53.

P. Davidson, Multimedia Finally Appears within Network's Reach, *Network World* **9**(14) (1992), pp. 39, 42, 51, 63.

D. Delisle, B-ISDN and How It Works, *IEEE Spectrum* **28**(8) (1991), pp. 39–42.

G. Doyle, Strange Bedfellows?, *Telephony* **218**(1) (1990), pp. 24–26.

J. Estrin, Hybrid Technologies Rewrite the Rules for Local Area Networks, *Mini-Micro Systems* **18**(1) (1985), pp. 195–196, 201–204.

M. Fahey, Carriers Target Data Market with Broadband Services, *Lightwave* **9**(5) (1992), pp. 1, 22, 24.

M. Fahey, Study Sees $4.5 Billion Loop Market in a Decade, *Lightwave* **8**(11) (1991), pp. 6, 8.

J. L. Fike and G. E. Friend, *Understanding Telephone Electronics*, Sams, Indianapolis (1984).

Future (The) of Television, National Cable Television Association, Washington, DC (1990).

J. Gantz, Bypass Is Not a Fatal Loss to Telcos, *TPT* **7**(2) (1989), pp. 26, 28, 30.

Growing (The) Importance of the PBX, Communications Products & Systems (May 1987), pp. 29.

Handbook and Buyers Guide: 1988 ISDN, IGI Publishing, Boston (1988).

High (The) Cost of Bypass, *Communications Week* (199) (1988), p. 16.

J. Iida, T-1 Paves the Way for ISDN's Arrival, *Information Week* (130) (1987), pp. 27–29.

Illinois Bell Files Tariff to Offer Non-Centrex Single Line ISDN, Illinois Bell, Chicago (June 20, 1990).

ISDN: An IDC White Paper for Information Systems Management, *Network World* 7(4) (1990).

G. C. Kessler, and R. K. Gojanovich, Before ISDN: CO LANs Offer Voice and Data Networking Now, *LAN Magazine* (February 1989), pp. 88, 90, 93, 96–97.

G. Kim, Studies Advance Fiber Penetration, *Lightwave* 9(2) (1992), pp. 1, 20.

L. Mantelman, Are Analog Local Loops Too Dirty for ISDN?, *Data Communications* 16(8) (1987), pp. 53–54, 56.

C. Mason, Telephone Costs Decline 5% in '89, *Telephony* 218(9) (1990), p. 11.

P. Meade, Welcome to the Land of the CO-LAN, *Communications Consultant* 3(8) (1987), pp. 58–59.

H. Morgan, Centrex Picks up Speed, *Telephony* 215(3) (1988), pp. 60, 64–65.

N. J. Muller, M. I. McLoughlin, and J. M. O'Neill, *A Management Briefing on Fractional T1 and the Transition to ISDN*, General Datacomm., Middlebury, CT (1991).

T H. Murray, Fractional T1: New Twists in Old Operations, *Telecommunications* 24(6) (1990), pp. 63, 65–66, 68, 70.

R. Orfali and D. Harkey, The BLOBs Are Coming, *LAN Technology* 8(1) (1992), pp. 61–62, 64, 66, 68, 70.

S. Pandhi, The Universal Data Connection, *IEEE Spectrum* 24(7) (1987), pp. 31–37.

N. Papadopoulos, Combining Data and Voice Network Management, *Data Communications* 15(13) (1986), pp. 121–122, 125–126, 128.

A. W. Paschal, Optical Publishing: Issues Facing Publishers and Subscribers Today and in the Future, *Proceedings of the Eighth National ON-LINE Meeting*, Learned Information, Medford, NJ (1987), pp. 387–395.

Perspective on PBX Systems, Perspective Telecommunications Group, Paramus, NJ (1984).

D. Powell, Will It Ever Get to Market? *Networking Management* 9(10) (1991), pp. 26–28, 30–32.

Public Telecommunications Technologies in the 1990s: Achieving Universal Service, Alliance for Public Technology, Washington, DC (June 1, 1990).

W. Reeve, The Subscriber's Loop, *Procomm Enterprise Magazine* 31 (1990), pp. 23–26, 57–58.

M. J. Richter, CO-LANs to Remain Part of RBOCs' Long-Term Strategy, *Networking Management* 7(5) (1989), pp. 50–53.

C. Roeckl, Fractional T1, *Communications Week* (252) (1989), pp. 34, 36, 40–41.

A. Scafidi, Intelligent Networks, *Communications Consultant* 6(2) (1990), pp. 25–27.

P. Schnaidt, Broad Horizon, *LAN* (November 1987), pp. 46, 48, 49–50.

R. Singh and V. Malhotra, Do You Know the Way to SMDS? *Telephony* 218(12) (1990), pp. 34–35, 38–40, 42.

J. Skene, The Last Mile: Alternative Ways to Implement ISDN Over the Local Loop, *Communications News* (December 1984), pp. 54–56.

W. Stallings, The Role of SONET in the Development of Broadband ISDN, *Telecommunications* 26(4) (1992), pp. 21–24.

T. Sweeny, Where Is Broadband ISDN? *Communications Week* (310) (1990), p. 26.

T-1 Applications Guide, Telco Systems Network Access, Fremont, CA (1989).

D. Tucker, N. King, and J. Sanders, Have Your Cake and Eat It Too! *Telephony* 216(20) (1989), pp. 26–28.

Chapter 7

The Path to Applications

It has been possible to access a computer from a distance of several hundred yards or several thousand miles since the 1950s; however the process has not always been easy. Usually the protocols that governed the communications were not standardized, and the equipment would work only across one vendor's product line. Access was possible but limited to highly specialized circumstances. The situation began to change with the advent of communications standards, and it will change even more with the development of public gateways and the introduction of an open network architecture. The former will enable users conveniently and easily to enter a nationwide system of interconnected communications services. The latter will enable a variety of equipment from a large number of vendors to communicate among one another through the nation's telecommunications network.

Tremendous structural adjustments are underway in the communications industry, especially involving the telephone companies. Most people are aware of these adjustments in only a peripheral way. Perhaps they subscribe to the services of a long-distance carrier that did not exist when they were children, or maybe they saw an article about proposed legislation or something about judicial proceedings in the courtroom of somebody named Judge Greene. There are few examples of events in the history of this country that have had such a profound impact on everyone but which are virtually unknown and certainly poorly understood. Chapter 7 attempts to fill in some of the holes by first examining the topic of public gateways—the justification for establishing them, how they operate, and the basic types. Because of the potential for abuse, a code of practice for public gateways is proposed.

The basis for ONA is the deregulation of the telecommunications industry that occurred in the 1980s. A history of regulatory activity leading up to the present is reviewed, and a model for introducing the ONA services offered by the telephone industry is presented. The chapter ends with a hypothetical implementation of the ONA model.

7.1. GATEWAYS

A gateway is the combination of hardware and software through which a user gains access to an information- or transaction-based system. For the purposes of the material that follows, there are three types of gateways: networked, host based, and distributed. A networked gateway joins together two networks, and its existence is transparent to the user. It may join two area networks, an area network with a PBX, or a myriad of other possibilities. It is a necessary element providing connectivity among components on two different networks. A host-based gateway resides in the main computing facility and provides a window through which a user can gain access to systems located inside and outside of the organization. Distributed gateways are sometimes also referred to as mass market gateways, and they serve as public switching centers through which systems available to the general public and paying customers may be accessed.

7.1.1. Why Gateways?

In most applications, the capability exists to connect a user to a system directly without a gateway; however there are several very good reasons for establishing gateways: Gateways reduce costs, make it more convenient and simple to access a particular system, contribute to the overall availability of the system, and enhance security.

Economics often dictates the existence of a gateway, because it may be impractical to maintain all the information that can be accessed on one system due to its sheer volume. The general rule of thumb is that when the cost of storing and maintaining the information on a centralized facility exceeds the storage and communication costs of accessing it at a second location, it should be relocated. In some systems, a dynamic balance is maintained by monitoring usage and moving frequently accessed information to a centralized site where it is kept. The information is relocated to the distribution site when it is no longer accessed enough to justify its retention. Another economic benefit results from communication services that often support gateway facilities: Packet-switched networks and specialized communication services priced for heavy traffic loads are likely to be developed. Without a gateway to focus traffic flow, users would access remotely located information through high-priced dial-up telephone services.

A commercial service provider who charges customers for accessing information or performing transactions may save money by having the gateway operator perform the billing function. This was the regional Bell operating companies principal motivation for creating gateways beginning in late 1988. Gateways removed the costly but necessary burden of operating a service from service providers and gave it to the telephone companies, which had efficient and well-oiled billing mechanisms in place.

A gateway is convenient and provides a common point of access; it is a one-stop store for reaching a multitude of services. Entry can be gained by dialing one telephone number or accessing a single menu and entering a user number and password once. After gaining access to a gateway, it is often possible to jump back and forth among services without having to initiate the dial-up and log-on routines. At the end of the month, the customer of a commercial service receives a single bill from the gateway provider, who then handles disbursements. A gateway gives the service provider visibility. It brings its existence to the attention of the viewer and like a listing in the yellow pages of the telephone directory, provides a convenient and single point of reference for a potential customer. The very existence of gateways will stimulate the development and use of electronic services. A local call is the entry to the gateway, and it opens a path to an information base or service that may be thousands of miles away.

An important advantage of a gateway is the simplicity of the user interface. By offering viewers a menu of service or information options, it is possible to avoid complex access routines. Cryptic system messages can also be avoided, and it is not necessary to access different computers.

A gateway provides a vertical interface for incompatible systems. In the absence of a gateway, each computer must have software for converting the code, transmission speeds, error-detecting routines, and so forth, used in systems communicating with each other. In an operating environment containing n different systems that have to communicate with m different computers, $n \times m$ conversion routines are necessary. For example, a company with users on two different systems that communicate with ten external computers may need to support as many as 20 conversion routines. In the ideal configuration, each system interfacing with the gateway is converted into the gateway's operating system. In the preceding example, this means that only $n + m$ or 12 conversion routines are needed. In fact if a commercial gateway, such as those operated by one of the telephone companies, is used, the company may not have to maintain any conversion routines.

A gateway increases the availability of third-party systems by providing a standard interface to a wide selection of choices that a single user can access. In some cases, a gateway increases security by adding another level of protection for systems that permit access by authorized users outside of the organization.

7.1.2. Gateway Operation

A gateway provides three general functions: (1) It establishes the user interface through which the users interact; (2) it connects the two ends of the communications stream by moving bits from point A to point B; and (3) it provides required administrative support (see the *Gateway 2000* reference at the end of the chapter for a more detailed explanation). These three functions are summarized in Table 7.1. It is unlikely that any gateway would incorporate all of these specific functions, and distributed gateways include more of them than host-based gateways.

One of the primary purposes of a gateway is to facilitate access with the least amount of delay, trouble, and cost; thus a directory listing the services, infobases, and other gateways is essential. For a user to move through a gateway or across gateways, consistency is important, especially in the commands needed to move around in the gateway; examples of navigation commands are next frame, previous frame, go to, keyword search, return to main menu. Supplementing these are commands for exiting the system and obtaining help. Where appropriate alternative gateway interfaces must be implemented; these are machine-to-machine interfaces, as opposed to a human-to-machine interface, designed to take advantage of the intelligence in the user device, such as a microcomputer.

Connectivity functions are normally transparent; this includes, protocol support to enable the interexchange of information between different sites, perhaps with different equipment. The gateway establishes the link automatically with the communications network and supports interactions between a user and the system or between two users; the latter activity may involve establishing a link to another gateway.

A host of activities are covered under the rubric of administrative support functions. Several important functions include user protection, billing, and policies for guarding the privacy of users.

The actual operation of gateways varies due to differences in system architecture and the business requirements of the gateway sponsor; a generalized description of what could occur follows:

Table 7.1.   Gateway Functions

User interface	Administrative support
Access	User preferences
Directory support	User protection
Consistency of operations	Service description
User support (Help!)	Service protection
Alternative gateway interface	System management
Connectivity	Billing services
Protocol support	User information
Information transport	User authentication
Session management	Privacy
Gateway to gateway interconnectivity	Testing services
	Optional management and support services

Source: *Gateway 2000: A Report of the Videotex Industry Association Study on North American Gateways*, Videotex Industry Association, Rosslyn, VA (October 1988).

1. A "call" from a user is routed to the gateway via the communications link.
2. The gateway obtains the destination (*e.g.*, phone number) of the call and the necessary parameters, such as the speed of the transmission and code (*e.g.*, ASCII) of the incoming data streams.
3. The gateway establishes a connection with the intended destination machine or the network serving it.
4. Parameters of the destination machine are obtained if appropriate.
5. The gateway establishes a data connection between the two parties and performs any conversions needed to support the dialogue.
6. The gateway monitors the exchange to the extent necessary to ensure the dialogue is proceeding normally and to collect charging data.
7. Appropriate accounts are billed or credited, and the statistical base maintained for performance analysis is updated.
8. The caller is released or a new session is initiated.

There are two ways that the session may occur via a gateway; the first and most common is the direct end-to-end transfer in which a menu or template is used to establish the session. After this is done, the data exchange occurs. The gateway performs such necessary conversions as speed and code conversions, but it does not interpose itself in the dialogue. The second transfer is template based, and it is found in some in-house systems and European videotex services. After the session is established, a set of preformatted screens called templates are presented. These templates reside at the gateway. The caller completes the template based on instructions accompanying the frame. This approach is satisfactory for querying an information base, but it is not suitable for applications involving the exchange of information between two people.

7.1.3. Host Based

The ITEX.25 is an example of a host-based corporate gateway; it is a product of Videodial, Inc., and it was introduced in 1987. The ITEX.25 was the first product of its type in the market. It operates in an IBM environment, giving users of IBM 3270 terminals access to external ASCII infobases via X.25 networks. Prior to the existence of this product, it was necessary for a user wishing to access internal and external sources of information to use a personal computer emulating a 3270 terminal and an asynchronous modem. This was an expensive solution and complicated the job of monitoring use and controlling costs.

Figure 7.1 shows how ITEX.25 is implemented. Any authorized 3270 terminal connected through IBM's Customer Information Control System (CICS) software can access ITEX.25. When access occurs, ITEX.25 presents a menu displaying all the sources available. Selecting one of the choices automatically initiates a connection to the corresponding service; no other action is taken. The next frame seen by the viewer is the normal welcome or sign-on display. At this

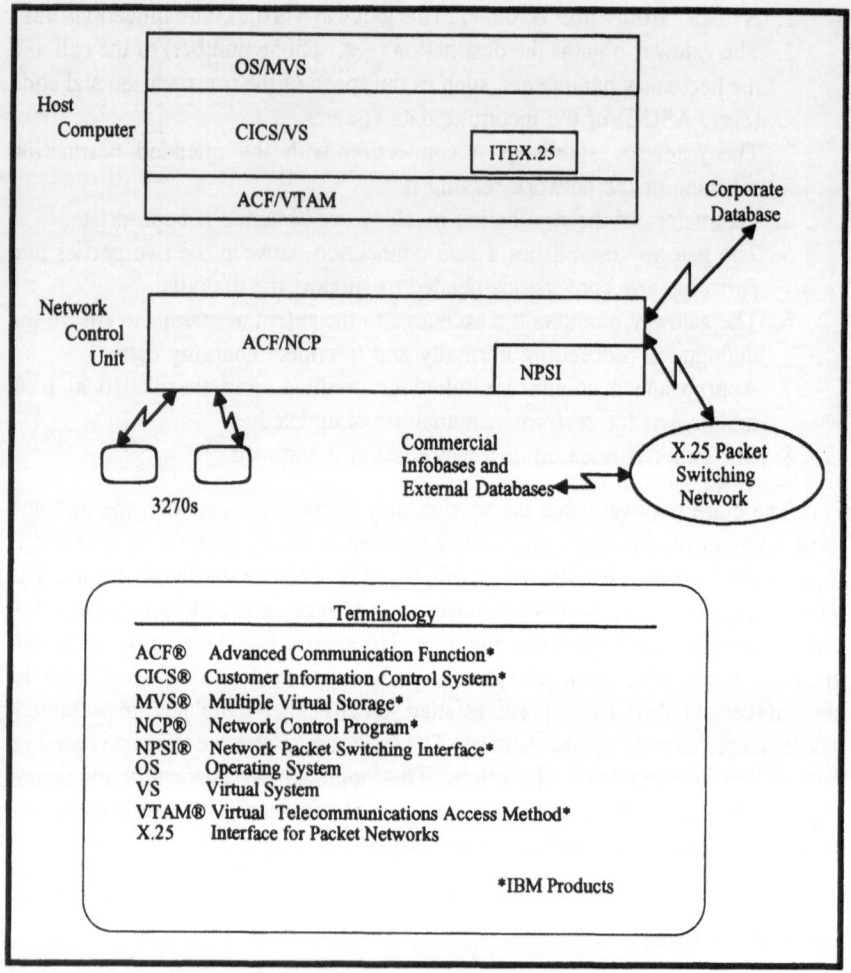

Figure 7.1. ITEX.25 systems environment: Videodial, Inc.

point, the user follows the procedure prescribed by that particular service, a feature for bypassing the ITEX.25 menu and directly select the service is also available. The software comes with predefined selection procedures for such popular infobases as Dow Jones News/Retrieval and CompuServe.

The ability of the software to monitor user activity is especially important. The ITEX.25 records the service accessed, the terminal involved, date, time of connection, and the time of disconnection. This information can be used for internal billing and to control access to outside services. An added advantage is the ability to negotiate volume discounts for services that are frequently accessed.

7.1.4. Distributed Gateways

The best-known distributed gateways are those operated by the RBOCs. These gateways are designed to facilitate the interface between a wide range of users and a wide range of services from a variety of market segments. The fact that the RBOCs are interested in hosting these gateways is not surprising, since almost 99% of U.S. telecommunications traffic involves RBOC networks at some point.

The forerunners of the RBOC gateways were those created in France for Teletel by the Postal Telegraph and Telecommunications Administration (PTT) in the early 1980s. Videotex services in other European countries were designed around a central computer that contained a large infobase. If the information being sought was not in the central infobase, the user would be switched to the appropriate site via a gateway feature. In Teletel the user directly accesses a concentrator called a Teletel Access Point (TAP) over the local loop. The TAP then passed the caller through the French packet data network, called TRANSPAC, to the intended destination.

The RBOC's were given permission to establish gateways in a judicial opinion by U.S. District Court Judge Harold Greene on March 7, 1988, the judge who oversaw the AT&T divestiture. One of the stipulations of the divestiture was a triennial review of the original agreement, which produced the following amendments to that agreement in U. S. v. Western Electric. et al., (1988):

1. The separated BOCs shall be permitted to engage in the transmission of information as part of a gateway to an information service, but not in the generation or manipulation of the content of information. Transmission shall mean the performance of the following functions: data transmission, address translation, protocol conversion, billing management, and introductory information content.
2. The separated BOCs shall be permitted to engage in voice storage and retrieval services, including voice messaging and electronic mail services.
3. In the performance of the services authorized herein, no BOC shall discriminate between and among providers of information or against other providers of information services or of voice storage and retrieval services.

The VICORP BETEX software is typical of commercial products available for establishing a distributed gateway (see Figure 7.2). It provides the basic gateway functions identified earlier and supports the three principal videotex standards implemented in North America: ASCII, Minitel, and NAPLPS.

Value-added network carriers, such as Telenet and Tymnet, provide nodes in hundreds of cities that are gateways to service providers. These carriers constitute the major purveyors of public gateways in North America, but CompuServe, Prodigy, and other service providers also operate their own networks with nodes in principal cities. Chapter 14 describes the Prodigy network architecture, which is a combination of communication services operated by Prodigy and Tymnet. Com-

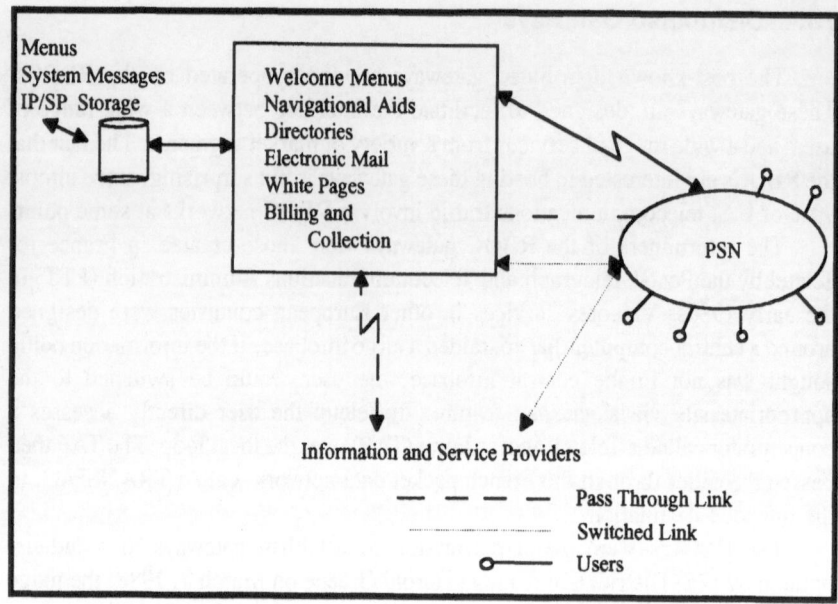

Figure 7.2. BETEX: Information Services gateway, VICORP Interactive Systems, Inc.

panies like DuPont and Digital Equipment Corporation operate worldwide videotex networks with private gateways, inaccessible by the public.

7.1.5. Gateway-to-Gateway Interconnection

Gateways are intended to facilitate the flow of communications between the
caller and the intended host; unfortunately gateways are often unable to access one
another, which creates islands of service. Attempts are underway to define a
technical means of overcoming this problem; however the mixture of public and
private gateways and the diversity of equipment and communications protocols
makes this a formidable undertaking.

Issues that must be resolved to interconnect public or private gateways.
How to exchange billing information
How to maintain the blocking features of the originating gateway while
     accessing information on the receiving gateway
How to allow the originating gateway to enforce credit limits
How to handle the selection of communication carriers based on user
     preferences
How to provide consistency in terminology and navigation commands

One of the most troublesome issues for both users of multiple systems and information providers operating on more than one system is the set of mnemonics on each system used to aid in renumbering services and options. Confusion reigns among users because the same terms may mean different things on different systems, and vice versa. Adding further confusion are information providers (IPs) who are forced to use different terms to provide their service on different systems. EAASY SABRE, the AMR reservation system described in Chapter 14, operates on more than two dozen systems; its mnemonic on three systems is EASY, EAASY, and SABRE. Confusion also occurs when an IP and a category of service are the same or similar. The Interactive Service Association has proposed establishing a voluntary registration of mnemonics and an on-line database to query common names. Until something is done, the problem will plague users and IPs alike.

7.2. CODE OF PRACTICE: GATEWAYS

There can be as many as four distinctly different participants to a gateway; these are the user, the gateway sponsor, the owner of the information or service, and possibly the system vendor where the information or service reside, if different from the owner. Each participant has his or her own views regarding what should and should not be allowed. The National Federation of Abstracting and Information Services (NFAIS) has developed a code of practice to address many of the issues surrounding gateway arrangements. The code is intended as a point of reference for establishing agreements among the participants. A description of the principal sections of the code is shown in Figure 7.3. The reference entitled *Code of Practice: Gateways* at the end of this chapter provides more information. This reference and the *Gateway 2000* reference mentioned earlier provide good coverage of the social, legal, ethical, and operational issues that should be included in a code of practice for gateway operators.

7.3. OPEN NETWORK ARCHITECTURE

In the past 35 years, the telecommunications industry has undergone a series of steps designed to increase competition. Since AT&T was and is the predominant firm in the industry, most of the actions undertaken have been directed at it.

7.3.1. A Short History of Telecom Deregulation

The number of court cases and their ramifications and the rule changes issued by the Federal Communications Commission (FCC) affecting deregulations are too numerous to review. However several of the major events that shaped the industry and are responsible for the creation of an ONA are listed in Figure 7.4.

I. IDENTIFICATION OF OWNERSHIP:
The ownership status of a database must be readily and publicly accessible throughout the Gateway arrangement to ensure copyright protection. The User has the right to know this information and the Owner has the right to expect that the Host System and Gateway will disseminate the information. Additionally, the Gateway has the right to receive the information from the Host System.

II. IDENTIFICATION OF USER(S):
User information must be recorded and conveyed along the Gateway chain. It is the obligation of the database User to supply sufficient identification information to both the primary Host and the Gateway operator. It is the obligation of the Host and Gateway operator to convey such user information to the database owner.

III. IDENTIFICATION OF HOST SYSTEM:
Any Host System entering into a Gateway arrangement must be identified to all other participants in that Gateway. It is the obligation of the primary Host to inform the Owner of a database it distributes that a Gateway arrangement exists. A Gateway operator has the obligation to inform users which Host System is being utilized and have the right to be notified if any Host in the Gateway arrangement enters into an arrangement with another Host System.

IV. IDENTIFICATION OF DATABASE(S) ACCESSED:
A Database accessed by any means in a Gateway arrangement must be identified by names that are approved by the database owner. Gateway Operators have the right to receive these names from the Host System and have the obligation to relay the name to the User.

V. DESCRIPTION OF GATEWAY COMPONENTS:
Any public description, by whatever means, of a Gateway or any component thereof must be accurate. Every participant in a Gateway arrangement has the right to receive such accurate data. To fulfill the right requires that each Database Owner, Host System and Gateway Operator in the arrangement assume the obligation of providing appropriate information to the other participants.

Figure 7.3. Description and rationale for a code of practice: Gateways. Source: *Code of Practice: Gateways*, National Federation of Abstracting and Information Services, Philadelphia (June 1987).

The recent history of deregulation began in 1949 when the United States sued Western Electric Company, Inc., and its parent, AT&T, alleging that they had monopolized and conspired to restrain trade in the manufacture and distribution of telephone equipment. After lengthy negotiations, the parties resolved the case before it came to trial. The consent decree or as it is more commonly known, the Final Judgement, was entered in 1956. The decree prohibited AT&T and its Bell operating companies from any business other than furnishing common carrier communications services. It prohibited Western Electric from manufacturing equipment other than that used by the Bell system.

At the same time that the 1956 decree was being hammered out, an unrelated but far-reaching case, Hush-A-Phone Corporation v. United States, was underway. For many years, AT&T had prohibited its customers from attaching anything to the network not supplied by the phone company. Hush-A-Phone Corporation

VI. CHANGES TO GATEWAY COMPONENT(S):
Any change to a Gateway component must be fully and promptly disclosed by the participant making the change. Every participant in a Gateway arrangement has the right to receive such information. To fulfill this right requires that each Database Owner, Host System and Gateway Operator in the arrangement assume the obligation of providing appropriate information to the other participants.

VII. ACCURACY AND QUALITY OF DATA/SERVICES:
The data and services provided by any Gateway participant must adhere to the standards for quality and accuracy which are publicly described by each. Each participant in a Gateway arrangement has the right to receive data and services which adhere to such publicly defined standards. Database Owner, Host System and Gateway Operator have the obligation to provide such data and services to one another and to the User.

VIII. UPDATING
The regular delivery of and timely access to information must be in accord with specific, predetermined agreements between participants in a Gateway arrangement. It is the obligation of both the Database Owner and the primary Host System to provide such timely access to the Gateway Operator and therefore to the user. It is the right of the Database owner and the primary Host System to expect that the other will perform the updates in accordance with their predetermined agreement.

IX. PRIVACY:
The nature of search queries and the corresponding results are privileged information and cannot be disclosed. It is the obligation of both the primary Host System and the Gateway Operator to maintain such privacy, which is the Right of the User.

X. USE OF INFORMATION:
The permitted uses of information are predefined by the Database Owner and cannot be extended beyond the scope of such definition by any other Gateway participant. The Primary Host System and Gateway Operator are obliged to disclose to the User the authorized use(s) of database information. The Primary Host System, Gateway Operator and User are each obliged to observe the Right of the Database Owner by complying with Owner's terms and conditions for the use of the database.

Figure 7.3. (*Continued*)

marketed a cuplike device by the same name that was attached to the telephone mouthpiece to help eliminate background noise. The Hush-A-Phone case was eventually remanded to the FCC. The FCC ordered the telephone companies to allow customer-attached devices that were not injurious; it did not permit the interconnection of non–telephone company equipment. However in 1968 the issue of attachment once again surfaced. A device called the Carterfone was being used in Texas oil fields to interconnect two-way mobile radio systems with the telephone network. The FCC in its Carterfone decision approved electronic devices, and not just the physical attachment of devices, to the telephone network as long as the device was not injurious. While the Hush-A-Phone decision wedged open the door to what until then had been the monolithic telephone

1956	Consent Decree	Barred AT&T from engaging in any business other than furnishing common carrier communications services
1956	Hush-A-Phone Decision	Allowed physical attachments
1968	Carterfone Decision	Allowed electronic attachments
1971	First Computer Inquiry, Final Decision	Federal Communications Commission confirmed data processing should not be subject to regulations; thus restricts AT&T from data-processing business
1980	Second Computer Inquiry, Final Decision	Deregulation of all offerings of customer premise equipment, *e.g.*, terminal equipment. Permitted enhanced services to be offered through separate subsidiaries
1984	Modification of Final Judgement	AT&T divests itself of RBOCs. Authorizes RBOCs to provide exchange telecommunications and access services; provide customer premises equipment; produce and distribute paper yellow pages Prohibits RBOCs from providing interexchange telecommunications services; information services; manufacturing or providing telecommunications equipment; manufacturing customer premises equipment
1986	Third Computer Inquiry	Permitted AT&T and RBOCs to provide enhanced services without structural separation if they complied with comparably efficient interconnection (CEI) and ONA. This was overruled on June 6, 1990, but reinstated on November 18, 1991.
1988	Triennial Review of MFJ	Permits RBOCs to engage in the transmission of information but not its generation; employ alternative transmission systems as and when they are able to propose methodology that will preserve content transmission dichotomy; enter the voice storage and retrieval markets Prohibits RBOCs from providing interexchange telecommunications, manufacturing
1989		AT&T given permission to engage in electronic publishing over its own transmission facilities.
1990		Federal appeals court requires judge overseeing the divestiture to apply different standard for restricting RBOC entry to information services.
1991		RBOCs are permitted to enter information services industry on July 25, 1991.

Figure 7.4. Major events in the deregulation of the telecommunications industry. Reported dates for these events vary slightly among sources depending on whether the decision date or effective date is reported.

network, the Carterfone decision threw open the door to manufacturers who wished to produce and market customer premise equipment (CPE).

In an attempt to provide some relief for AT&T and to further clarify what the company could and could not do, the FCC issued a series of rules, commonly referred to as Computer I, II, and III. The 1956 decree prohibited competition between telephone and nontelephone companies. The position proved unsustainable due to changes in technology, and in 1966, the FCC initiated the First Computer Inquiry or Computer I. In Computer I, the FCC distinguished between using computers to perform switching operations (which is a regulated communications service) and using computers to process information. The FCC confirmed regulation of the former, and not the later. Furthermore carriers other than the Bell operating companies subject to the 1956 decree were permitted to participate in the unregulated data-processing industry market through a separate corporate entity; this is known in the industry as structural separation.

In Computer II the imposition of structured separation rules were relaxed because it was believed competition in telecommunications had increased. Communication services were divided into basic services, enhanced services, and CPE. Carriers other than AT&T were allowed to enter the enhanced services market and sell unbundled CPE without structural separation. However, due to its size and dominant position, AT&T was required to separate its enhanced services and CPE. This meant that an AT&T subsidiary providing unregulated goods or services had to have separate offices; maintain separate books; employ separate personnel for operations, installation, and maintenance; undertake its own marketing; and use separate computer facilities for providing enhanced services.

Between Computer II and Computer III, an event occurred that turned the whole communications industry upside down: It turned out the Final Judgement was not so final as everyone assumed. In 1974 the Justice Department had once again initiated a suite against AT&T, contending it has used its monopoly power in the local exchange market to exclude competition in the equipment and intercity markets. After extensive hearings, the trial began in January 1981. One year later, on January 8, 1982, the Justice Department and AT&T announced a proposed modification of the 1956 Consent Decree that made further litigation of the 1974 case unnecessary.

The Modified Final Judgement (MFJ) required AT&T to divest itself of its 22 wholly owned Bell operating companies. AT&T kept its imbedded CPE, its enhanced services business, Western Electric, and Bell Labs. Perhaps most important to AT&T, it freed the company from the restrictions of the 1956 Consent Decree, including restrictions on engaging in unregulated business. The Bell operating companies (BOCs) were limited to providing exchange services within geographic areas called LATAs, information service access, yellow pages, directory services, and CPE. The BOCs could not offer any interLATA services or information services; manufacture or provide telecommunications products or CPE; nor provide any other product or service other than exchange telecommunications and access services that is not a natural monopoly service regulated by tariff. A BOC had to obtain a waiver to offer any service that was prohibited. Within a few years, so many waivers were requested and granted that the MFJ had been significantly altered.

News of the AT&T breakup was greeted with shock and some dismay. A January 9, 1982, eight-column banner headline announced "AT&T Split up, Transforming Industry"; the *New York Post* proclaimed in its headline, "Ma Bell

Gets Divorce, We'll Pay the Alimony." It took a long time for the clamor to subside and for everyone to become accustomed to the change. In the intervening ten years since divestiture, subscribers, stockholders, and employees have reviewed the results with mixed feelings. Users have witnessed about a 50% increase in local rates and a 33% decline in long-distance charges; overall telephone service costs have risen about 17%. The AT&T stock has just about doubled, and stock of the baby Bells has nearly tripled. Employees have been hit the hardest: There has been a combined reduction of about 200,000 employees, and massive exchanges of employees among the companies has taken place, resulting in disrupted careers and uprooted families.

On the heals of the MFJ, the FCC once again changed the rules of the game with the third Computer Inquiry in 1986. The FCC believed that changes in the telecommunications markets eliminated the need for structural separation and that nonstructured safeguards would be adequate to prevent anticompetitive conduct by AT&T and the RBOCs. As a condition for allowing AT&T and the RBOCs to provide enhanced services without structural separation, both parties were required to comply with two new requirements: CEI and ONA. These requirements were to ensure that all enhanced service providers have equal access to basic network facilities, or as it is referred to in the industry, to ensure a level playing field for everyone.

Equal access is not a new concept, although it is being considerably extended in Computer III. Historically equal access has been used in equipment manufacturing and interexchange services for a long time. Standardized plugs and telephone jacks enable all manufacturers to make equipment for sale to the general public. One of the most important provisions of the MFJ was the requirement that BOCs provide exchange access, information access, and exchange services for such access that is equal in type, quality, and price to its own offerings. It was this requirement that sparked the long-distance telephone wars of the mid-1980s between AT&T, MCI, Sprint, and dozens of smaller companies. This led to a period when every telephone subscriber was bombarded with literature, telephone calls, and door-to-door salespersons.

In Computer III ONA is an overall design of basic network facilities and services to permit all users of the basic network, including the enhanced service operations of the carrier and its competitors, to interconnect to specific network functions and interfaces on an unbundled and equal access basis.

Two events brought the judicial and FCC developments of the 1980s to a close. One of the MFJ stipulations was that a review be conducted at three year intervals. The results of this review permitted the RBOCs to transmit information but not to generate it. A little more than a year later, AT&T filed a motion requesting the court to remove the prohibitions against its own transmission facilities engaging in electronic publishing. The motion was granted effective August 24, 1989, opening the door for AT&T to participate in videotex and ancillary services.

The story continued to unfold as a new decade dawned. In April 1990, a federal appeals court ruled that the judge overseeing the divestiture agreement, Judge Harold Greene, had applied a legal standard that was too strict and required the judge to reconsider the issue based on whether it was in the public interest to allow the RBOCs to participate in the information services arena. In the same opinion, the court upheld Greene's decision to maintain the manufacturing and long-distance restrictions but narrowed the criteria on which these restrictions were based.

The final domino fell in July 1991, when Judge Greene, with considerable reluctance, ruled that the baby Bells could enter the information services market. Although the judge placed a stay on his decision pending appeals, the die was cast. A few weeks later, the stay was lifted. The doors of the information services market have been thrown wide open, and permission for the RBOCs to manufacture and sell long-distance service may not be far behind.

7.3.2. ONA Model

The BOCs worked together to develop a common ONA model. Although each BOC may choose to implement ONA differently, the model depicts the generic elements of any ONA connection to the telephone network. The ONA model consists of three parts: the basic serving arrangement, the basic service elements, and complementary network services (see Figure 7.5). A basic service arrangement (BSA) is the connection an enhanced service provider makes to and through the BOC's network. It consists of the service provider's physical access

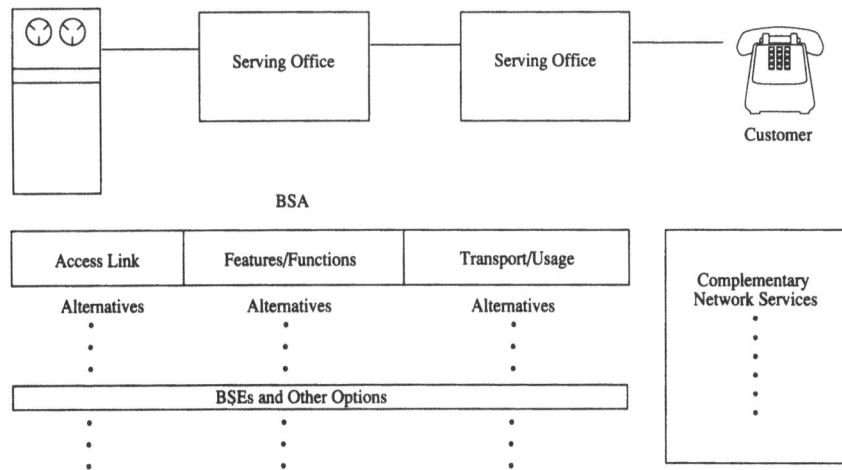

Figure 7.5. ONA model. *ONA Special Report Number 4: Common ONA Model*, Bell Operating Companies, Washington, DC (1987).

link, the features/functions associated with the access link, and the transport (packet, switched, or dedicated) that connects the service provider to its customer.

A BSE is an optional network capability associated with a BSA, *e.g.*, the capability to deliver the calling party's telephone number to the enhanced service provider when the call is made. Complementary network services (CNS) describe how a customer connects to the network; *e.g.*, call forward busy line or do not answer, enables the enhanced service provider to receive calls intended for the end-user when the end-user is unable to answer the telephone. In addition to the BSA, BSE, and CNS components, the model also permits the incorporation of auxiliary services that may not fall within the defined ONA model; examples of such services are billing and collection.

7.3.3. Basic Serving Arrangements

Before a service provider can order a specific package of BSEs, some form of working telephone service is necessary. As an illustration, a service provider could not order calling number identification if the service provider did not subscribe to some type of telephone service through which the information could be delivered. After selecting a BSA, several types of access link may be possible, for example, a choice between a 2- or 4-wire access link.

The features/functions are capabilities located at the CO serving a particular service provider; there are four broad categories: circuit switched, packet switched, dedicated, and dedicated network access link. Within these categories further refinement is possible: A circuit switched connection may be a voice-grade line or a voice-grade trunk. These categories may be even further refined: A circuit-switched voice-grade line may be coupled with a feature/functions option that provides one-way in, one-way out, or two-way communications.

The transport portion of the model completes the connection from the service provider's office to its customers' serving offices. Based on the BSA selected, the transport element may be usage sensitive (call setup plus minutes of use for circuit switched and per one thousand characters for packet switched) or distance sensitive; other possibilities also exist.

To determine the composition of the BSAs, each RBOC surveyed service providers and conducted forums. Since the input each BOC received was different, the BSAs vary among the BOCs. Table 7.2 lists several examples from the 1992 BOCs' *Services Descriptions ONA Services User Guide*. These examples and others from the BSEs and CNSs that appear in later tables are not offered by all the BOCs, which creates problems for a service provider operating within the service areas of different BOCs.

7.3.4. Basic Service Elements

A BSE is an optional network function; it is not required, but when incorporated with a BSA, it provides additional utility. The BSEs enable a service

Table 7.2.   Examples of Basic Serving Arrangements

BSA Category/Type	Title	Description
1/A	Circuit switched	Provides an enhanced service provider (ESP) with a connection to the circuit switched network; examples of connections are a standard telephone line or a PBX trunk; examples of signaling are dial-pulse and dual-tone multifrequency.
1/B	Circuit-switched trunk	Provides an ESP with a trunk connection to the circuit switched network.
2/A	X.25 packet switched	Provides an ESP with an X.25 access to the public packet-switching network; data rates of 1.2, 2.4, 4.8, 9.6, and 56 Kbits/sec are supported.
2/B	X.75 packet switched	Provides an ESP with X.75 access to the public packet-switching network; data rates of 9.6 Kbits/sec are supported over analog and digital facilities, and rates of 56 Kbits/sec are supported over digital facilities.
3/F	Dedicated digital (<64 Kbits/sec)	A dedicated service that provides an ESP with a 4-wire digital channel to the ESP's client at speeds of 2.4, 4.8, 9.6 or 56 Kbits/sec
3/G	Dedicated high-capacity digital (1.544 Mbits/sec)	A dedicated service that provides an ESP with a dedicated channel at speeds of 1.544 Mbits/sec

Source: *Services Descriptions ONA Services User Guide*, Bell Operating Companies, Washington DC, (January 31, 1992).

provider to tailor a particular product to meet the customer's specific need. Since not all BSEs and BSAs are compatible, a service provider must approach the selection process on two levels; for example, since some BSEs work with circuit-switched lines and some work with dedicated digital lines, requirements to support a particular application must be defined in terms of both the service elements and serving arrangements needed. Then the particular BSEs that form the minimum desired level of service must be selected. After that BSAs compatible with this minimum essential set are identified. As additional BSEs are included, the set of compatible BSAs is reduced. The decision point is reached when the last BSE added eliminates the most desirable BSA remaining. This incremental procedure then requires the decision maker to go back one iteration to that combination of BSEs and BSAs before the most desirable BSA was eliminated. Table 7.3 lists several examples of BSEs.

7.3.5. Complementary Network and Ancillary Services

Complementary network services help a service provider offer enhanced services to end-users. A CNS differs from a BSE because it is offered to the end-user, and not the service provider; Consequently a CNS is under the control of the

Table 7.3. Examples of BSE line switching

Element	Description
Call detail recording reports	A paper or magnetic tape report containing originating and terminating telephone numbers, time of day call mode, duration of the call, and date of call
Carrier selection on reverse charges	Charges for calls are paid by called party; can be user to encourage calls from parties of choice in particular geographic areas
Multiline hunt group—overflow	A software-defined search for an idle terminal to which a call can be completed; when all terminations are busy, the calls to be routed to another telephone number within the same switching machine, but outside the hunt group
Uniform 7-digit access number via overlay networking	Provides the ESP with a uniform 7-digit directory number for use across a LATA, state, or regional company; this enables a client to dial one number from all locations, and the calls will be routed to a specified ESP location within each LATA.

Source: *Services Descriptions ONA Services User Guide*, Bell Operating Companies, Washington, DC (January 31, 1992).

user, and not the service provider. Giving control to the end-user prevents the service provider from activating or deactivating critical capabilities related to providing certain services. For example, there are a set of CNS options for Call Forwarding. If the service provider had the ability to deactivate the call-forwarding set by users, it would create confusion and perhaps cost sales, among other problems; Table 7.4 lists a few examples of CNSs.

Ancillary services vary widely among the RBOCs because they are not a part of the ONA model; they are optional services intended to support or complement a service provider's products. Because of their features, they cannot be included

Table 7.4. Examples of Complementary Network Services

Title	Description
Call forwarding—busy line or don't answer—custom control of activation/deactivation	Enables subscriber to activate feature by dialing an access code
Call waiting—cancel	Enables subscriber to suspend call waiting for the duration of a single call and thus prevent interruptions of data traffic.
Message waiting indicator (MWI) ability to receive visual message waiting	MWI is a device with an illuminating lamp controlled by signals from the telephone company's CO.
Speed calling	Enables a subscriber to establish a connection to certain directory numbers by dialing as few as one digit

Source: *Services Descriptions ONA Services User Guide*, Bell Operating Companies, Washington, DC (January 31, 1992).

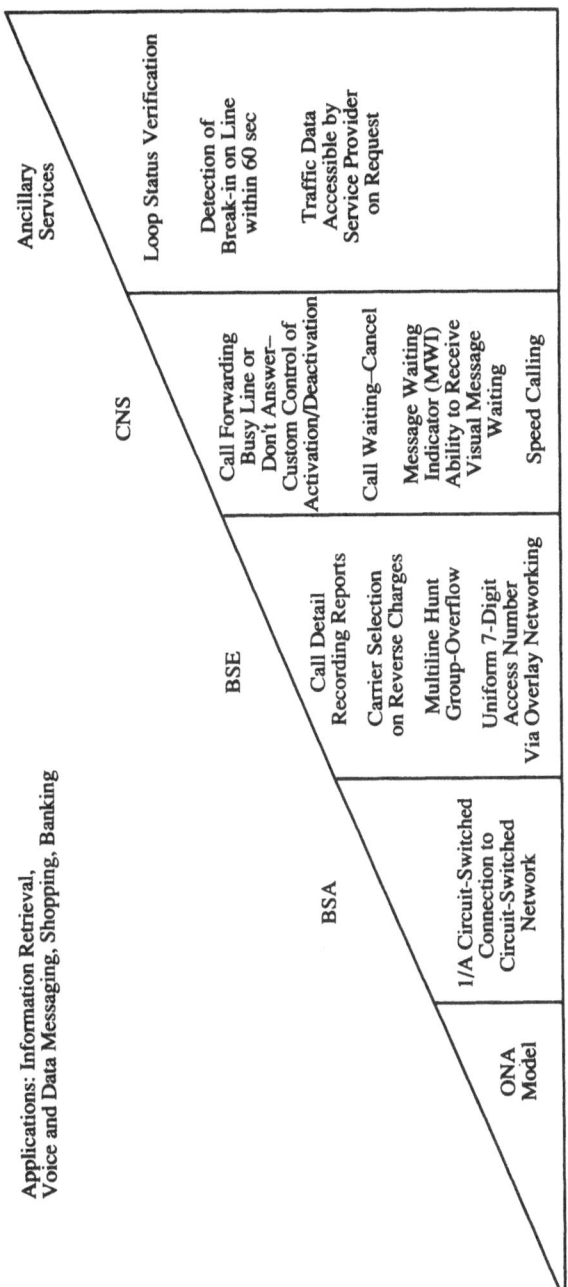

Figure 7.6. An enhanced service offering based on the ONA model.

in the BSE category; examples are the previously mentioned billing and collection activities conducted by the RBOC on behalf of a service provider. Other possibilities are protocol conversion, statistics on such traffic as the number of attempted calls, the number of completed calls, the origin of calls, call duration, and the ability to have status reports on whether communications lines are operating.

7.4. HYPOTHETICAL ENHANCED SERVICE OFFERING

A commercial service built around the ONA is illustrated in Figure 7.6. The service provider in this example is offering information, voice and electronic messaging, shopping, and banking. The user's data access is through a modem-equipped personal computer or a dumb terminal capable of limited graphics. The basic serving arrangement is the circuit-switched voice-grade network connection described in Table 7.2. The BSEs and CNSs correspond to those described in Tables 7.3 and 7.4, respectively.

7.5. SUMMARY

A gateway is the combination of hardware and software through which a user can gain access to a computer system. Gateways have existed in some form almost from the advent of computers, but it is only since the early 1980s that communications standards have evolved to the point where access has become relatively seamless. The most important benefits of a gateway are reduced costs, ease of use, and increased availability of multiple systems. Gateways can be conveniently classified as either host based or distributed. The most widely known distributed gateways are those operated by the telephone industry.

The telecommunications industry has undergone a tremendous upheaval during the 1980s. The decade started with the announced breakup of AT&T and ended on the threshold of almost total deregulation. Part of the deregulatory process was the creation of an ONA. An ONA is intended to provide a level playing field for every company wishing to provide goods and services in or through the telecommunications industry. An ONA varies at different locations in the United States. The ONA model consists of three major parts—the BSA, BSEs, and CNS. The implementation of ONA is changing the way the United States communicates.

REFERENCES

C. Clarke, The Strategic Implications of Open Network Architecture, *Telecommunications* 22(3) (1988), pp. 41–42, 44, 47, 76.
Code of Practice: Gateways, National Federation of Abstracting and Information Services, Philadelphia (June 1987).

ONA Special Report Number 4: Common ONA Model, Bell Operating Companies, Washington, DC (November 1987).

M. S. Fowler, A. Halprin, and J. D. Schlicting, "Back to the Future": A Model for Telecommunications, *Federal Communication Law Journal* **38**(2) (1986), pp. 145–199.

D. Gabel, Gateways: From Here to There, *Network World* **5**(28) (1988), pp. 33–35, 38.

Gateway 2000: A Report of the Videotex Industry Association Study on North American Gateways, Videotex Industry Association, Rosslyn, VA (October 1988).

P. W. Huber, *The Geodesic Network: 1987 Report on Competition in the Telephone Industry*, Superintendent of Documents, U.S. Government Printing Office, Washington, DC (January 1987).

C. Mason, Appeals Court Reverses Greene on Information Services Ban, *Telephony* **218**(15) (1990), pp. 8–9.

ONA Planning, Materials from Information Industry Seminar, Southwestern Bell Telephone Company, St. Louis (March 1 & 2, 1988).

D. Powell, What You Should Know About ONA, *Network Management* **7**(8) (1989), pp. 36–41.

RBOCs Get the Go Ahead on Information Services, *Telecommunications* **25**(9) (1991), pp. 9, 11.

Recommended Practice for Gateway-to-Gateway Interconnection—Preliminary Draft, Videotex Industry Association, Silver Spring, MD (1990).

Recommendations on Information and Service Provider Mnemonics by the Service Location Subcommittee of the Business Consistency Practices Committee, Videotex Industry Association, Silver Spring, MD (1990).

Report and Recommendations of the United States Concerning the Line of Business Restrictions Imposed on the Bell Operating Companies by the Modification of Final Judgement, Civil Action No. 82-1092 (D.D.C. February 2, 1987).

Services Descriptions ONA Services User Guide, Bell Operating Companies, Washington, DC (January 31, 1992).

Services Descriptions ONA Services User Guide, Bell Operating Companies, Washington, DC (July 31, 1991).

D. M. Simons, Romancing the Telephone, *Information Times* (Fall 1984), pp. 43–53.

D. G. Smith, and D. C. Pitt, Open Network Architecture: Journey to an Unknown Destination? *Telecommunication Policy* **15**(5) (1991), pp. 379–394.

Southwestern Bell Telephone Company's Open Network Architecture Plan, Southwestern Bell, St. Louis (January 25, 1988), pp. 158.

A. Taff, FCC to Let Regulated BOCs Offer Enhanced Services, *Network World* **8**(46) (1991), pp. 6, 8.

W. B. Tunstall, Disconnected Parties: Divestiture in Retrospect—Part I, *Telecommunications* **26**(3) (1992), pp. 41, 43–44.

In re United States of America v. Western Electric Company, Inc., et al., Civil Action No. 82-0192, Triennial Review and Amendment to the Modification of Final Judgement (Initial), (D.D.C. September 10, 1987).

In re United States of America v. Western Electric Company, Inc., et al., Civil Action No. 82-0192, Triennial Review and Amendment to the Modification of Final Judgement (Final), (D.D.C. March 7, 1988).

In re United States of America v. Western Electric Company, Inc., et al., Civil Action 82-0192, (D.D.C. July 28, 1989).

Chapter 8

Software

Software is much like the mystical ether of the Middle Ages from which charlatans tried to conjure solutions to life's desires; like that ether, software does not have substance. It is largely the manifestation of ideas. It is difficult to measure, since it has neither weight nor physical dimension, and it can be effectively evaluated only through use, a sad lesson many have learned.

The term software is synonymous with program. The 1980 amendment to the copyright law defines a computer program as "a set of statements or instructions to be used directly or indirectly in a computer in order to bring about a certain result." The development of software has passed through three phases. The first phase, which took place in the 1950s and early 1960s, was primarily concerned with an empirical approach to the study of programming concepts. This was followed by a more mathematical approach to develop theories for the empirical work done earlier. The third phase has emphasized an engineering approach to harness these theories.

As software development has passed through these phases, the orientation of application software has shifted from a program-centered view to a data-centered view. A program-centered view focuses on program development and considers data piecemeal, such as when it is the object of program transformation. This view is appropriate when computers are solving numerical problems. The data-centered view focuses on the database and considers programs as bugs that crawl around the database and perform necessary functions, such as querying and updating.

As computing power has migrated from the computer center to the desktop, the relative importance of the data-centered view has increased. Most end-users of a computer want to receive or send information and perform relatively simple transactions, such as transfer funds or create a spreadsheet analysis. The shift from the program- to the data-centered point of view has been compared with the shift from the earth-centered to the sun-centered model of the universe introduced by Copernicus. As computing power continues to migrate toward the user, software development will revolve around the end-user's needs.

Chapter 8 begins with a model for classifying different types of software and then briefly explores the basic functions found in software that support the creation and delivery of information. Brief overviews of Windows® 3.1 (Microsoft Corporation), OS\2® 2.0 (International Business Machines, Inc.) and QuickTime® (Apple Computer, Inc.) are presented. These overviews are followed by the description of a videotex server called UNITRAX® (American Telephone and Telegraph Company). Chapter 8 concludes with several important points to consider in selecting and acquiring software.

8.1. CLASSIFYING SOFTWARE

Figure 8.1 shows a model for classifying software. All software can be divided into one of two broad categories—operating-systems software and applications software. Each of these may then be further subdivided.

The operating system provides for the fundamental operation of the hardware and software; it is subdivided into routines for controlling how various parts of the system function and routines for performing various processes. Control software is normally acquired from the hardware vendor. In this taxonomy, it consists of the supervisor, input-output, and communications programs. The supervisor or executive controls the operation of other routines, and it is often thought of as part of the machine itself. Among its responsibilities is sequencing, set-up, and executing all jobs entering the system. The input-output or I/O software moves data between the processing unit and either storage or I/O equipment. Closely allied with I/O routines is the communications software, which monitors communications activity and performs a variety of functions whose complexity depends on the sophistication of communications traffic. The complexity may range from a one-way limited flow of data to a full video and audio two-way exchange.

Processing software provides the underlying support needed to run various application packages. In this taxonomy, it is subdivided into service, language translation, and database management routines. Service or utility software accomplish the everyday tasks incidental to operating a computer; examples are sorting, merging, and moving data between storage devices. A language translator converts one computer language into another; normally the process involves converting from a form understandable by humans to one understood by the particular machine being used. Over one thousand programming languages have been developed. The infobase/database management system provides a systematic approach for storing, updating, manipulating, and retrieving what is presented

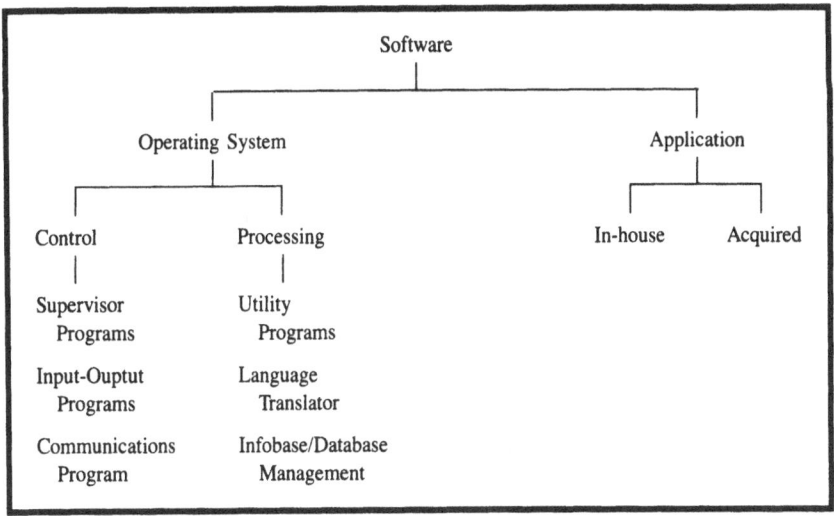

Figure 8.1. Software Functions in a Multiuser System

through the computer. There is a large selection of packages available, and cost varies tremendously from several hundred to tens of thousands of dollars.

Application software is the other main branch of the taxonomy presented in Figure 8.1. This category contains software for performing activities that are an integral part of a particular organization's business; traditionally this encompasses personnel management, inventory control, accounting and financial needs, market analysis, order entry, product distribution, and so forth. Most of the features a user associates with a particular business activity resides in this software.

8.2. FUNCTIONS

An enhanced system has value because of its ability to meet users' needs. What it can do is determined by the functions the software performs. The major functions of a multiuser system are listed in Table 8.1 under the four headings of information base management, administration, information retrieval, and user services. In some instances, these features are contained in a single program, but in most cases they are the result of several programs working in concert.

8.2.1. Information Base Management

Until recently the information base consisted of data that was displayed or printed as text; today digitized audio, sophisticated graphics, and picture-quality images are part of the information base. It is a much more formidable task to

Table 8.1. Software Functions in a Multiuser System

Feature	Function	Feature	Function
Infobase management	Create (text, audio, video, graphic)	Information retrieval	Access
			Browsing
	Editing		Path functions
	Structure		Scrolling
	Contents (automatic,		Update registry
	manual, bulk)	User services	Transaction support
	Utilities		Clipping
	Archive and backup		Forms Filling out
	Compaction		Interest matching
	Copy		Opinion polling
	Orphan check		Response
	Overview		Document interexchange
	Recovery and undo		Electronic messaging
	User profile		Static
	Verification		Dynamic
Administration	User management		Distributed processing and
	Accounting and billing		information transfer
	Bulletins		Electronic audiovisual
	Configuration change		presentations
	Enabling/disabling		Closed user groups
	Monitoring		Guided tours
	Multiple gateways		
	Security		
	Statistical analysis		
	User control		

create, maintain, modify, and present information that contains a mix of these technologies. Adding to the complexity is the need to integrate different storage media and storage devices; for example, a basic frame containing text may be stored on a floppy disk and a picture-quality image that overlays part of the text frame may be stored on an optical disc device. Mixing text and images leads to the object-oriented infobases discussed in Chapter 3. The emphasis shifts from storing bytes to what one company refers to as BLOBS. As images complement text in the 1990s, BLOBS or their equivalent will become the basic building blocks of an infobase.

The functions found in the infobase management system can be divided into three groups: creating, editing, and maintenance or utility routines. The creation function can assume three distinct forms; the first involves creating the overall structure of the infobase, and it occurs when the infobase is initialized. It involves such activities as creating the files that will contain user records, routing information, and the contents of the frames. It defines such basic features as security and the links that will be permitted among various parts of the infobase.

After the structure of the infobase has been established, the structure of each frame and its contents must be created. Frame structure consists of deciding on the general layout of the frame and its links. Establishing the links involves identifying what frames point to it and what it points to, which can become a complex undertaking when text and video or other display combinations are being merged from different storage media.

Creating the contents of the frame is considerably more involved than meets the eye, and it probably entails using a special authoring software package. What one sees is the last step of a lengthy and expensive process that is more fully discussed in Chapter 10. The software that assists in the mechanics of frame creation may support free form creation and the use of templates. In this context, a template is a fill-in-the-blank display designed to assist the person creating the frame.

Content creation software often includes routines for enhancing the text; for example, there may be a clip art library with pull-down graphical menus for inserting images in the text. There may be charting capability which transports business charts produced on popular packages, such as Lotus 1-2-3, translates charts into the appropriate format, and inserts them into the infobase. Routines for converting voice to text, processing and storing input from cameras and scanners, and accepting images created with drawing tablets and spatial digitizers may also exist. Such software uses a spatial digitizer to determine the physical dimensions of an object, then converts these dimensions into a computer-generated image.

The edit function enables the structure and the contents of the infobase to be modified. Structural changes include inserting and deleting frames, changing links, archiving and restoring frames, and automatically renumbering or reindexing frames. Because of relationships among frames, the software is more sophisticated than first imagined; for example, well-written software would prohibit deleting an index frame or index entry that is not empty or for which there exists a cross-reference.

Content editing changes the composition of a specific frame. The software performs many of the same activities as a word-processing package—cursor manipulation, inserting and deleting rows and specific characters within a row, highlighting, color changes, content searches, reformatting text after it has been input, repositioning items in a frame, and so forth. In addition it may have functions to support bulk updating in which whole sections of the infobase are changed automatically.

Bulk updating may occur off-line or on-line. Magnetic tape or floppy disks are the common media for off-line updating. The system does not have to be closed to users, although performing the updating at night or during a period of relative inactivity will lessen the impact on user response time. On-line updating is trickier because large segments of the infobase may be impacted by changes at the same time they are being accessed by users. An example is the process involved in updating an infobase from a news or financial services wire. The software must capture the information on a periodic basis, perhaps hourly, reformat and segregate it, and then insert it in place of the currently stored information. Bulk updating is an efficient and economic way of maintaining an infobase. The alternative is the time-consuming and labor-intensive process of rekeying and individually moving text and imagery to its proper location.

In addition to the functions just described, the traditional type of recovery, archiving, and backup software found in most data-processing installations must exist. Since enhanced systems are often transaction oriented and involve integrating information from a variety of media, these chores of good system management are especially important as well as more difficult to provide.

A number of utility programs are included to perform routine maintenance functions. A composition utility reorganizes the physical storage of the infobase to assimilate additions and reuse space freed by deletions. Typically compaction software creates a new and empty infobase with the same name as the old one. Information is then copied into the new infobase. The process is complex and may be very time consuming.

A copy utility is often available to copy existing information from one location to another. This utility is useful for replicating the infobase in other locations, and it normally provides the option for copying individual frames and images.

A user profile routine is needed to create and modify each user's stored profile which is a record created for an authorized user. The profile normally contains the user's name, user type, address, telephone number, identification code, password, account information, security clearance, and sometimes usage statistics.

An overview utility may be provided to assist in infobase management and verification. The routine will point out a list of frames (but not their contents) and their direct access numbers, keywords, titles, and security level. A companion feature provides information about the date and time of last access or update, the number of accesses, information about links to other information, and how many bytes of storage are needed.

A verification feature checks for consistency and looks for paths that lead to nonexistent frames and down paths from which the only return is to abort the session and log on again. On systems that support keyword access, the keywords are checked for duplication, and the administrator is notified of potential problems. A related function found in some commercial software prevents the intentional or unintentional creation of orphan frames. An orphan frame is one not linked to other frames and not reached through a menu. Orphans are unintentionally created when menus or other frames through which they are reached are deleted.

8.2.2. Administration

Table 8.1 lists a number of administrative functions. The user control function provides the ability to log onto and off of the system; the latter feature may permit the log off to occur with or without a line disconnection. The user control function also enables a user to change a password, skip forward and backward, retrieve help displays, and perhaps refresh a currently displayed frame if it contains a transmission error or has information that may have changed since it was first retrieved.

Universal connectivity amplifies the need for security; the traditional importance of securing data, the communications exchange, and the hardware and software have not changed; issues of security have become more important. Establishing gateways for users to access other systems is a two-way road that heightens the possibility of unauthorized users gaining access to an organization's information system. In environments with multiple infobases, software must be able to control which infobases each user has access to, whether or not passwords are required, how use will be accounted for and billed, and the ability to switch among the infobases in the same session.

Financial transactions, electronic messaging, and the electronic exchange of documents introduce new issues of security and authorization. As an illustration, a paper contract is legally executed when the signed document is postmarked; the legality of an electronic message is not so clear. Is it legally executed when transmitted or received? Answers to such questions do not exist, which make software to facilitate the authenticity and security of information very important. The types of features that are helpful are the ability to cancel transactions, the power to recall and verify receipt, and time stamping.

Software must be able to monitor the communications flow and signal strength in order to detect problems and unreasonable levels of errors. Network management software must perform such traditional activities as enabling and disabling equipment, reconfiguring the network, and supporting access to internal and external gateways.

Statistics gathering is important for determining system activity and balancing the resources to support periods of peak demand. Accounting and billing software must exist to bill users or charge budgets properly. The ability to integrate voice and data in some systems means voice toll calls, facsimile transmission, electronic mail, and costs for accessing commercial infobases must be recorded and charged to the correct user.

If the company sells its information or provides services to outsiders, perhaps through one of the gateways operated by the telephone industry, the accounting data collected must be fairly detailed. Each time a user begins a session, the total costs must be tracked. A record must be generated that contains the user's name; the time and date; the length of access; the number of the frame viewed or some other form of identification; the price of each frame viewed or service performed; charges for resources consumed, such as printing and mailing information viewed; and perhaps the session sequence number; port number; terminal number; and inward-calling information if the company must charge the user and rebate the communications carrier.

Accounting information must also be maintained for information providers who store information on the company's computer for access by closed user groups or the public. This accounting information must track resources consumed by the information providers as well as maintain information about what the information providers are owed. Accounting is a very important aspect of any computer system, and it is crucial to the justification and continued operation of a system providing special services.

8.2.3. Information Retrieval

These functions may be divided among the routines for accessing the infobase, navigating through it, and transforming information that is retrieved. A variety of access techniques are available, but the most common for systems of the type being discussed are menu based.

A menu-based approach involves returning to a preceding frame, returning to the main frame, going to the next frame using a function by the same name if it exists, or proceeding to one of the choices in the menu. In business applications, access is often most conveniently handled with a direct access made possible by specifying a frame address. Another popular method of access in business applications is keyword based in which a text string is provided and a frame is sought with the specified string. A more sophisticated search is possible when keywords are combined using the Boolean operators + and or. When the complexity of the application exceeds the possibility of question-and-answer or keyword methods, a query language may be needed. In the coming years, artificial intelligence will facilitate the access process and enable sophisticated searches by unsophisticated users.

An update registry if present will record major changes to the content since the user last accessed the system. This can save time, since an update registry prevents looking at information that has not changed. A path function enables the user to insert a selection of frames to be viewed each time the system is accessed. This permits faster access to information of recurring interest. Some software products also create a chronological list of the frames that have been accessed so the user can backtrack during the same session.

Browsing and scrolling are two approaches for traversing the infobase once access has occurred. Browsing might be thought of as moving horizontally rather than vertically through the infobase; for example, in a personnel application, all frames relating to managers at the same level in the organizational chart may be linked together so that a viewer can browse through them. Scrolling is the process of viewing a set of related frames arranged in a hierarchical fashion or moving text up or down that will not all fit on a screen at the same time.

8.2.4. User Services

A transaction was defined earlier as the collection of related actions whose completion identifies a discrete unit of work. Typical examples are a funds transfer and ordering goods or services. Transaction software can be categorized according to how the user interacts with the system performing the transaction. The software may involve filling out forms via a formatted screen and transmitting the information. It may be response based, so that predetermined choices are presented for selecting; many audiotext services used with a touch-tone telephone operate this way. It may be document based, so that information is transmitted in standardized formats; the ANSI X12 standards for electronic document interchange is an example of this type of transaction software.

Other user services are clipping, interest matching, and opinion polling. Clipping enables a user to select a set of keywords that are then compared to content when it is added. If a match occurs, either a copy is made and stored in the user's mailbox or a record of the contents is made. Interest matching enables users to submit a set of key words that are then matched with a database of keywords submitted by other users, so that an individual or individuals with similar interests are identified. Opinion polling supports surveys, market research, and public opinion sampling.

User services software incorporates voice-, text-, and image-based messaging. The simplest form of messaging is response based. Predetermined choices are presented, and the listener or viewer selects one and routes it to the intended recipient; birthday and special meeting announcements are sometimes sent this way.

Conferencing is the second form of messaging. It may be static or dynamic and varies considerably in the level of sophistication possible. Static conferencing is the electronic mail service found throughout business. One user composes a letter and transmits it to one or more people who may then respond in kind. The input may be voice or text based, and the output may be either the same form as the input or translated into the other form.

Voice-based dynamic conferencing is, of course, a telephone conversation. Text-based dynamic conferencing is widely referred to as an electronic forum. In some systems, the speaker has the floor until relinquishing it, while in other systems, the speakers can interrupt each other just as in a telephone call. The most sophisticated form of conferencing is video based, which ranges from freeze frame video images to full-motion televisionlike conferencing.

Distributed processing functions and file transfers are common features found in most computer systems. Some software packages simplify the procedure with form-filling displays for entering the data and command menus for sending it for processing or storage. In some systems, libraries of programs are stored in a file server and accessed via a LAN, or they are stored in a central computer facility and accessed via the telephone or other public communication services.

When accessed by phone the program transferred is called telesoftware. Once the program has been downloaded, the user can go off-line and execute it; when the program is no longer needed, it is deleted. Software provided through file servers or from public libraries relieves the user of maintenance responsibility, and it is a cost-effective way of giving everyone access to the latest version.

Electronic audiovisual software is a unique feature not found on traditional systems. It enables an audiovisual presentation to be created for display on other users' desktop units, public access terminals, or the originator's machine as part of a presentation. This software is often directed at educational, directory, and advertising applications.

A closed user group (CUG) is a group of users that have a designated part of the information base that only they are able to access. This is a feature popular with trade and professional associations. The guided tour feature enables a scrolling set of frames to be displayed until an interruption occurs, such as someone entering a command via a touch screen. This feature is popular in software designed for kiosks.

8.3. SOFTWARE PRODUCTS

There are literally scores of software packages that support enhanced system; unfortunately many of these products are islands unto themselves; incapable of supporting content developed for another system. However in late 1991 and the first several months of 1992 order began to emerge from the chaos as large companies like Apple Computer and Microsoft Corporation introduced extensions to their operating systems that supported multimedia applications. These product introductions lent order and reduced some of the chaos but in no way eliminated it.

This section briefly describes four software products: Windows® 3.1 (Microsoft Corporation), OS\2® 2.0 (International Business Machines, Inc.), Quick-Time® (Apple Computer, Inc.), and UNITRAX® (American Telephone and Telegraph Company). The first three are classified as either operating environments or operating systems; however none of the three is designed to serve a large population of users. To do so requires a special product called a server. The videotex server UNITRAX illustrates the type of capability needed in a system with many simultaneous users. Like the first three software products, UNITRAX is not completely compatible with any of the others. Much more extensive descriptions of these four products can be found in books and manuals specifically devoted to one of the products. The discussion here provides only a basic understanding of what each does and how it fits into an interactive environment.

8.3.1. Windows

Windows has been referred to as an overnight success that took ten years to occur; the chronology of *Windows* appears in Table 8.2. Microsoft announced development plans for it in November 1983, and shipped *Windows* 1.0 in November 1985; the original release and the one that followed in December 1987 were not very successful. The programs were slow; they consumed prodigious amounts of

Table 8.2.   A Short History of *Windows*

Date	Event
November 1983	Intent to develop announced
November 1985	*Windows* 1.0 shipped
October 1987	Excel for *Windows* shipped
December 1987	*Windows* 2.0 shipped
March 1988	Apple Computer files "look and feel" suit
November 1989	IBM and Microsoft announce agreement to limit *Windows* to low-end machines
December 1989	Word for *Windows* shipped
May 1990	*Windows* 3.0 shipped
September 1990	IBM and Microsoft rewrite the terms of earlier agreement
August 1991	Lotus 1-2-3 for *Windows* is shipped
November 1991	WordPerfect for *Windows* ships after missing ship date three times
April 1992	*Windows* 3.1 is shipped

memory considering the relative amounts of memory on machines of that era; and *Windows* had an unforgivable habit of catastrophic failure. But the software attracted adherents because it offered an easy-to-use mouse interface built around icons and pull-down menus; in fact its resemblance to Apple Computer's Macintosh interface led to an unsuccessful look and feel suit initiated by Apple in March 1988. *Windows* really gained acceptability with release 3.0 in 1990: Over nine million copies of this version were sold; release 3.1, which followed almost exactly two years later, cemented *Windows'* position as the predominant graphical user interface. Technically *Windows* is an operating environment, and not an operating system, because it runs on top of DOS and depends on many DOS functions to operate.

Windows 3.1 made multimedia available to millions of people. The product includes many features that were part of a separate product called Multimedia Extensions for *Windows* that had been shipped in 1991. By incorporating multimedia functionally with *Windows*, Microsoft managed to sneak onto millions of desktops a whole new set of capabilities for a nominal cost; for example, the introductory cost to upgrade from release 3.0 to 3.1 was $49.99. Having an imbedded market of this size made it possible for developers to assume that their users have at least the software if not the hardware needed to run multimedia applications. The availability of software to support multimedia applications is a big step in combating the chicken-or-the-egg problem that confronts every new technology that depends on a large enough market to justify the cost of creating new products. The basic multimedia features provided in release 3.1 were animation, audio, and drivers for controlling CD-ROM and videodisc players. These features are being supplemented as time passes, but the basic features, such as sound can support many creative applications—for example, adding voice or music to charts and graphs; a moving video playback feature called Audio-Video Interleaved (AVI) will be one of the first additions.

Windows/NT is the newest manifestation of *Windows*. It is an operating system designed to replace DOS, Unix, and OS\2 on high-end workstations and network servers. It is a 32-bit product compared to the basic 16-bit *Windows* products previously discussed. *Windows*/NT is designed to overcome some of the limitations of its predecessors, such as memory management, and it has the ability to run on computer chips other than those of the Intel family and its clones.

8.3.2. OS\2 2.0

IBM's answer to *Windows* is OS\2, the most advanced operating systems for PCs on the market. Although the April 1992 release of OS\2 2.0 did not have integrated multimedia drivers like *Windows* 3.1, it had just about everything else. The minimum requirements for OS\2 2.0 are a PC based on the Intel 80386SX, 3 Mbytes of main memory (preferably 8 MB), and a 40 Mbytes (preferably 80MB) hard disk. The software supports 32- and 16-bit architectures, and it can run both DOS and *Windows* 3.0 applications as seamlessly as if they had been written for OS\2. The magnitude of the operating system is suggested just by virtue of the 14 high-density 3.5 disks on which it is shipped.

The biggest benefit of OS\2 2.0 is its 32-bit architecture, which enables it to address large amounts of memory. A 16-bit application common to DOS can access memory only in 64-Kbyte segments ($2^{16} = 64,000$). Since files larger than 64 Kbytes must be collected individually and pieced together, considerable time is wasted, and the speed at which such applications are processed is adversely affected. The OS\2 2.0 does not have to worry about this problem; it can address as much memory as it needs, which can be up to 4 Gbytes; other advantages are multitasking, multithreading, and virtual memory support. Multitasking is the concurrent execution of two or more tasks (*i.e.*, units of work). A thread is a set of activities, such as reading a file or scanning the text as it is loaded into buffer memory. Multithreading enables a thread that is delayed, perhaps due to a search for additional data, to have its execution suspended and the execution of another thread started. Virtual memory support allows a job too large to fit into available memory to be broken into pieces and the pieces moved in and out of the processor as each piece is needed.

The one major drawback of OS\2 had been the limited number of applications that were programmed for it. When OS\2 Presentation Manager was introduced in 1988, only the IBM ProPrinter would work with it, and there were few software application packages available. This shortcoming has been corrected by incorporating the ability to support DOS and *Windows*, not to mention the over 800 developers creating new applications specifically for OS\2. The current version of OS\2 is competitively priced with *Windows*, but the former is much more powerful. The questions are whether it can overcome *Windows*' juggernaut and how well it can promote multimedia capability through the multimedia extensions which are acquired by adding the IBM Multimedia Presentation Manager/2 software?

8.3.3. QuickTime

QuickTime is Apple Computer's multimedia software for incorporating graphics, video stills, motion, animation, and sound into presentations. It was introduced late in the fourth quarter of 1991. Initially QuickTime was offered as an extension to System 7, the Macintosh operating system, but versions have been released or are underdevelopment for other operating environments, including *Windows* and OS\2. Essentially QuickTime works as a software compression tool. In its initial release, it was build around three compression schemes—JPEG (Joint Photographic Experts Group) for still images, an Apple-developed compression algorithm for animation, and a video compression algorithm capable of reductions of 25 to 1.

QuickTime consists of three major software modules: a Movie Toolbox providing standardized movie controls for applications; an Image Compression Manager for compressing and decompressing various file types; and the Component Manager for controlling external devices, such as a CD-ROM. The philosophy behind the software is to enable QuickTime to assume responsibility for controlling devices and compressing files, which will enable applications to import movies, capture photographs, and so forth, by virtue of being QuickTime compatible. This philosophy is similar to the one adopted for *Windows* 3.1.

The CD-ROMs are the predominant storage medium due to their price and capacity, and this is likely to continue for the foreseeable future. This ensures a large pool of material but inhibits the ability of the system to support full-motion video because of the data transfer rate for CD-ROMs. In fact QuickTime can support only moving images smaller than full size on the screen. The fastest these images can be repeated is at a rate of about 15 frames per second (fps), whereas full motion is generally considered to be 30 fps.

8.3.4. UNITRAX

UNITRAX is a member of the software family known as videotex servers. It is a multipurpose, multipresentation Unix-based product from AT&T written to support text, graphics, and digitized images. The product supports the commonly used videotex presentation standards: NAPLPS, ASCII, PRESTEL, and CEPT. The meaning of these acronyms is contained in the appendices. UNITRAX operates on UNIX minicomputers, and it can support as few as 8 and as many as 64 simultaneous users.

Like most servers, UNITRAX defines multiple classes of users. The different classes represented in a server give a hint of the system's capability. In the case of UNITRAX, there are five classes:

Retrieval user: Allowed to view all unprotected frames; may be given
    permission to access protected frames belonging to as many as ten
    information providers
Bulletin board publisher: Reviews and publishes notices on a publicly
    accessible bulletin board
Editing user: Creates, amends, copies, and deletes frames
Registration user: Registers other users; has the ability to determine what
    various users may do
System administrator: Has overall system control

The software supports three control philosophies. It can operate in a centralized fashion and assign full control to a single administrator; this approach is
consistent in a small organization. The software will allow control to be dele-
gated to a set of registrars, or each user can grant edit privileges and register users;
this is appropriate for large organizations. The third control philosophy is a
combination of the first two and permits the administrator to hold onto some
functions and distribute others.

The software supports three broad categories of applications: information
retrieval, form filling, and personal facilities. The basic unit of information is
called a page. Each page has a unique number ranging from 10 to 999,999,999;
each of the pages may in turn have up to 26 frames lettered A–Z associated with
it, e.g., 647C, 1524Q. This arrangement can accommodate over 25 billion
displays. There are four ways of accessing a frame: by menu, direct retrieval with a
frame number, or by a keyword associated with the desired frame; if a microcomputer or a terminal with function keys is being used, function keys F1–F8 are able
to access a frame relative to the currently displayed frame (e.g., next, previous).

The form filling facility is used for collecting information; two types of form
filling are data collection and action initiation. The data collection form filler is
used for ordering goods and answering questions. Examples of action initiation
frames are the login and change password frames.

The personal facilities software supports several separate applications. There
is an extensive mail facility with a variety of options. A personal-filing facility
allows a user to create up to five files in a personal file drawer and there is a bulletin
board option for creating and sending public announcements. A path or routing
facility also exists for selecting a set of frames that are frequently accessed by
the user. The frames can then be viewed each session by simply entering a single
key to initiate display of the sequence.

The UNITRAX is used internally by AT&T, but it has not been widely
adopted by other organizations. The software is well documented and designed
with flexibility and compatibility in mind. An application development package is
available for tailoring specialized applications for combining text, color, graphics
and photographic-quality images on the same frame.

8.4. AUTHORING SOFTWARE

This section discusses software for creating and maintaining the infobase. There are a variety of ways of undertaking these important activities, and there are many software programs available for accomplishing them.

8.4.1. Configurations for Content Creation

A system for creating content can be either on-line or off-line, and it can be dedicated to the task of making frames or shared with other applications. The capital investment per employee for on-line systems is generally smaller than for off-line systems, primarily because on-line systems support multiple users and because work stations are often little more than dumb terminals. On-line systems make it easier to centralize control and monitor what is taking place on an hour-by-hour basis.

Off-line systems are usually built around a simple user microcomputer. The infobase is updated periodically by logging onto the host and transmitting the frames as a file or by physically transporting a floppy disk or cassette tape. The major advantages of off-line creation are (1) it does not tie up the resources of an expensive mainframe; (2) the number of work stations is not constrained by the capability of the host computer; (3) and there is greater local control over the development cycle. This last point deserves amplification.

Off-line production increases system availability. If the host is down for repairs or preventive maintenance or if communications between the host and frame creation system are disrupted, frame production is not impacted. Off-line development enables sophisticated equipment to be integrated with the production process, such as cameras for capturing photographic-quality images. Perhaps most importantly, off-line production enables units to manage the production process without being monitored by a central authority. Creativity can be given freer reign until the frames are transported to the host.

A creation system is dedicated or shared; micro-based off-line systems are often dedicated to just the creation process. There is a wide assortment of commercial software packages available for MS-DOS personal computers. These packages are relatively inexpensive and easy to use. Their major shortcoming is a limited set of features and failure to integrate still video and graphics of high quality. And there is the ever-present danger that the packages may produce frames containing options that are incompatible with the service operator's host software.

Another set of micro-based systems exists that is built just for frame creation. These systems are relatively expensive, but they are designed for volume levels of production. They integrate a variety of equipment, and most conform to standard reference models issued by standards-making bodies. Consequently compatibility within the same vendor's product line and even across vendors is usually not a problem.

Shared frame creation systems are found on mainframes and minicomputers. Systems operated on mainframes may run in the background, allowing the system to perform other work with a higher priority. Packages sharing the mainframe with other applications are usually part of the host's enhanced software package; therefore compatibility is assured, and portability of the frames to other systems with the same software is possible. Shared systems on mainframes are relatively inexpensive per work station, but they generally do not support the development of frames with sophisticated graphical content or photographic images. Shared systems on minicomputers are similar to the stand-alone units just described. The number of work stations is usually limited to a handful. Other than sharing the processor and storage, these units resemble the stand-alones and may support the same basic set of peripheral devices that can be attached to stand-alone devices.

8.4.2. Multimedia Authoring

There are at least five basic types of authoring facilities for creating and organizing multimedia content; these are authoring systems, application shells, hypermedia application generators, time line processors, and programming languages. An authoring system is designed to simplify the creation process. It provides preformatted screens in which the author fills in the blanks. Application shells simplify the process even further by supplying the structure of the presentation. The author then supplies information about the details of the presentation by filling in the blanks.

Hypermedia generators allow cross references to be set among related items in the infobase. One item may be text with links to a graphic that is linked to an animation, and so on. Time line processors present the author with a list of media forms available and provide a way of invoking the various items at any point on a time line that spans the duration of the presentation. Programming languages provide the greatest flexibility and require the most work. The languages range from special-purpose code to general-purpose languages used for authoring. The widespread availability of inexpensive authoring facilities obviates the need for programming in most applications.

The particular choice of an authoring facility depends on the application and the sophistication of the author among other things. The Bergman reference at the end of this chapter describes authoring facilities and their limitations.

8.4.3. Authorware

Authorware Professional for Windows® is an example of an authoring program for multimedia applications. It is a product of Authorware, Inc., a company formed in 1987 to develop products for the multimedia market. The underlying strategy behind the software is the creation of an authoring system that does not require any programming in the conventional sense. It operates under

Microsoft *Windows* and has the ability of integrating text, animation, sound, still pictures, graphs and video into a seamless presentation by moving icons with a mouse. The result can then be run with *Windows* or a run-time-only version of the software.

8.4.3.1. Authoring Environment

The authoring environment consists of command icons on a flowchart and a set of menus containing commands for various authoring functions. The icon set used for building the sequence of actions are display, animation, erasure, pause, decision, interaction, calculation, organization logic flow, start and stop flags, movie, sound, and video. These icons are selected with a mouse and moved into an ordering that is similar to a flowchart but represents the sequence in which the action is to occur. A "try it" feature permits the creator to preview the flowchart during the creation stage to make editorial changes. This is a very powerful feature not found on most authoring systems. Various design aids are provided to permit considerable flexibility in the design; for example, flowcharts are easy to reorganize and nine different interaction types are built in, such as key presses, click/touch areas, time limits, and tries limits. A productivity library is also included to facilitate the developer's work. The library contains clip media for animations, graphics, movies, and sound.

8.4.3.2. Multimedia Tools

A tool set is provided for working in the following media forms: text, graphics, sound, animation, and still or motion video. Each of these tool sets offers an extensive number of features for designing and editing the content, which may either be developed with the assistance of Authorware or with other software and then imported by the clipboard or one of several file options. A special effects option provides transitions between displays and further customizes the content, especially if it were created on other software and imported.

8.4.3.3. Multiplatform Architecture and System Requirements

Software is also available for the Macintosh. To use applications developed in one environment for the other, conversion routines exist for translating applications from Macintosh to *Windows* platforms with nearly 100% accuracy. It is possible to jump between applications or files to share data. Authorware is compatible with a raft of complementary products from other vendors, which provides a much larger base of possible content than would otherwise exist; for example, Adobe Type 1 fonts, ZSoft SoftType fonts and other *Windows*-compatible screen fonts may be used. The recommended system requirements for authoring are a 20+MHz 386 machine with 4 Mbytes of random access memory and a 16-color VGA card.

Authorware is widely regarded as one of the best packages for multimedia authoring on the market, and it is also one of the most expensive.

8.5. SELECTING SOFTWARE AND LIVING WITH THE DECISION

Selecting the right software is difficult, time consuming, costly, and fraught with pitfalls. Most of the software available for the types of systems described in this book are relatively new and untried; some of what is advertised is "vaporware"—it does not yet exist, and until a few customers materialize, it will not be developed. Even when all the right steps are followed for selecting and acquiring a piece of software, the relatively few installations and the rapidly changing technology within the field may result in a poor choice.

Many books and hundreds of articles have been written about selecting software. Many provide lengthy checklists. Even when the lists are filled in and the homework has been completed, there is still a lot of risk involved in selecting software. You really cannot be sure the choice was correct until it has been used for several months, but by then it is yours! Five major factors to consider in selecting software for an enhanced system are listed in Figure 8.2. Cost often ranks near the top of anyone's list, but it usually proves not to be the most important factor in the long run; operational factors should be considered first and foremost. Unfortunately operational issues are the least clear cut of the five factors.

In Chapter 2 the general features that characterize an enhanced system were described as functional simplicity; the ability to integrate equipment, media, and transmission services; a reliance on a graphical user interface; and a focus on information and transactions. Functional simplicity consists of features that make the system easy to use and that enable the system operator to implement new applications or modify existing ones. The first set of features imply user involve-

```
Cost

Operational
    Functional simplicity
    Information and transaction focus
    Integrative
    Types of user interface

Installation and maintenance

Documentation

Contractual and licensing
```

Figure 8.2.   Major factors in selecting software.

ment in the selection process—something that is often not done. The second set of features suggests a software package written in one of the more popular languages, a copy of the source code, and an application development module.

Another principal feature of a software package for providing enhanced services is its ability to integrate different technologies, which implies conformance to standards; for example, there are software products on the market that profess to be based on the graphical standard known as the North American Presentation Level Protocol Syntax (NAPLPS), but based on does not necessarily mean conforming to. However conformity to the NAPLPS standard can be determined with the NAPLPS test package, which consists of documentation, test slides, and test package data. Testing covers geometric-drawing instructions, the attributes, various code sets, and a check for compliance with the reference model developed for the standard. The package tests hardware and software to uncover areas of divergence from the standard.

Users work with information, not data; if the system delivers data, the user must still convert it into information. Some software products make the task of information retrieval difficult; it is very important for the appropriate retrieval option to exist for a particular application. The two following paragraphs outline some of the characteristics to consider when selecting retrieval software; the reference by Lochovsky and Tsichritzis cited at the end of the chapter goes into greater detail.

The query process can be divided into three parts: request, reply, and dynamics. Characteristics related to requests are the number of keystrokes required, the type and number of commands available, the latitude given the user in formulating a request, and the selectivity of requests. Characteristics related to replies include the form in which the reply is presented, whether the reply can be presented via more than one medium, and control over the reply. Control refers to the ability to control the speed of the reply or to change the form of the reply from text to voice, among other things.

The dynamics of interaction concerns issues of bandwidth, the amount of gamesmanship involved in the interaction, protocols supporting interaction, the responsiveness of the system, and user's control over the system. A similar approach could be followed for examining characteristics related to performing transactions. The value of these exercises is the insight into the user's needs gained by the person selecting the software. All of these operational factors are affected by the user interface provided. In theory the more graphically based the interface, the easier it is to use and the less technical competence required.

A third major factor to consider when selecting software is the level of support provided for installation and maintenance. Installation support can vary from providing a person on site to help, to mailing a box of tapes with instruction of how to proceed. After the system has been installed, it is important to have a prior understanding about the cost and level of maintenance support and any major modification to the software package in process or planned for the future.

Documentation is very important. Software packages described earlier in this chapter all come with a relatively well-written and complete set of documentation. Since a disproportionate amount of responsibility for selecting software is often given to the systems side of the house, users sometimes suffer because their need for well-written and easily understood documentation is not given enough weight. Well-written information is needed by everyone who will be involved with the system including the users.

The contractual agreements that accompany most software products are standard boiler plate and not subject to modification. As any owner of a personal computer knows, simply breaking the cellophane on the manual binds the user to the terms of the license. Despite the relatively inflexible nature of most contractual agreements, obtaining and reviewing a copy of the terms may uncover issues that must be clarified or resolved prior to acquiring the software.

8.5.1. Escrowing Software

Software vendors are under tremendous pressure to ship products before they are fully tested. Entry to the industry is relatively easy, especially for developers of software for micro- and minicomputers: Very little fixed investment is needed and most of the expenses are tied up in labor and marketing. Consequently competition is intense, and products, with a few noticeable exceptions, such as Lotus 1-2-3, seldom achieve a dominant position before a raft of copycats appear.

Software defects or "bugs" are common because software is not adequately tested. According to one study, in large programs, defects in coding, documentation, and incorrect bug fixes result in an average of 300 serious errors per 1000 lines of code; another source reports between 40–100 defects per 1000 lines of source code. Whatever the right number is, bad software abounds.

Even more potentially dangerous to the unwary buyer is the short history and tenuous existence of many software vendors, especially vendors with only one product. The death of a key person, threat of a copyright infringement suit, introduction of new technology by a major hardware vendor, and many other factors can destroy the fragile balance between the solvency and insolvency of a software developer.

A firm must protect its software investment; one solution is a well-structured escrow or software deposit arrangement. The concept is simple: A negotiated set of instructions is worked out among the parties involved, then if a set of agreed to events occurs, the escrowed information is released to the user of the software. The information might be source code for the software, embedded algorithms, flowcharts, test and diagnostic information, and other information needed to modify and maintain the software. A deposit agreement can provide a cushion of protection for the buyer, and it can also provide an audit trail if ownership issues arise as well as test the integrity of the vendor. The agent could be an attorney, a jointly-held safe deposit box, or even a bank; Figure 8.3 lists several responsibilities of the escrow agent.

8.5.2. Bill of Rights

This is the era of end-user computing; hardware and software have been moving away from a centralized data-processing environment and toward the user. The end of the journey is at the user's fingertips or within the sound of his/her voice. End-users have been conditioned by the expectations and demands of the consumer rights forces. It may be considered as radical to suggest that software users have rights, especially to vendors, but it is going to be increasingly difficult to foist poorly developed software and bad documentation on users. Whether or not data-processing personnel and software vendors believe users have rights, users believe they have rights. One example of those rights is shown in Figure 8.4. A wise person will remember them when designing and supporting an enhanced system.

8.6. SUMMARY

Software is an amorphous entity: It is difficult to describe and difficult to shop for. Software can be divided into two categories—systems software and applications software. Systems software can be further divided into control and processing software, and applications software can be divided into software produced in-house and software acquired externally.

A system has value because of its ability to meet user needs. The functions it may be called on to perform fall into four categories: infobase management, administration, information retrieval, and user services; each of these may be further subdivided. Four software products were discussed briefly: *Windows*, OS\2 2.0, QuickTime, and UNITRAX. The first three are classified as either operating environments or operating systems, UNITRAX is an example of a videotex server.

Be a neutral third party
Understand the business and technical details on which the escrowed material is to be released
Arrange for physical security and retention of materials
Certify deposit of materials
Be prepared to provide verification of deposit, contents, or changes

---

Figure 8.3. Responsibilities of the escrow agent.

Right to product quality
  Perform advertised functions without error
  Operate without damaging data and other software
Right of functionality
  Work on machine advertised
  Perform functions advertised
  Perform a minimum set of functions associated with type of software; e.g., all word-
    processing software should "write" to a disk or a printer.
Right of continuous service
  Right to make a backup copy
  Opportunity to upgrade to new releases without relinquishing backup copies until new
    release arrives
Right to program support
  Notification of major software defects
  Given a method of avoiding or working around known errors
  Timely notification of upgrades and fixes to known errors
Right of system integration
  Ability to integrate and operate in conjuction with other products

---

Figure 8.4. A bill of rights for software users. Extracted from *Bill of Rights for Software Users*, Capital PC User Group, Inc., Gaithersberg, MD (1985).

Important considerations in developing the content for a system are the philosophy regarding how equipment will be configured and the choice of software that will be used. Hardware may be configured on-line or off-line, it may be used just for the purpose of creating content, or it may be shared with other applications. Authoring software may appear in a variety of forms; the five forms discussed were authoring systems, application shells, hypermedia application generators, time line processors, and programming languages. The Authorware software was examined briefly.

Acquiring the right software is a time-consuming and difficult job; unfortunately one can never be sure if the right decision was made until the software is used. Consequently software acquisition is often little more than an act of faith in the final analysis. To help the wary buyer keep the faith, it is sometimes worthwhile to escrow the software and speak up for one's rights.

REFERENCES

C. Andres, Authoring Software Offers Multiple Choices, *MacWeek Inc* 5(22) (1991), p. 30.
AT&T UNITRAX Software System: Description and Features Manual, AT&T Information Systems, Issue 1 (1987).
AT&T UNITRAX Software System: Editor's Manual, AT&T Information Systems, Issue 1 (1987).

Authorware Professional, Authorware, Inc., Redwood City, CA (1990–92).

R. E. Bergman and T. V. Moore, *Managing Interactive Video/Multimedia Projects*, Educational Technology Publications, Englewood Cliffs, NJ (1990).

Bill of Rights for Software Users, Capital PC User Group, Inc., Gaithersberg, MD (1985).

P. Bonner, *Windows* 3.1: Faster and Smarter, *PC Computing* 5(5) (1992), pp. 128–132, 137–143.

D. Buxbaum, The Latest in Compression Software Solutions: From QuickTime to *Windows*, *Advanced Imaging* 7(5) (1992), pp. 50, 52, 54.

J. Canning, Multimedia Authoring Tools, *InfoWorld* 14(10) (1992), pp. 76–77.

D. Hamilton, OS\2: A 32-Bit Tutorial, *Corporate Computing* 1(1) (1992), pp. 216, 218, 220–222.

J. King, OS\2: The Big Blue Route, *Corporate Computing* 1(1) (1992), pp. 192, 194, 196, 198, 200.

F. H. Lochovsky and D. C. Tsichritzis, *Telidon Behavioral Research 5: Interactive Query Languages for External Data Bases*, Department of Communications, Ottawa, Canada (1981).

Protecting Software with Escrow Services, *Data Processing & Communications Security* 8(5) (1984), pp. 22–24.

S. Rosenthal, *Windows* Does Video, *MPC World* (April/May 1992), p. 29.

J. Sanders and K. Coakley, Setting the Stage for *Windows* 3.1, *PC Week* 9(14) (1992), pp. s/4, s/6–s/8, s/11, s/13.

L. Seltzer, Microsoft Emphasizes the "New" in *Windows* NT, *PC Week* 9(14) (1992), pp. s/6, s/19.

J. W. Semich, Inside the New OS\2, *Datamation* 37(24) (1991), pp. 26–30.

G. Smith, OS\2 2.0, *PC Computing* 5(5) (1992), pp. 48–55.

Testing *Windows*, Macintosh Multimedia Authoring Tools, *InfoWorld* 14(10) (1992), pp. 77, 80, 82–85, 88–89, 92–93.

D. Todd, QuickTime Ships, Cross-Platform Player Coming, *New Media* (February 1992), pp. 10–11.

D. Todd, *Windows* 3.1 Saves the Day for MPC Developers, *New Media* (March 1992), pp. 12–13, 16–17.

P. D. Varhol, OS\2 2.0 Challenges *Windows* 3.1, *PC World* 10(6) (1992), pp. 192–195.

B. Waring, QuickTime off to a Fast Start, *New Media* (June 1992), pp. 20–21.

P. Wegner, Programming Languages—The First 25 Years, *IEEE Transactions on Computers* C.25(12) (1976), pp. 1207–1225.

Chapter 9

Laying the Groundwork

Every idea needs a champion, and every champion needs his/her supporters; it is not enough to come up with the solution to a problem. The world is full of solutions for every problem imaginable. What is usually needed is a plan to sell the solution, which is easier said than done. It is often difficult to envision all the ramifications of an innovation. In his welcoming speech to the 1990 International Videotex Industry Association (IVIA), Heiko Falk, president of the IVIA, provided a brief list of doubting Thomases. In 1908 the director of the London Patent Office proposed closing the office, since all essential inventions had been made. In 1927 the founder of Warner Brothers, Harry Warner, was reported to have asked; "Who in the hell wants to hear actors talk?" Thomas Watson, founder of IBM, supposedly estimated in 1943 the global need for computers to be five! More recently Ken Olson, founder of Digital Equipment Corporation, reportedly said in 1977, "I can see no reason why anyone would want to have a computer at home." The types of information systems discussed in this book have their doubting Thomases, Kens, and Harrys, too; it is human nature for some people to pose these questions and for others to answer them.

Perhaps the most accurate picture of the situation confronting system developers is summed up in a story quoted in Jerome Aumente's book *New Electronic Pathways*. When Robert Fulton demonstrated his steam boat, a group of people stood on the bank and yelled, "It won't go, it won't go." When it did go, they began yelling, "It won't stop, it won't stop." Once the critics become supporters, a system will often sell itself; until then some hard work lies ahead for the innovator.

This chapter discusses the idea of selling the system to users, management, and data processing and nurturing their expectations. The reader should pay particular attention to the points that are raised. As experienced developers of systems know, most of the time spent in developing a system is spent on technical matters, but it is the human element that is usually the reason why a system fails. Developing an enhanced system moves beyond the traditional confines of data processing and telecommunications. A new set of experiences and expenses are encountered. This chapter describes what some of these will be and concludes with a look at some of the human skills that will be required to make the services possible.

9.1. SELLING THE SYSTEM

Selling the system is a two-tier process; first it is necessary to convince users and management of the need for, and value in, having the capability of the proposed system. Once that has been accomplished and the system built, it is necessary to convince the appropriate people to use it. The first condition suggests a very important principle: Sell the functionality made possible, not the technology. In marketing campaigns, this is referred to as selling the steak rather than the sizzle. The sizzle is temporal and not of much use; technology, like the sizzle, has no value in and of itself. Its value is in the applications it makes possible; therefore it is the applications that should be emphasized, and not the speed of the system nor the color and sound that may be possible.

The second condition is also important; once a system has been implemented, it may not be used, which is something technicians rarely understand. The prevailing belief among the technically educated is that once the system is in place the users will swarm to it like bees to honey. People accept change but seldom embrace it, and in many cases, they may even resist change, if it alters social relationships—a move to a new office or new reporting responsibilities. An effort must be made to educate and encourage use and not to assume that a system will be used because it is available. The dust covered microcomputers on the desks of executives throughout corporate America are testimonials to how difficult it is to foster new ways of working when the power of choice resides solely in the hands of the user.

What is salesmanship? Unfortunately the word conjures up images of checkered coats, wide ties, and a foot in the door. However putting the word in its proper perspective makes the idea of selling the system much more palatable. Salesmanship is an interpersonal skill that strives to put the right person in touch with the right idea at the right time to obtain a commitment for action. Selling involves communication: Information must be presented clearly and concisely, and it must involve the right people, those who will approve the system, use it, and support it. They are managers, employees, or customers of the company and typically data processing. Each person has his/her own agenda and may or may not be in favor of the proposed system. The proposal has to involve the right idea and it cannot be presented as a technology in search of a justification. Timing is crucial: The proposed system must be a solution to a problem that exists or is anticipated. And last, the commitment of everyone involved with the system should be sought; especially the users.

9.1.1. The User

A customer votes with money, an employee votes with time. The best way of ensuring that the users vote the way you want them to is to solicit their involvement. Companies often form customer focus groups and conduct surveys prior to launching a new product as a standard operating procedure, but few companies form employee user groups, the in-house equivalent to the corporate focus group. In-house groups can function as a sounding board for ideas; they can provide input and help secure support, and they can be the forum for helping cope with the technology. User groups can be a source of feedback about how well data processing is responding to the needs of the user population.

The success of an enhanced system is partly a matter of understanding what information users need and how they want to receive it, but considering only information needs and the vehicle for delivery is a myopic view of end-user computing. It is also important to at least consider the following advice, since some people have a genuine fear of computers regardless of how friendly you make them.

Consider social and behavioral factors, not just technical factors.
Do not assume a solution from the start; work toward the solution based on the evidence.
Make sure coworkers are users; its difficult to resist using a system if one's peers are users.
Transform the boss into a user.

9.1.2. Management

Managers do not think the same way as users; managers think in terms of investments and returns, so the approach to selling them is different. The process begins by identifying a problem and setting a clearly defined objective. The objective should be a practical solution to the problem, not necessarily an economic one. Chapter 12 contains a section on the objectives companies have cited for implementing enhanced systems and what some of the benefits are as reported in an extensive mail survey. This is the type of information management needs when you are trying to close the sale.

Many computer-based applications are not economic in the sense that a flowing back of dollars will pay for the investment. Such an expectation has caused problems for many a data-processing manager. Since managers tend to think in terms of investment and returns, it is important to qualify the kind of return. In many instances, it is desirable to shift from a cost displacement type of mentality to one where the payoff is adding value or extending the personal effectiveness of professional and managerial workers or improving customer relations. These are the types of returns discussed in Chapter 1 and summarized in Figure 1.4.

The next step is formulate a strategy for developing and implementing the solution that incorporates an acceptable level of risk. If it is practical, establish a demonstration of the proposed system. Provide a way for managers to support the program without jeopardizing their reputations and the economic viability of the unit they manage. Learn how to deal with rejection, but remember that the word no often does not mean no; it means not now. Create a grass-roots effort within the organization by finding other spokespersons to champion the cause, since a well-orchestrated chorus can be more convincing than a solo.

Last advertise success by making management aware of favorable outcomes as well as the potential risks. It may make sense to keep information from business competitors, but not from others within the organization. Do not be reticent about sharing what competitors are doing with regard to similar technology: Most people do not like to be outdone, and no manager worth his or her salt will play second fiddle to a competitor. And do not be shy about taking someone else's money in the organization. The funds may not be in your budget, but there may be colleagues who can afford to help underwrite the investment for a share of the benefits.

9.1.3. Data Processing

People in data processing may be the most difficult to convince unless the idea and proposal for upgrading the system originates there, since data processing often will have responsibility for designing and implementing the system along with the headaches of trying to live up to user expectations and finding the resources needed. Selling the system to data processing should be approached in three stages, the first of which involves developing a new way of thinking within data processing. Information processing is no longer confined to number crunching and file dumps that produce stacks of printout; it is transaction oriented and display based. That is to say, users want answers to specific questions on demand, and they want those answers presented in a way that effectively communicates the information. This implies graphics, color, perhaps sound, and other enhanced features. Sensitizing personnel to users' expectations and current trends in information technology is a necessary beginning.

The second stage is changing that department's perceived mission. Few data-processing departments have a mission statement but if asked to write one, it would emphasize a central role in providing computer power to the organization. In the 1980s conflict often existed between data processing and the rest of the organization over the proliferation of microcomputers. Data-processing personnel regarded the spread of micros as an erosion of control and a reduction of data-processing importance in the organization. Frequently users acquired machines despite protestations from the data-processing manager. Data processing's mission should be to facilitate the introduction of computer-based technology to support the organization's computer-based applications regardless of how these applications are performed.

The third stage is explaining to data processing the benefits it may realize: a reduction in training requirements and less user hand holding; media transformations, such as shifting information from paper to computer storage and back to paper, may be eliminated; data processing will be able to distribute its output quickly, effectively, and virtually automatically; new applications can be implemented more easily and time wasted attending meetings or searching for information can be reduced. These and other benefits are described in greater detail in Chapter 2. The key is helping data processing understand that it, too, can benefit directly from assuming an active role in enhancing computer-based services.

9.2. NURTURING EXPECTATIONS

"The check is in the mail." "I'm from the federal government and I'm here to help you." "Of course I'll respect you in the morning." There is a considerable difference between nurturing expectations and building false expectations. It is important to avoid adding to the list of false expectations: "Enhanced computer technology will solve all of your problems."

Since their introduction in the 1950s, computers have been maligned for building false expectations. Today managers are smarter and users more experienced in the worldly matters of computer technology, but hyperbole still exists. The expectations of everyone who will be affected by the system should be nurtured as a part of the selling process, but there are ways of doing this without inadvertently misleading anyone such as:

Emphasize the application and not the technology. Some people put an over abiding faith in the technology and lose sight of the application and how it will affect performance. It's easy to be enamored with the "bells and whistles" and forget about the impact on employees and customers. Talk in terms of application; not technology.

Present a realistic picture. Recognize that nothing ever goes according to schedule and that unanticipated costs always occur. Many benefits cannot be quantified, and no effort should be made to do so; some changes in how work is performed can lead to changes in employee attitudes and customer satisfaction that ultimately but indirectly affect the bottom line.

If selling the system is the seduction, implementation is the consummation. To avoid the more vulgar aspects that can develop in the relationship

between everyone involved, communicate often and honestly with the parties involved; share problems and seek help in developing solutions.
Make sure the system provides the value intended. The job is not over when the hardware and software are installed, and training is completed; monitor, adjust, and redo if necessary.
Remain flexible. Implementing a new system is a learning experience for everyone. Benefit from the learning that takes place and avoid rigidity.

Introducing technological innovation does not have to generate conflict, although it sometimes does. Technological change is measured in weeks and months; its impact on human elements, which are manifested through the social and organizational fabric of a company, are measured in years and perhaps decades. Recognizing that technology and human behavior and belief do not change at the same pace is the first step in dealing with the misunderstandings and false expectations that often occur.

9.3. ECONOMIC REALITIES

The Law of Increasing Realization: the recognition that what you thought could be done as an adjunct to normal business operations will cost more and take longer than you ever realized. This may be a facetious way to express the impact of enhancing a system but it probably has a ring of truth to it for several reasons.
Different technologies are assembled in an enhanced system. These technologies may not be well understood, and due to their inherent characteristics, it may be difficult to integrate them. The infrastructure needed to support the system may not be in place. The project personnel may not have the training and skills necessary to implement the system properly. Last the enhanced system is complex, and as the complexity of any system increases, the risk of various components not working together increases too. This leads to unforeseen problems that require time and money to resolve. The amount of time and money invested is an economic issue.
Economic issues are an important element in planning a proposed system and budgets and cost control procedures are the foundation of economic planning. They assume increased importance during periods of major change, and while they may not prevent problems from occurring, they can signal the beginning of a problem and lessen its impact on other areas of the organization.

9.3.1. Traditional Budgeting Methodology

There are two fundamental differences between budgeting for information processing and budgeting for most other activities in an organization. The first difference is that items are almost always expensed and rarely capitalized. The second difference is that many expenditures for handling information are in the users' budget rather than the information-processing department's budget.

It is not unusual in some organizations for information processing to consume as much as 6–10% of the organization's total budget. Nor is it unusual to find the annual expenditure for a large project to exceed several million dollars. Expensing information-processing projects grew out of a decision in the early 1960s, when the Internal Revenue Service ruled that data-processing projects should be capitalized. A group of major corporations banded together and successfully fought the ruling on the basis that computer projects were no different from any other administrative project that is expensed. Since computer projects are not capitalized, they do not appear on the books as an asset. Consequently they often do not receive the same long-term managerial scrutiny and control that is afforded an investment depreciated over a number of years.

The second difference is the inability of the information-processing manager to exercise direct control over expenditures for information processing. The proliferation of microcomputers and departmental computers has spread expenditures across departmental budgets. Consequently the information-processing department's budget may account for only a small part of an organization's total expenditures in this area, which complicates the task of planning for and implementing new systems.

With these differences in mind, the items that comprise the budget for information processing are shown in Figure 9.1. A typical budget will contain line items for personnel, hardware, software, training and education, acquiring information (e.g., census tapes), user support services, communications, office equipment, supplies, facilities, and the all-encompassing category called other.

In general the finer the detail, the tighter the control possible. The line items in Figure 9.1 have different degrees of importance depending on the company and the situation. A variety of studies are reported by the popular press each year that express various components of a budget as a percentage of the whole budget. There is a modest degree of variation among the studies according to the size of the company and the type of industry. A typical breakdown is personnel (45%), hardware (30%), software (10%); communications (10%), facilities (3%), outside services (2%).

9.3.2. Significantly Impacted Line Items

Adding enhancements to a traditional data-processing environment will affect most areas of the budget, but the three line items that will be impacted the most are expenditures for personnel, hardware, and communications. Enhanced services can vary tremendously in size and the features they contain. A service may consist of a text-only interactive information service for the employees at one site, or it may consist of an interactive system containing pictorial-quality displays serving employees throughout the country. The size of the service and the

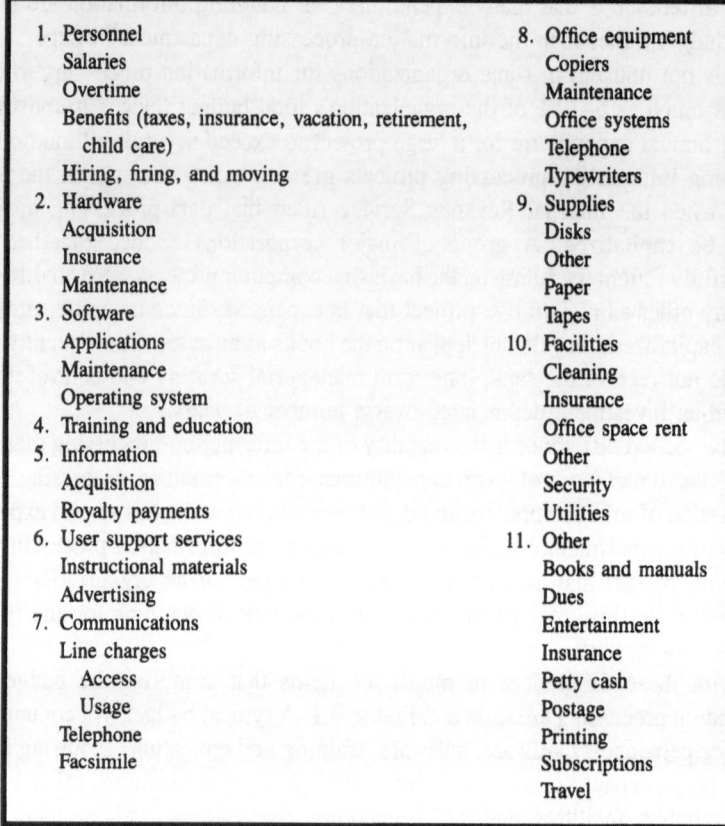

1. Personnel
   Salaries
   Overtime
   Benefits (taxes, insurance, vacation, retirement,
     child care)
   Hiring, firing, and moving
2. Hardware
   Acquisition
   Insurance
   Maintenance
3. Software
   Applications
   Maintenance
   Operating system
4. Training and education
5. Information
   Acquisition
   Royalty payments
6. User support services
   Instructional materials
   Advertising
7. Communications
   Line charges
     Access
     Usage
   Telephone
   Facsimile

8. Office equipment
   Copiers
   Maintenance
   Office systems
   Telephone
   Typewriters
9. Supplies
   Disks
   Other
   Paper
   Tapes
10. Facilities
   Cleaning
   Insurance
   Office space rent
   Other
   Security
   Utilities
11. Other
   Books and manuals
   Dues
   Entertainment
   Insurance
   Petty cash
   Postage
   Printing
   Subscriptions
   Travel

Figure 9.1. Identifying Budget Items.

features it contains affects the type and number of people needed to create and maintain it, the kind of equipment necessary for producing the information and receiving it, and the costs associated with delivering it. A text-only system is normally the least expensive to produce and deliver, and full-motion video is the most expensive; for example, the final cost to produce the latter may easily exceed $2000 a minute.

9.3.2.1. Personnel Costs

Personnel costs are increasing more rapidly than any other element in the budget, and these account for approximately 41% of all money allocated. Staff salary costs rose 5.4% in 1989 from 1988. In specialized fields where shortages exist, increases were considerably higher.

When data processing collected data and had it entered on cards or magnetic media, relatively unskilled labor was used; however as data are replaced by information, additional skills are needed. For example, in a study from the University of Nebraska, it is estimated that an all-graphics sequence requires nine times the production and development work of a sequence with text alone. Adding video to text and graphics requires more time, since people are needed to research the topic, develop a storyboard, create the content, and then edit it. Job descriptions for these positions are presented in the next section, and the process involved in creating and maintaining the contents is described in Chapter 10.

The following example illustrates the personnel resources involved in creating and maintaining the contents of a small electronic magazine. Assume the company is interested in developing an electronic version of its annual report, but since the report will be updated at least quarterly, it is more appropriately called the Corporate Report. It will be available to employees through the in-house network. Others such as stockholders, investment analysts, and the general public will be able to access it by dialing a toll-free phone number. It will also be distributed by mail on a floppy disk to the media, career centers on college campuses, and other targeted groups.

The primary objective of the report is to improve stockholder, employee, customer, and public relations. The secondary objective is to market the firm's stock to the general public and the financial service industry to the extent possible under existing federal and state regulations. The simplest form that the electronic Corporate Report could take is a monochrome text-based format, which virtually everyone's equipment would accept. A monochrome text-based format would not require a color monitor, and there would be no need to worry about graphics standards, extensive transmission needs, and extensive user storage requirements. On the other hand, a monochrome text-based format limits the ability to illustrate important concepts and enliven material.

The most complex form the report could assume incorporates a combination of text, graphics, photographic-quality images, and video overlays. This format might be compatible with a network of public access terminals and a limited number of executive work stations. From a technical perspective, such a format would not be an effective way of disseminating information outside the organization because most equipment would not be able to display the report. Clearly an acceptable trade-off must be found between meeting the objectives for the electronic report and the technical sophistication employed.

Based on the objectives just stated for creating the report, photographic-quality images and video overlays are inappropriate, but graphics are desirable. Table 9.1 illustrates the time it would take to create a frame depending on the extent to which colors and graphics are incorporated. Time estimates are based on interviews with people who create frames and on statistics reported in the literature. In any particular situation, the actual time to create a frame depends on the complexity of the display, the skill of the frame creator, and the features of the frame creation equipment.

Table 9.1. Time Estimates for Frame Creation

Type of Frame	Description	Frames/hr	Time/Frame (min)	Operator Skill Needed
Text	Alphanumeric characters, preformatted and monochrome, 500 characters maximum per frame	10	6	Low
Text	Alphanumeric characters, two colors, 500 characters maximum per frame	4	15	Low
Text with simple graphs	Alphanumeric characters, three colors, straight-line graphics, charts	2	30	Medium
Graphics	Multicolored charts and simple outlines	1	60	Medium
Complex	Detailed multicolored illustrations	0.1–0.5	120 and more	High

Source: Adapted from A. Alber, *Videotex/Teletext: Principles and Practices*, McGraw-Hill, (1985), p. 273.

The Corporate Report is closely patterned after, but not identical to, the firm's annual report. Differences between the paper and electronic versions are significantly greater than first imagined. For example, text on the electronic version will be limited to a maximum of 100 words per frame; less if a graph or table is included, photographs must be replaced with renderings, and the use of colors must be much more carefully considered because there is considerable variation in how the same color appears on different types of equipment. The overall length of the report is shortened considerably, since it will be distributed on-line. To illustrate the necessary changes, Table 9.2 summarizes the difference between AT&T's 1991 annual report and a hypothetical electronic version.

Table 9.2. Features of the 1991 AT&T Annual Report and Features of a Hypothetical AT&T Electronic Corporate Report

Features	1991 AT&T Annual Report	Hypothetical Corporate Report
Number of pages (frames)	40	30
Typical number of words/page (words/frame)	600	100
Number of tables	35	5
Number of graphs and charts	4	15
Number of photographs and illustrations	15	None
Highlighting	Underlining, color	Underlining, color, flashing
Pseudo animation	No	Yes
Downloading for analyzing of financial information	No	Yes
Updating	No	Quarterly[a]
Question/answer support	No	Yes[a]

[a]Possible with on-line version.

Despite its brevity and lack of photographs, the electronic version has several unique features: Items can be highlighted by color, underlining, and by either flashing (blinking) an image or the background surrounding the image; pseudo animation can be incorporated to enliven the display and to focus attention on particular aspects of the report. One of the most useful features is the ability to transfer tables, such as the balance sheet and income statement, to a spreadsheet software package for ratio, liquidity, and other forms of financial analysis. Anyone who has a question or who would like more information can leave a note in the electronic mailbox that accompanies the on-line version of the report. The response can be provided electronically, by phone, or mail.

The estimated time required to create the initial version of the electronic report is shown in Table 9.3. These estimates represent only the time someone would spend seated at a frame creation terminal; it does not include the time required to convert the annual report into a form suitable for electronic presentation.

It will take about 21 hours to create the electronic version; the development work necessary before someone sits down and begins to enter the information is much more variable and takes considerably longer. It is the development work that precedes the step of entering the information that ultimately decides what the viewer will see. Working from the annual report, it could take as few as 42 hours and as many as 210 hours. Whatever the investment in time to develop the report, it will exceed the time to create the frames. The steps involved in the development process are described in the next chapter.

A major advantage of having an electronic version of the annual report is the ability to update frequently and distribute throughout the year. The electronic version can reflect recent developments of a financial nature, information about new products, recent contracts, key management changes, and a host of other newsworthy items that are normally made known through press releases.

It takes more time and money to develop and create a frame than it does to

Table 9.3.   Time Estimates to Create Initial Electronic Version of Corporate Report

Type of Frame	Description	Number of Frames	Time/Frame (min.)	Time/Type (min.)
Text[a]	Alphanumeric characters, two colors, 500 characters maximum per frame	8	15	120
Text with simple graphics	Alphanumeric characters, three colors, straight-line graphics, charts	10	30	300
Graphics	Multicolored charts and simple outlines	10	60	600
Complex	Detailed multicolored illustrations	2	120	240
	Totals	30		1260 (21 hr)

[a]Assumes an average word size of 5 characters.

update it, but in a relatively short time, the accumulated cost of modifications can exceed the original cost of production. The following analysis demonstrates the importance of adopting a plan for updating the electronic report at the time of its initial creation. For purposes of illustration, assume that the Corporate Report will have to be updated throughout the year. The extent of the changes can not be determined beforehand precisely, however it is anticipated that at least 25% and perhaps as much as 50% of the report will have to be replaced. The balance of the report will simply be updated with minor changes.

Table 9.4 illustrates the time required to replace portions of the report and update the rest each period, assuming it will take only 20% of the time to make minor changes to a frame than it took to create the frame. As shown in Table 9.4, it will take 444 minutes (7.4 hours) if approximately 25% of the report is replaced and the balance updated, which increases to 612 minutes (10.2 hours) if approximately 50% of the report is replaced. These estimates may seem excessive, but they may in fact be optimistic. Such factors as fatigue, illness, interruptions, equipment breakdowns, and mistakes contribute to lost time and affect productivity.

Table 9.4. The Estimated Time Required to Revise Completely a Portion of the Frames and Modify the Rest Each Period in the Corporate Report

Type of Frame	25% Completely Replaced			
	Number of Frames Replaced	Time	Number of Frames Modified	Time (min)
Text	2	30	6	18
Text with simple graphics	3	90	7	42
Graphics	3	180	7	84
Complex[a]	0	0	0	0
Total	8	300	20	144
Grand total			444	

Type of Frame	50% Completely Replaced			
	Number of Frames Replaced	Time	Number of Frames Modified	Time
Text	4	60	4	12
Text with simple graphics	5	150	5	30
Graphics	5	300	5	60
Complex[a]	0	0	0	0
Total	14	510	14	102
Grand Total			612	

[a]The two detailed multicolored illustrations are not changed.

The labor required to update the Corporate Report for several different update frequencies is shown in Table 9.5. An employee working a 35-hour work week would spend 3.50 weeks a year updating the report if it were updated monthly and 50% of the contents were replaced each time; this is 5.83 times more labor than was required to create the Corporate Report. The amount of time required annually to update the report can be reduced by updating less frequently or reducing the number of frames replaced each time.

This simple example indicates just how labor intensive it is to maintain the electronic version of the report. Clearly it is important to consider the question of updating when the project is being planned, not after it has been created. It is also important for the reader to remember that the time estimates are for creating the physical content; development work that precedes this phase will require additional time, and how much additional time will depend on what must be done to develop the material. If information is being extracted from a paper version of the annual report, the effort will be minor; if the process starts with an idea and must progress from that point with little or no content already available, the investment in time may be significant.

9.3.2.2. Hardware

Predicting the cost of computer hardware is a difficult task. The rate of innovation and the speed at which products are rendered obsolete is almost beyond comprehension. There is probably no other industry in the history of mankind that has year-after-year consistently succeeded in delivering more performance for fewer dollars; in fact the unit cost of executing computer instructions declines by 20% or more a year, which does not mean that an organization's computer costs decline each year. The opposite is probably true, but it does mean that for each dollar spent, more is received in return.

There is a clear trade-off between what a company can afford in terms of computer technology and employees' ability to take advantage of the technology at their disposal. If an organization were to place a full video computer on every employee's desktop, some would not garner any greater value from it than from a dumb terminal. The trade-off is illustrated in Figure 9.2. The value of equipment

Table 9.5. Labor Requirements for Updating the Corporate Report

Update Rate	25% Completely Replaced		50% Completely Replaced	
	Minutes/Year	Employee Weeks	Minutes/Year	Employee Weeks
Semiannually	888	0.42	1224	0.58
Quarterly	1776	0.85	2448	1.17
Monthly	5328	2.54	7344	3.50

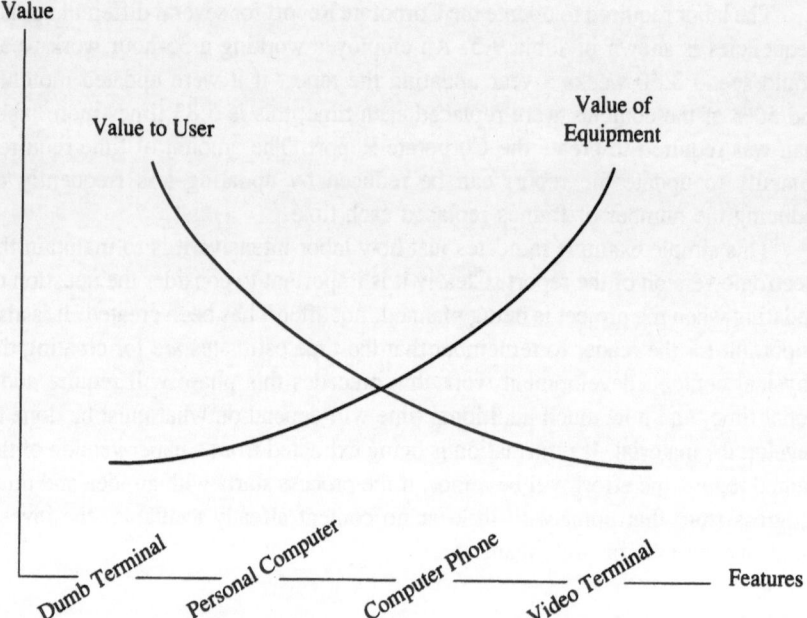

Figure 9.2. The conceptual task of providing the right equipment to meet the user's real needs.

increases as more features are added; however the value to individuals varies according to the use they can make of the features. For example, if the Corporate Report discussed in the preceding section were the most sophisticated type of application to be rendered, there would be little value in having a video terminal. The task facing every systems manager is balancing technology with the need.

9.3.2.3. Communications

After costs for personnel, the most rapidly increasing item in the information budget is the cost for communications. The proliferation of computer equipment supporting remote access, the growth of transaction-oriented applications, and the distribution of databases have combined to create a tremendous demand for communication services. This demand has increased the amount of money spent on communications relative to other items in the budget.

Figure 9.3 illustrates the time required to transmit the Corporate Report at different transmission rates and the features that can be supported effectively at these various rates. The Corporate Report is estimated to be 23,400 bytes large, based on an analysis of the different types of frames, as described in Table 9.3 and the estimated number of bytes contained in each type. For example, a text frame of 500 characters equals about 100 words; considering the space separat-

Time   Features

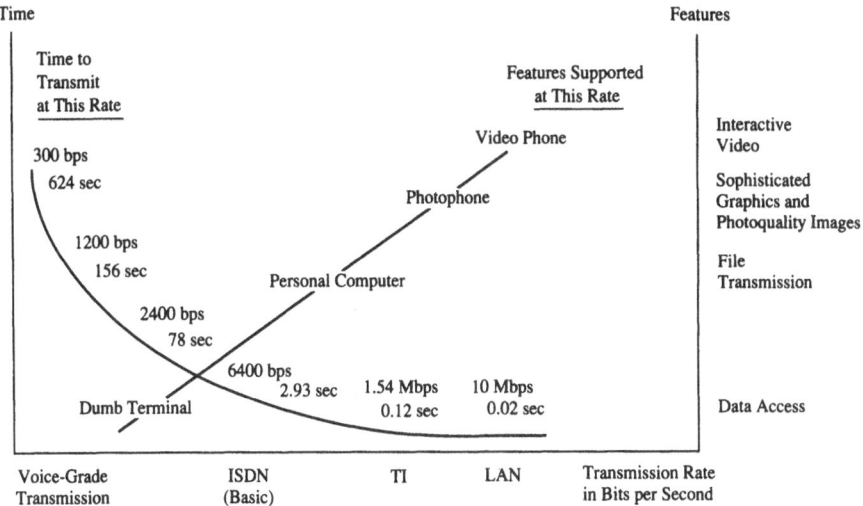

Figure 9.3. The time to transmit the Corporate Report at different transmission rates and the features supported at these rates.

ing the words, the total number of bytes is approximately 600. A detailed multicolored illustration, referred to as a complex frame in Table 9.3 may require 1800 bytes. As a rule of thumb, 1800 bytes is considered the upper limit in the frame creation industry for access via telephone.

Viewers can either access the Corporate Report and examine it frame by frame or download it to their own machine and read it off-line at their leisure. Accessing the report on-line and reading it in its entirety are impractical because of the time involved; however it could be browsed in this manner.

In Figure 9.3 the time to transmit the Corporate Report at various transmission rates is illustrated by the curved line. The relative speed of voice-grade transmission services in comparison to ISDN and faster services is dramatically illustrated. The line sloping upward to the right shows the features supported and the appropriate type of terminal for different transmission rates. Naturally it is possible to use a device at the lower end of the line with a higher transmission rate (*e.g.*, a dumb terminal on ISDN), but the opposite is not necessarily true (*e.g.*, a videophone on a voice-grade transmission).

9.4. STAFFING CONSIDERATIONS

People who work in the computer area of most organizations are divided into two groups: those who work in development and those who work in operations. The development group normally consists of programmers, systems analysts, and

quantitative specialists. The operations segment is composed of machine operators, data entry, and a section responsible for control, security, and database
administration. Neither group contains individuals with the skills needed to create
and maintain the contents of an enhanced information system. The greater the
incorporation of graphics, text, and high-quality imaging, the greater the need for
a new category of employee. This point is readily apparent in the survey results for
companies using enhanced systems that are discussed in Chapter 12.

Figure 9.4 states job descriptions for employees in this new category. They
are an amalgamation of those skills found among people in the print-publishing,
graphics, and computer-based training industries. Actual staffing will be determined by the content that must be created; the greater the disparity in content, the
broader the range of skills needed.

The individual who manages the development and creation unit must have
strong communication skills. In addition this individual must have a solid back-

---

Manager of development and creation

Directs and coordinates the planning, creation, and production of the contents for the
infobase. Coordinates the activities of personnel who are meeting with users, determining
needs, and developing the contents. Establishes the maintenance policy and audits the
accuracy and timeliness of the contents. Works closely with the existing development and
operational groups.

Specifically: Consults with management and users to determine new operational and
strategic applications of information resources. Manages personnel involved in all aspects of
planning, developing, creating, and maintaining the infobase. Develops budgets and
integrates the activities of the creation and production group within the organization's overall
computer operations. Introduces technological innovations of a computer nature to the
organization. Educates staff on technical issues, authoring systems, and equipment.

Writer/designer

Establishes purpose for information. Researches the subject and meets with information
providers and representatives of the user population. Creates initial specifications and
coordinates development of the product.

Specifically: Writes goals for information. Develops the script for the "magazine" and
creates a storyboard in cooperation with other members of the development and creation unit.
Revises script based on editorial, technical, and client comments. Develops instructions
regarding branching and such screen features as flashing and color.

Graphics designer

Designs and creates the visual layout for the display.

Specifically: Works closely with the writer to design and create graphics to complement the
application. Consults with writers and production personnel about the storage requirements
and level of complexity required to store and generate the electronic version. Reviews initial
displays in their entirety to make sure colors, images, and so forth, are in harmony.

Figure 9.4. Job descriptions for personnel in the development and creation units.

Input specialist
Produces the contents of the information base from source documents and the storyboard. Operates the authoring system and the composition equipment Interacts with the writer and graphics designer to translate information into electronic form.

Specifically: Fully trained to read the storyboard and create the frames linking them in the correct order. Merges text and graphics. Draws relatively complex graphical relationships. Operates equipment efficiently and effectively with special emphasis on the ability to create graphic images. Reviews frames while they are being created and makes appropriate adjustment for problems not previously identified, such as color that clashes or is not legible. Types accurately at a high rate of speed.

Editor
Reviews storyboards and the finished product for clarity, completeness, and ability to communicate the message. Consults with others in the development and creation department on technical requirements and corrects errors and notes deficiencies.

Specifically: Alters, adapts, and refines scripts and finished product to assure clarity, conciseness, and accuracy. Conducts a technical review of product and compares it with source documents and storyboard. Ensures that documentation developed to support the project is accurate, up-to-date, and widely available. Recommends alternative design approaches when initial approaches are not feasible or affordable. Designs the overall structure of the infobase, enforces conformance to standards governing content and oversees staff training.

Analyst/programmer
Analyzes, codes and tests special routines for processing data.

Specifically: Interviews and surveys users and researches methodologies for performing specific functions. Develops logic for performing the work and writes computer programs. Thoroughly tests routines prior to installing them.

Figure 9.4. (*Continued*)

ground in the computer field, be aware of innovative ways of using computer technology, and understand the need to provide users with information in the appropriate format.

Storyboards are created by the writer, who researches the subject and develops the material. The writer plans the logical flow and presentation of information consistent with the application in which it is to be used. The writer works in close cooperation with the graphics designer, who designs and may actually create the visual layout of the presentation. The input specialist is the person seated at the frame creation equipment who creates or copies from source documentation what the viewer will see on the screen. Last come the editor, who is responsible for editing and final preparation of the material, and the analyst/ programmer, who develops special routines for processing data.

These people constitute the backbone of the team that will create the contents of the system. As the presentation format evolves more and more toward audio and video, these employees will be supplemented by people with audio, photographic, and video camera skills. However many organizations will chose to use outside sources for audio and video production because of the expensive equipment and special skills involved in producing this form of content.

9.5. SUMMARY

To sell the system, emphasize functionality and application, not the technology. Once the system is implemented, users must be encouraged to integrate it into their work routine. Employing in-house user groups, demonstration projects, and convincing the data-processing personnel that enhancing the systems will make their jobs easier are three of the many measures that can be undertaken to sell the system. In the process of selling the system, nurture realistic expectations and avoid building false expectations.

The Law of Increasing Realization states that what you thought could be done as an adjunct to normal business operations will cost more and take longer than you realized; this is especially true when adding enhancements to a traditional data-processing environment. Line items most severely impacted will be personnel, hardware, and communications.

The labor needed to develop and create the electronic version of a report can be significant; however the time required to maintain it can be even greater if it is updated frequently. The unit cost of executing computer instructions declines by 20% or more a year, but as a system is enhanced, additional hardware features are needed that increase the overall cost of equipment. The breakup of AT&T, increased competition in the telecommunications industry, and technological improvement have reduced the per unit cost of transmitting information, but this decrease has been more than offset by the increased bandwidth needed to support information system enhancements.

As systems are enhanced, a new set of employee skills are needed that are an amalgamation of those skills found among people working in the print-publishing, graphics, and computer-based training industries.

REFERENCES

A. F. Alber, *Videotex/Teletext: Principles and Practices*, McGraw-Hill, New York (1985).
J. Aumente, *New Electronic Pathways: Videotex, Teletext, and Online Databases*, Sage, Newbury Park, CA (1987).
J. Carey, *An Economic Assessment of Electronic Text*, Report N. 6 of the Electronic Text Report Series, An Annenberg/CPB Project, San Diego State University, CA (1984).
C. Currid, Tips for Selling Technology to Corporate America, *PC Week* 8(17) (1991), p. 66.
H. Falk, Welcome: International Videotex Industry Association Conference, *VIA Update* 9(7) (1990), pp. 1–2.

G. Gery, *Making CBT Happen*, Weingarten Publications, Boston, MA (1987).

G. Gery, The CBT authoring team, *Data Training* **6**(3) (1987), pp. 48–51.

S. C. Mott, P. G. W. Keen and J. D. Sulser, *Electronic Information Services: Looking Ahead by Looking Back*, International Center for Information Technologies, Washington, DC (1988).

W. E. Perry, *Data Processing Budgets: How to Develop and Use Budgets Effectively*, Prentice-Hall, Englewood Cliffs, NJ (1985).

Selling Telecommunications Concepts to Management, *Telecommunications*, Datapro Research Corporation, Delran, NJ (1983).

Strategies for Growth: 1991 Annual Report, AT&T, New York (1991).

Chapter 10

Planning and Producing the Infobase: Part 1

Planning and producing the infobase are complex and time-consuming processes. Many factors must be considered, not the least of which is determining exactly what information is needed, how to produce it, and the form in which to present it. A variety of skills are needed, and many different people may be involved at this stage.

To be successful, a methodology is needed to lead the team systematically through the process. In a sense, methodology is a road map that will indicate the direction to be followed and the series of steps to be performed to traverse the path successfully from beginning to end. A good methodology contributes to good communication among the parties involved and creates a natural division of the work to be done. It will be an aid for estimating the amount of work necessary and assist in managing the processes involved in getting it done. A methodology will define roles and responsibilities and offer guidelines for selecting the tools and techniques to be used. In short a good methodology delivers a product of high quality that is on time and within budget.

By following the methodology proposed in this chapter and Chapter 11, the development and creation processes will be better understood; however the material presented is only an introduction to the subject. An organization planning to become deeply involved in these processes must go beyond the information presented here to build in-house expertise, and the more complex the content, the more special the preparations needed. In fact when video and audio are included, most organizations will find it more practical to have the work performed by companies in that business. Chapter 12 reports on the extent to which outside expertise is sought, based on results from a nationwide survey.

Chapter 10 introduces an infobase that might be found in the corporate relations department of any large business. The importance of quasi standards is reviewed, followed by a 10-step methodology for planning and producing the information (see Figure 10.1). The process begins by identifying the need for information and setting goals that the system is expected to satisfy; after this has been done the content and design of the material should be tackled. Once this has been completed, the information source can be located and the economies examined in presenting the information. In Chapter 11 the remaining steps in the process are covered.

Both Chapters 10 and 11 concentrate on the process of creating and managing the infobase, and not on the specifics of organizing the information. In systems where there are large blocks of information accessed linearly, traditional menus and lengthy files may prevail. In applications where information must be accessed in a nonlinear manner, techniques based on hypertext are appropriate. Hypertext refers to information that is organized in chunks and linked together; for example, a page of text may have bold-faced words that are linked to other parts of the infobase containing information about them. Selecting the bold-faced word initiates the link to the additional information.

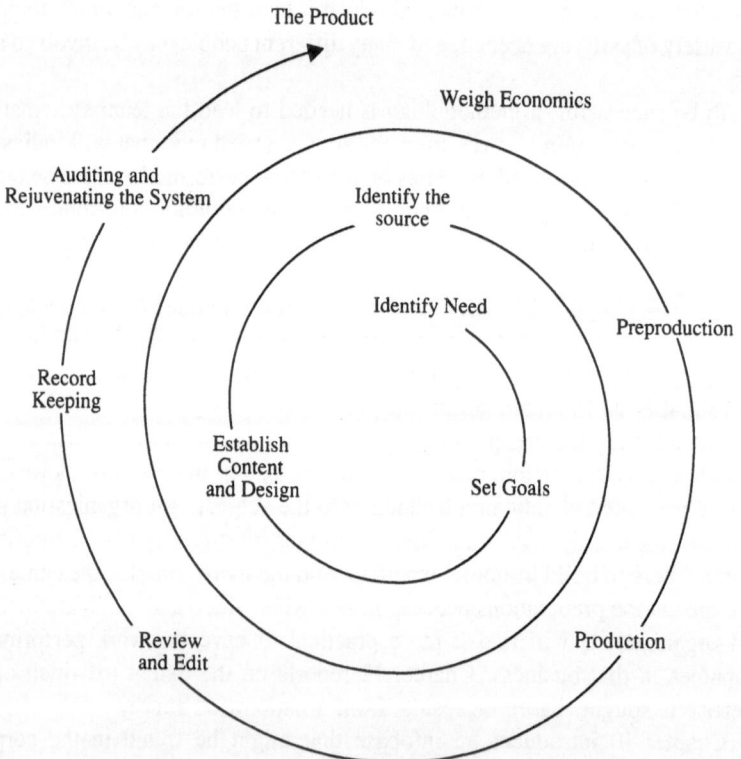

Figure 10.1.   A model for planning and producing an infobase.

Authoring and the retrieval software determines how linear or nonlinear access is to the content. Documentation for the software that is chosen will introduce the appropriate vocabulary and explain the technique for organizing the content so these topics are not explored in detail here. Instead the focus in Chapters 10 and 11 is on creating and maintaining the infobase.

10.1. AN EXAMPLE OF A FOCUSED INFOBASE

To illustrate how this methodology can be implemented, creating a small infobase for the Corporate and Stockholder Relations Department of a hypothetical company will be discussed. This departmental infobase may be only a small portion of the infobase that exists within the company.

It is assumed that information resides on a computer with information from other areas of the company and it can be accessed by the general public. It is further assumed that the contents of the infobase were created under the auspices of the Development and Creation department for which job descriptions were provided in Figure 9.4. This ensures a level of consistency across the corporation's computerized infobases and also focuses the skills of the development group on the task of designing and creating the initial structure, which is especially important if graphics and video are a part of the system.

Once the infobase has been created, it becomes the responsibility of the department to maintain it. This is consistent with the prevailing notion of distributing computer resources and responsibility for managing them downward rather than centralizing everything at one location. The main menu for the infobase, which has been named THE WINDOW is shown in Figure 10.2.

The primary purpose of THE WINDOW is to improve relations with the general public and stockholders. In the lexicon of Chapter 1, it is an illustration of using computer technology strategically for a public (external) application. Access to the information is free and anyone may browse the contents.

Menu items 1–5 provide information. Item 1 contains a brief synopsis of company news with an emphasis on current and nonfinancial subjects. Item 2 is the Corporate Report described in Chapter 9; it contains financial information. Item 3 is a directory of middle- and upper-level managers and instructions for communicating with them. Item 4 is a compilation of recent press releases and articles about the company. Item 5 contains announcements and display ads for selected products.

Item 6 permits stockholders to ask questions and express viewpoints. Stockholders of record are issued a user number, password, and mailbox. Individuals listed in the Management Directory can be queried, or general questions can be left in the departmental mailbox, which are then answered by a designated individual. In addition stockholders can express opinions and leave suggestions. The comments left by stockholders are accessible only by authorized indivi-

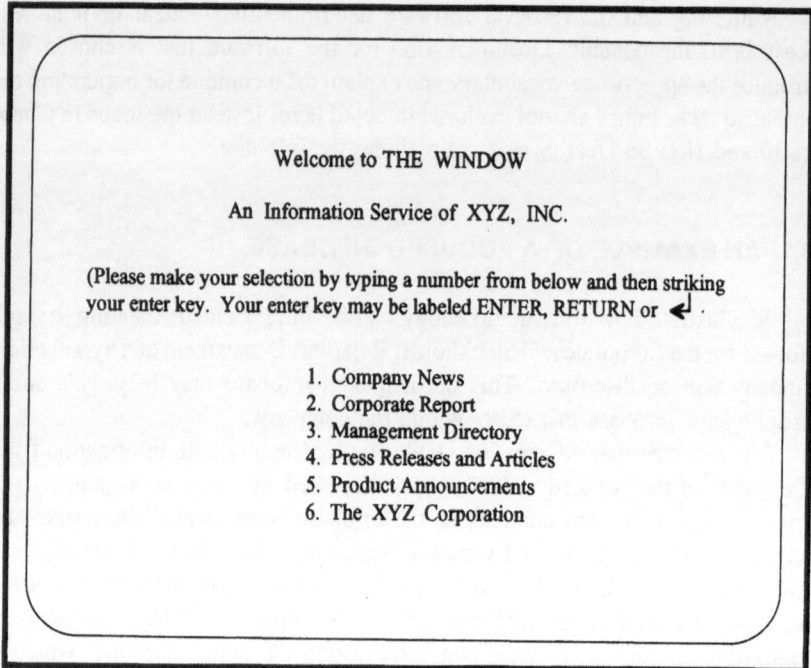

Figure 10.2. The main menu for the hypothetical infobase of a corporate and stockholder relations department.

duals in the company. Item 6 constitutes a public forum for stockholders and provides upper management an innovative and explicit way to determine the sentiment of the owners.

THE WINDOW is hypothetical, but companies are implementing various on-line information services for customers and employees; this is especially true of hardware and software vendors who use the technology on which their business rests in this way. The E-mail features on corporate systems provide an additional method of customer support; this feature can also be an in-house message center to help employees keep in touch with one another, especially when employees are traveling.

Corporate on-line services are used for order entry, catalog and pricing information, polling customers, and classified ads among customers. Examples of these and other applications are described in the cases discussed in Chapters 13–15. Some companies implement internal on-line information services that draw heavily on outside sources; Glaxo, Inc., a pharmaceutical research company, makes extensive use of commercial on-line infobases to mold a daily newsletter that is delivered electronically in-house. Another example is the Town Crier by General Motors, which is described in Chapter 15. In 1991 Digital Equipment

Corporation (DEC) implemented an on-line version of its annual report for employee access only. The company later added proxy material and prospectuses. The two caveats for an organization interested in developing and implementing its own version of THE WINDOW are (1) be aware of copyright issues and (2) formulate a written policy. The former is a reminder that material distributed by the media is often copyrighted. It is often permissible to redistribute the information internally but providing it to the general public may violate copyright restrictions. The written word is a powerful tool that is sometimes misused. An organization contemplating its own edition of THE WINDOW should have a policy statement governing the type of information that may be presented, rules of etiquette, guidelines concerning proprietary information, and so forth.

10.2. DEVELOPING QUASI STANDARDS

An infobase composed of the various menu selections shown in Figure 10.2 can be compared to a paper magazine containing a number of articles. Typically articles in a magazine are on different subjects written by different people. Some authors write in the first person, others perhaps in the third person; some articles may be lavishly illustrated, and some may be all text. One article may fill exactly three pages, and the next article might occupy ten pages; on some pages, an article may occupy all the space and on other pages it may be juxtaposed with part of another article or an advertisement. Although the articles differ in content, the magazine conforms to a particular style that differentiates it from another magazine. Each magazine has different editorial standards regarding spacing, words on a page, the use of photographs and so on. The same basic principles apply when information is stored electronically. Quasi standards must be established for the infobase as a whole and for the appearance of each screen. These quasi standards are partly determined by the software being used and partly by the conventions adopted by the unit that designed and created the frames.

The Savoy reference at the end of this chapter gives the reader a sense of the role of standards in electronic publishing. The particular project Savoy describes is called Ebook3, an electronic book, parts of which are animated. A book is normally regarded as static and linear, but it does not have to be when it is electronic. Static images can be inserted in the text, and dynamic graphics can support the visualization of animated sequences. The reader can employ the full power of the medium by inserting data that enable items being discussed to demonstrate certain features before the readers' eyes. Underlying the technology is a prescribed set of rules serving as a style guide—rules for color, linking different subjects, organizing the contents and other aspects of the creation process.

If quasi standards are not developed, three problems result: The unit responsible for updating the contents does not have guidelines necessary for performing its job; second the different portions of the infobase maintained by different units become editorially inconsistent, which eventually leads to a mishmash of colors, fonts, graphic and pictorial styles that confuse the viewer and create a bad impression; third readers become confused by the inconsistency among screens. Therefore quasi standards are needed at two levels; they are needed to shape the infobase structure, and they are needed for design and creation at the frame level.

10.2.1. Infobase Level

The infobase can be thought of as an inverted tree. The base of the tree is the main menu, and the submenus are the tree's branches. The leaves of the tree then become the information that is sought. Thinking of the infobase in this way provides a logical structure to what would otherwise be a very difficult concept to picture mentally (see Figure 10.3.)

The software used to create the infobase imposes certain *de facto* standards; for example, the AT&T UNITRAX software described in Chapter 8 divides the address space of the infobase into 90 page ranges numbered 10–99. Several of the

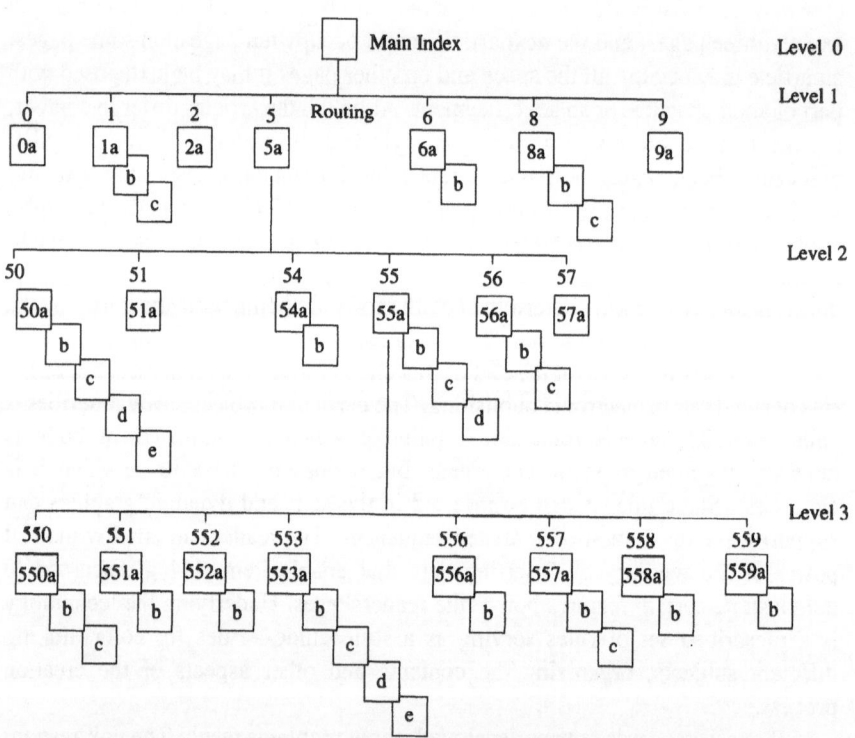

Figure 10.3. Conceptual view of a portion of THE WINDOW infobase.

two-digit pairs are reserved for special purposes. Eight levels are permitted in the infobase (in Figure 10.3, there are four levels); therefore within any two-digit pair XX, page numbers can range from XX–XX9999999. Furthermore each page in the UNITRAX system consists of one or more subpages or frames, up to a maximum of 26. The maximum size of a frame in UNITRAX is 16K bytes, including frame control information.

The types of *de facto* standards that result include the address size and the combination of numbers and letters for naming the address. Size limitation constrains the amount of graphical detail allowed and prohibits incorporating video frames into the infobase itself, although there are ways of merging UNITRAX frames and video images. There is a requirement that keywords begin with an alphabetic character and consist of 18 or fewer characters; similarly for direct access, a frame number must consist of 2–10 characters. The characters are all numeric except for the last, which may be a letter ranging from a–z. All software packages have naming conventions that create a basic set of rules common to that package.

An organization may establish categorization schemes for segmenting the infobase among different units comparable to the organization of a company telephone directory. Divisions can be given branches beginning at the two-digit level and departments within divisions a three-digit level, and so on; for example, personnel can be assigned the number 20 and the benefits department within personnel 201, the payroll department 202. Within each branch, specific sets of frames can be designated for information about that branch, such as an index or keywords relevant to that segment of the tree.

The information required to determine the general design of the infobase includes the following:

Approximate size of the infobase
Number of frames in each branch
Categories of information and required links among frames and between branches
Categories in need of frequent updating and categories relatively static
Areas likely to require more space
Answers to these types of questions lead to general recommendations governing how information will be collected, prepared, stored, and presented.

General recommendations include the following: Incorporate available data whenever possible but do not try to transfer a print-based product into an electronic version without editing it according to the guidelines presented later in this chapter. Be selective in what is added to the contents; do not fall into the trap of filling the available memory with information just because the information is available and the computer memory is not being used. Group information and services into logical categories based on relationships among the content, what information the user is likely to access next, and what information the user needs to back up to. And perhaps most important, lay out a replica on paper of the general structure of the infobase; this process is important because it is the basis for establishing frame links and providing the infobase with a logical structure.

10.2.2. Frame Level

Whether supporting text, graphic, pictorial, or video displays, most systems place constraints on particular areas of the display. For example, the UNITRAX system reserves the first and last lines of the screen image. On the first line to the left is the main keyword or name of the information provider, in the center is the price, and on the right is the frame number; any of these fields can be left blank. The last line is called the command line, and it may repeat several of the fields from the first line as well as system prompts and error messages. The remaining lines on the screen are available for information.

Different systems support different display parameters. A common variable among systems is the number of lines available for displaying text. In fact in some packages, such as UNITRAX, the number of lines varies according to the type of presentation standard. Since different standards often require different types of terminals, the UNITRAX software prevents a user from accessing a frame whose type, in terms of presentation standards, does not match the user's terminal. This creates two issues that must be planned for. The first is the number of available lines. Some software supports a variety of terminals with different numbers of columns and rows, such as 40 columns by 20 rows, 40 columns by 24 rows, 39 columns by 20 rows, and 80 columns by 24 rows. The rows in each case often include several reserved at the top and bottom of the display for system-generated messages. The second issue to consider when the same system supports multiple-line displays is how to prepare the same information for users with different terminals. Obviously a frame with 22 rows of text will not fit on a screen capable of displaying 18 rows. Consequently the creation unit must either design to the lowest common denominator, *i.e.*, the terminal with the fewest rows, lowest resolution, fewest colors, *etc.*, or modify information during the creation stage to match each terminal type. Designing for a multi-terminal environment may be accomplished with a common frame type. A frame could be created using the ASCII GO character set with default foreground and background colors and a 39-column by 20-row configuration. The common frame would not conform to all terminal types, but it could capture many of those likely to be found in North America.

Given the fact that different terminals and even presentation standards may be supported by the same system, it is necessary to develop general frame-level standards to accommodate these variations. Such standards may pertain to page headings, logos and graphical designs, menu layouts, terminology, navigation sequences for exiting a frame or the system, and even the maximum number of choices permitted on menu frames. To ensure that everyone creating and maintaining frames follows the general standards, an editor is needed to develop and enforce standards, oversee training information providers, and enforce the standards.

It is important to achieve a balance between maintaining consistency among different segments of the information base and giving sufficient leeway to each unit to develop a creative and effective set of frames. Figure 10.4 contains some general guidelines.

Each frame should satisfy three basic functions. First it should identify itself as the intended frame or the path to the intended frame. Second it should carry as much as possible a whole unit of information; in the case of a video where the display is continuous, the sequence of frames treating a particular subject should make a complete thought. Third each frame should contain some index information to guide the user to the next frame. Good design does not confuse the user; it provides what is sought; it is easy on the eyes and mind, and it uses the resources of the system efficiently; the guidelines in Figure 10.4 will lead to this outcome.

The overriding rule when creating text is to be concise. Use abbreviations and avoid superfluous information. The viewer should be able to find information quickly, then move on. Do not indent paragraphs nor leave blank lines unneces-

```
Presentation Medium
    Text
        Be concise
        Limit fonts
        Establish text/background contrast
        Use complete thoughts
        Keep menus simple
    Graphics
        Balance text and graphics
        Do not over design
    Video
        Make technology unobtrusive
        Consider viewer's ability to absorb information
        Design with updating in mind
    Features
        Color
        The fewer the better for text
        Be aware of the psychological implications of particular colors
    Sound
        Use sound discriminately
        Provide alternative cues
        Do not rely exclusively on audio outputs
    Flashing and Animation
        Use judiciously and sparingly
```

Figure 10.4.   General guidelines for frame design.

sarily; however these rules have to be balanced by the need for breathing space in the text. This can be accomplished by numbering or lettering information, using bullets, and highlighting. Choose a set of fonts for the entire infobase, perhaps one for the header, one for subtitles, and one for the body of the text. Text is more legible if there is a significant contrast with its background; dark text on a light background is usually easier to read. Try to end sentences on the same frame as they began. Keep menus simple and place prompts consistently on the left side of the screen. Locate helps, returns, and exit commands in the same place on every frame where these are appropriate.

Balance text and graphics to enhance each other; it is generally a good idea to have graphics and text separated on the frame, so that one can be changed without impacting the other in a design sense. Do not over design: Users who can not receive graphics will receive garbage or a system message. Avoid trying to create "still television" by building elaborate graphic displays that take a long time to transmit and paint on the user's screen, yet do not contribute significantly to the information being conveyed.

Video is the most complex display medium, as such, it can be a little overwhelming at times. Keep the technology as unobtrusive as possible—a goal that has been achieved in video games. Video displays can convey a tremendous amount of visual information in a brief time, therefore during sequences when important information is being transmitted, minimize hand and eye movements and other actions that may divert the viewer's attention from the screen. When text, graphics, and video are being integrated into the same series of displays it is important to provide a way of updating them without affecting the series design. Because video is expensive and time consuming to produce, it should be created with longevity as a goal.

Some specific rules for using color are stated in the section on content and design; in general the basic rule is the fewer colors the better, especially for text. Colors convey subtle meaning and attract attention to a particular part of the screen, but when done inadvertently by the designer, color distracts the viewer and may even convey an unintentional message.

Sound is a useful enhancement, but it can become annoying. Sound is often used as a negative or "punishing" response to an incorrect answer. Avoid unnecessary beeping and incorporate a convenient way of turning off the sound and switching to an alternative cue. Care must also be taken not to rely exclusively on audio outputs. Some users, especially those using public access systems, may be hearing impaired. When sound is incorporated, provide a clear and complete message.

Flashing text and animation can be very effective but use them sparingly, since flashing text on a screen can be as distracting as blinking lights on a Christmas tree. Both flashing and animation can highlight information and retain the viewer's attention.

10.2.3. User's Response to Poor Design

Users respond to poor design in several ways, but the most general response is confusion over the intended message and what to do next. This is especially true when a series of displays have been presented, and the last display simply hangs there waiting for a response of some kind. Frustration occurs from the inability to convey one's intention to the computer, and frustration is heightened if an unintentional response cannot be undone. Confusion and frustration are often followed by a sense of panic—the feeling that you should be doing something but not knowing what it is or how to make the system stop and wait. The opposite response to panic is boredom, which may be caused by slow response times or lengthy menu searches.

The outcome to these responses varies: The user may reject the system and turn to other solutions or perhaps use only a portion of the system. Sometimes the task or role for which the system is used will be modified, which is comparable to changing the question to fit the answer. The worst consequence is misusing the system intentionally or unintentionally—bending the rules to fit the needs of the moment and thus compromising the integrity of the system.

10.3. PLANNING AND DESIGNING THE INFOBASE

A good design helps prevent unwelcome user responses. It shapes how information will be used and users' perceptions of the system. The ten steps leading to a good design are discussed in the following sections.

10.3.1. Identify the Need

There are two philosophies that prevail as the basis for creating an enhanced system. The first is to accept the existence of the system and attempt to find applications that fit it. This is comparable in business to developing a product and then searching for a market. The usually cliché is a product in search of a market. Many successful products, including copy machines and personal computers were developed in this way. The advantage to this approach is simple: It is direct and immediate. The risk is that a product, or in this case a need for presenting information electronically, may not exist.

A more rational, albeit more time-consuming and costly, approach is to identify a set of needs that can be most effectively met by an enhanced system. In Chapter 9 (Figure 9.4), the manager of the development and creation unit and the writer/designer are charged with responsibility for identifying the need for particular information. The identification process may involve interviewing and surveying potential users to learn what is desired or analyzing what competitors are doing. It can be derived from an investigation of the market place and from developing market trends. It may be the outgrowth of a strategic initiative or the result of responding to a problem. Whatever method is used, the outcome is the conscious identification of a need that can be satisfactorily addressed by an enhanced information system.

10.3.2. Set Goals

Identifying a need helps identify the goals that justify an investment, and there must be an expected return on the investment. It may be possible to quantify the expected outcome, but the return will likely be more subjective than objective. The corporate report described in the previous chapter has primary and secondary goals. The primary goal is to improve stockholder, employee, customer, and public relations. The secondary goal is to market the firm's stock to the extent permitted by federal and state laws.

In this case, the primary goal is more important because it represents a new and unique way for the company to communicate with its constituencies. The Corporate Report will single out the company as innovative and caring. If the company happens to be in the communications industry, such as AT&T, it also tacitly creates a model for other companies to follow, which may advance the business interests of AT&T, e.g., selling communications services to users. However it is virtually impossible to attach a dollar figure to the expected returns, although it may be possible to measure some aspects of improved corporate relations quantitatively.

The existence of clearly stated and measurable goals makes three important contributions to the success of the system:

They help sell the system

They set the tone and direction

They play an important role in establishing the content and form the system will assume.

Goals make selling the system to decision makers easier. Achievable goals clearly define the value to the organization and put trade-offs in perspective. Although it may be difficult to measure subjective goals, it may still be possible to establish measures that indicate the system's value. For example, the Corporate Report may be partly judged on the basis of how many times it is accessed, the amount of free press it receives, the change in public and stockholder perception of the company as determined by pre- and postsurveys, reductions in the frequency of telephone and mail solicitations for information, and so forth. Clearly stated goals set the tone and direction of the system. A goal to generate improved corporate relations will reduce the chance the project will be redirected so that its resources can be used to achieve a different outcome.

Perhaps most importantly, goals establish the content and form the system will assume. If the goal of the project is to increase distribution of the annual report to nonstockholders, there are many ways of accomplishing this. But if the goal is to improve corporate relations, then such features as periodically updated information from the chief executive office, a question-and-answer section, an opinion poll, and similar features are likely to attract interest and present an image of innovation. If the Corporate Report is also distributed on a floppy disk to the media, college career centers, and other targeted groups, interactive features are disabled or excluded.

Interactive features are a key ingredient in the process; these establish the perceived one-to-one relationship with the user that is missing from the glossy report mailed out annually and increase the likelihood that corporate relations will be improved.

Primary and secondary goals for the hypothetical Corporate Report and several quantifiers to measure the attainment of the goals are listed in Table 10.1.

10.3.3. Establish Content and Design

This step is comparable to the type of development work found in data processing, where a systems analyst examines user needs and develops the general design to be followed. Here the manager of development and creation, a writer/designer, a graphics/designer, and perhaps a video specialist perform a similar function.

10.3.3.1. Establish Content

Content is shaped principally by need, but availability of information, economics, and technology also play a roll. To determine what should be included, the need that is being addressed must be carefully analyzed. Users must be contacted and interviews and surveys performed. Such factors as age, occupation,

Table 10.1.   Corporate Report: Goals

Goals		Quantifiers (12 months)
Primary		
Improve corporate relations	10%	Improved stockholder relations pre- and postsurvey
	10%	More articles, extensive reports in trade journals
	10%	Improved customer relations pre- and postsurvey
	5%	Stockholders accessing service
Secondary		
Market stock	5000	Downloads requested
	2000	Requests for full annual report
	5000	Transactions executed

environment, and whether the users are internal or external must be weighed, for example, information structured for company executives might require a higher level of reading comprehension than information intended for the general public. Material viewed at work might require greater bandwidth than material viewed at home, content designed to educate may require greater detail than material designed to inform. Studying the intended user group is an important step in determining the appropriate content.

If the need has been established internally and the users are external to the system, such as those intended for the Corporate Report described earlier, it is necessary to put the two ends of the service together; at least figuratively. For the Corporate Report, this may involve creating a suitable representation of the corporate report and testing it with a set of stockholders to learn if it produces the intended results.

Content is also shaped by what information and services are available or can be developed in a timely and economic fashion; there are usually more ideas than available resources to implement them. Information creation or the acquisition of information created elsewhere may be expensive. Technology may also play an important role in determining what can be included. A need that can be fulfilled only by video, for example, may not be feasible to develop because of the cost and the inability of the intended audience to receive and display it. In most instances, the lowest common denominator technically is a simple text-based service, with perhaps a limited graphics content.

10.3.3.2. Design Considerations

The medium will affect the message! A medium is most effective when its use emphasizes its inherent advantages; therefore a paper-based product can seldom be, nor should it be, converted without change into an electronic medium. Information and services to be presented electronically must be designed for the appropriate electronic medium, that is, text, graphics, audio, or video presentation. It is unlikely that the same design would be effective for all four media.

When designing the form the presentation will take, decisions must be made about screen layout; menu; the amount and style of text; color patterns; whether to use graphics and the appropriate style; the need for audio or video; and decisions about such transient indicators as blinking text.

10.3.3.1.1. Formatting. The term format refers to the predictable pattern of screen design that is consistently applied to the way that information is entered and displayed. For text this refers to the number of characters per row and to the number of rows per frame, among other things. Formatting is important for a variety of reasons. It helps ensure compatibility among information providers. Without a specified format the following type of problem will occur: A system with a format of 24 rows of 40 characters can be made to accept information from a system with a format of 20 rows and 39 characters; however the reverse is not the case. A smaller frame format does not easily accept information from a larger frame format. If only text is involved, the problems are solvable; if graphics or video is involved, solutions are much more difficult.

A well-designed screen format can favorably impact human comprehension and performance, reduce errors, and speed computer-processing time. A poorly designed display may have the opposite effects on performance, error rates, and computer operations. Screen design has also been linked to visual fatigue due to unnecessary and excessive eye movement.

A display screen has two sets of information associated with it. One set is the information that will be seen, and it consists of a menu, a form to be filled out, or information to be read. The second set of information controls the first set. This set probably consists of a creation date, keywords, color coding, and security authorizations. Formatting involves information viewed by the user.

The relatively small size of the screen and the limited amount of information that can be displayed at one time present some unique problems. The basis for dealing with these problems is grounded in what is known about paper-based forms of communication. In Western culture, reading takes place from left to right and from top to bottom. This creates certain expectations in the viewers' mind and causes them to focus on certain areas of the printed page. For example, readers look at the top for guidance and placement information and to the left for placement and cueing information. They look to the left side for paragraph headings and starts and to the bottom for footnotes and explanations. Significant changes to these conventions in a screen format will increase the chance for confusion.

The first step in determining the format is determining the portion of the screen that should be used. Work done by a novice is easily recognized, because the entire screen is filled, which usually overpowers the viewer. In a paper-based medium, text is prepared with a width-to-depth ratio of about 1:1.414 (see the reference by Rubens at the end of the chapter for a more complete discussion). This ratio is based on the International Standards Organization's standard for an AO sheet of paper; in the United States, this equals approximately an 8½ by 11 inch sheet of paper. Photographic dimensions are the inverse: 35-mm film produces slides with a width-to-depth ratio of 1.41:1; television screens have approximately a 1.33:1 ratio. Computer monitors vary in size, particularly because of the case that holds the screen, but their ratios come closer to those for photographs and television screens than to the printed page. Thus the screen designer starts with a world that has been turned topsy-turvy. The trick is to present information in a format that compensates for what the user is accustomed to seeing.

The amount of room allotted to the screen designer may not be large; for example, if a decision has been made to conform to a common frame type, *e.g.*, 39 columns by 20 rows, only 780 character spaces are available. Assuming an average word size of five characters and a space between each word, this permits only about 130 words to be displayed on each screen. When allowances are made for indenting, separating paragraphs with a blank line to improve legibility, and providing borders around images, not much space is left; consequently formatting becomes even more important.

The Galitz reference provides a number of formatting tips. Displays should be orderly and clutter-free. Plain, simple English should be used. There should be a clear indication of what relates to what, *e.g.*, headings, field captions, instructions, and so forth. A simple way must exist for learning what is in the system and how to access it. An attempt must be made to avoid overloading the viewer with information: Only what is needed should be provided, and that information should be presented on a single screen if possible. With all of these points in mind, the designer has one more goal—to make the arrangement on the screen pleasant to look at. An example of unaesthetic screen design is the data channel on cable television networks, which are divided into two or three horizontally scrolling windows.

Specific suggestions by Galitz and others referenced at the end of the chapter include the following. Indicate an obvious starting point in the top-left corner of the screen. Arrange information in a logical manner, with the most frequently requested information first, using a top-to-bottom, left-to-right orientation. Develop a pattern so that specific areas of the screen are reserved for the same kind of information, such as helpful tips, transfers to the main menu, and user common sequences. Group elements and units of information with blank lines, framing lines, and different intensity levels. Except where lists are presented, avoid two or more columns of text, since that line length is shorter than the viewer is accustomed to in paper-based text.

In concluding this section on formatting, it is important to remember that a single format will probably not be suitable for all the different kinds of display: inquiry, data entry, menu, question and answer, and transaction. Differences exist in how each of these screens is treated by the viewer; for example, a menu screen is quickly scanned from top to bottom, while a data entry screen is slowly read from left to right as data are input. Arranging each screen depends to a large degree on how it will be used; again form should follow function.

10.3.3.1.2. Menu Design. A menu is a set of choices presented to a viewer. The viewer's selection may be another frame or the command to exit the system. The menu frame may contain a graphical illustration, an advertisement, information, or instructions to help the user, as illustrated in Figure 10.2.

A well-designed menu frame has a number of benefits (see Figure 10.5), the most important of which is simple and unsophisticated access to the system's contents. Anyone who can read and enter a keystroke can make full use of the system. Instructions on how to navigate through the infobase can even be given in a few words on the opening frame, as illustrated in Figure 10.2. Menu choices

Simple and unsophisticated access
Indicates contents
Facilitates classification of contents
Helps develop search strategies
Guides user logically through information
Provides structure for planning and allocating types of information

Figure 10.5. Benefits of well-designed menus.

indicate what the system contains; during the design and creation of the contents, menu choices facilitate content classification and organization.

Menu items can help guide the reader in searching for particular information, since the arrangement of items suggests a certain logic for proceeding. The visual arrangement on the screen does not however imply the same arrangement for the information in the system, since two adjacent items on a menu may be on different disk drives.

Menu items can also form a structure for allocating the amount of information that will exist for each topic. This is an essential step in the planning process and one often overlooked when storing information electronically. This is in contrast to newspapers, magazines and book publishers who have guidelines on the size of the end product. Table 10.2 indicates how much information will be targeted for each selection and what percentage it comprises of the public access information base, THE WINDOW, discussed earlier.

The overall structure of THE WINDOW is similar to the electronic version of the Corporate Report examined in detail in Chapter 9. THE WINDOW will consist of frames containing 100 words or less; there will be some graphic displays; a liberal use of color; some pseudo animation; and support features for downloading particular information, sending electronic mail, asking questions, and requesting

Table 10.2. Information Allocation
for THE WINDOW in Frames[a]

Menu Item	Frame Allocation	Percentage
1. Company news	45	13
2. Corporate report	30	8
3. Management directory	45	13
4. Press releases and articles	30	8
5. Product announcements	30	8
6. XYZ Corporation	180	50
Total	360	100

[a]Does not include required submenus.

additional information. After several meetings, considerable discussions, a few arguments, and an agreement to comprise, the allocations in Table 10.2 are made.

One last benefit of a well-designed menu is the opportunity to incorporate information, advertisements, instructions for on-line help, and navigation commands. A menu frame is a breaking point in movement through a system. Since it commands the user's attention, it is a logical as well as an ideal place to insert supplemental information that can be presented succinctly. From a system perspective, it is easy to create, and the amount of computer overhead required for support is relatively low. Once created a menu frame affords a reasonable response time for a large number of users.

Menus are effective because they depend on user recognition rather than recall, but they are not without problems. In general the larger the infobase, the more menus needed. The main menu for THE WINDOW will contain several submenus; for example, the selection labeled Company News could be divided into Personnel News, Corporate Newsletter, Union News and Views, the Ecology Report, and so on. Several of these may have submenus.

Menus are helpful, but they can be time consuming and boring. Even in a small infobase, such as THE WINDOW, it may be necessary to access three or more menu frames to reach the target and its possible to access the wrong branch. Each menu is viewed in isolation, and relationships between menus are usually not clear. Once in the wrong branch, a user may become frustrated and leave the system or worse yet, assume the information is unavailable; similarly poorly worded or ambiguously worded menu entries may lead to incorrect assumptions about what is contained in a branch of the infobase. Such problems can never be eliminated entirely, but their frequency and severity can be reduced by observing several rules when creating menus.

Although it may at first seem antithetical to suggest providing a paper-based guide to the user, it is helpful. A graphic portrayal of the first two or three menu levels showing their entries and connections gives the user an idea of the relationships among different branches of the infobase, what each contains, and the best way of reaching them. Figure 10.6 illustrates relationships among menus in THE WINDOW infobase.

Clear, logical, and complete choices reduce uncertainty and the chance of proceeding in the wrong direction. Generalizations and ambiguous words and phrases, such as miscellaneous or others, do not indicate contents and are likely to frustrate the user. A balance must be struck between conciseness and completeness.

Develop a logical hierarchical relationship among the frames. Main categories should be listed on the main menu, with subcategories on submenus. If the system supports keyword searching or direct access through go to commands, provide identifiers for critical menus, so that intervening levels of the hierarchy can be omitted.

Organize choices within the menu according to how the user is likely to search, which may involve putting the most frequently accessed items at the top; if there is equal likelihood of access, then alphabetical ordering should be considered. If there is a geographical relationship contained within the information, then choices ordered according to points on the compass may be appropriate, *e.g.*, North America, Central America, South America. Common sense is the governing rule when organizing choices.

Provide consistency among sets of menus. Numbers or letters work best on the left. An even margin on the left side makes scanning easier. Resist the temptation to arrange menu items in the form of clever designs, such as a Christmas tree. There are no hard and fast rules about how many items to list on a frame, but several research studies have suggested that five to eight items are best. Two or three items per frame lengthen the search too much, while too many items at once overwhelm the viewer. Menu labels should appear in the same location. Incorporate navigational prompts on each menu. As indicated before, menus are logical dividers in a system and are therefore appropriate locations for prompts indicating how to obtain help, exit, return to the preceding menu or return to the main menu.

Group information into logical categories. A menu should reveal what is in the branch about to be entered; where appropriate provide additional information about the contents. On menus with only a few choices, consider incorporating a small section of information, perhaps a generic description of the branch about to be entered.

Try to anticipate paths a user may incorrectly take when searching for particular information, then at critical menu junctions, insert statements directing the user elsewhere for certain information or indicate what information is not contained in the system.

By following the ideas described and summarized in Figure 10.7 the developer will be ready for the most critical part of creating menus—testing the system by monitoring use and interviewing users.

10.3.3.1.3. Text Issues. Although text has been presented on computer screens for decades, very little attention has been paid to adapting it to the medium on which it appears. Because text composition and presentation are not taught to data-processing personnel, the user is forced to read text that is not only poorly written from the standpoint of English grammar, but text presented in a format that impedes comprehension. Text preparation involves the same attention to detail as creating a graphic display. Text should be prepared for the medium on which it will be displayed, and not for a paper-based medium, which is often the case. Figure 10.8 adds to the general guidelines for preparing text given in Figure 10.4.

It is important to try and capture as complete a thought as possible on one frame, because unlike reading a newspaper or magazine, one frame is all the viewer can see at a time. The reader can scan the first several sentences and then

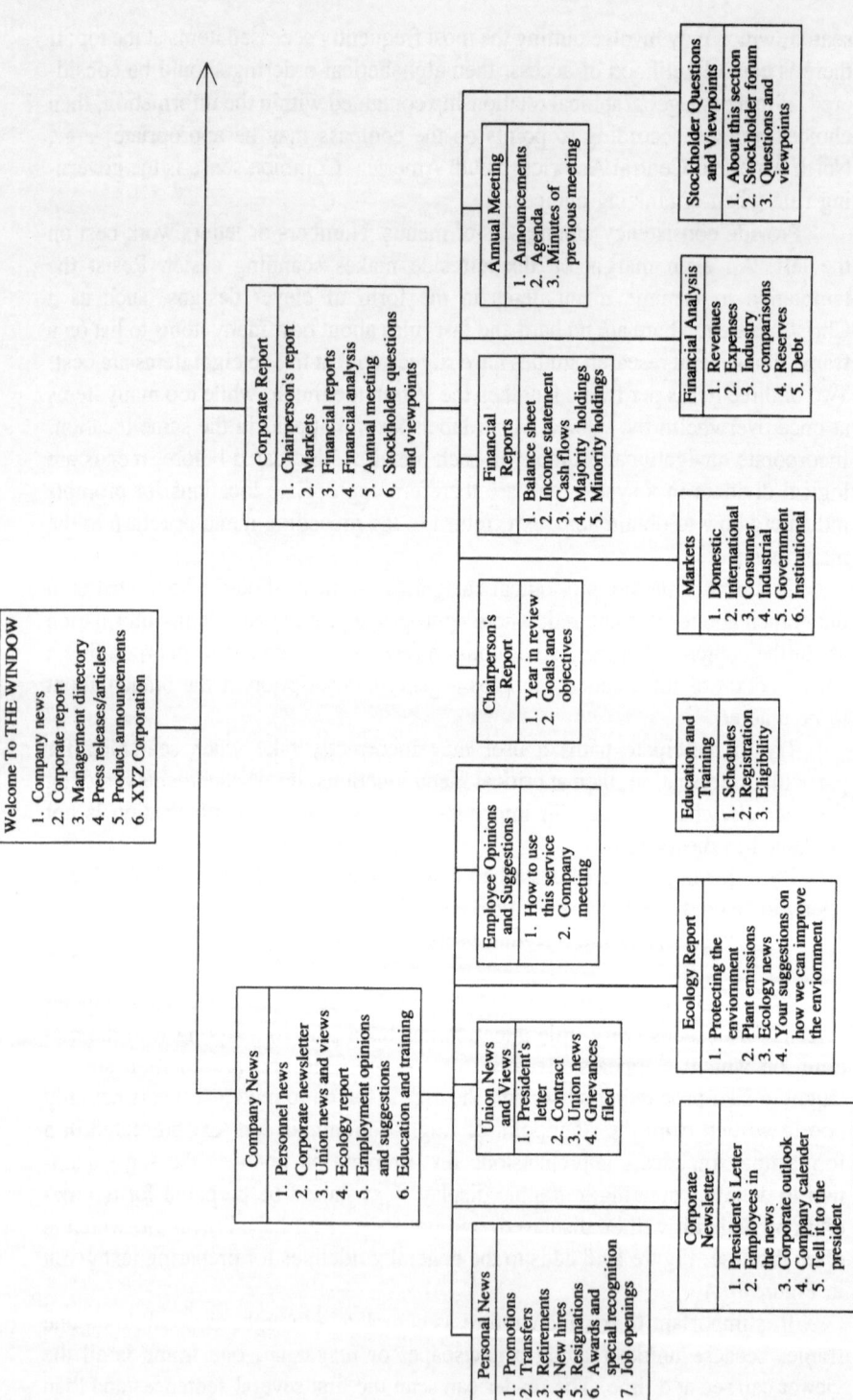

Figure 10.6. Menu structure for THE WINDOW.

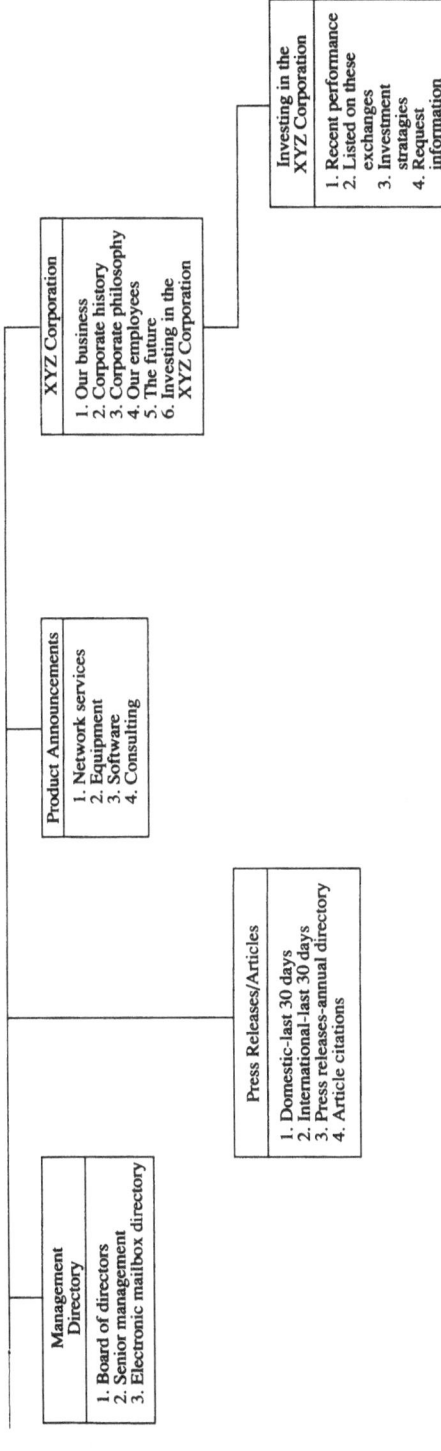

Figure 10.6. (*Continued*)

Illustrate the first two to three levels of menus and their interrelationships.
Make choices clear, logical, and complete.
Avoid such generalizations as miscellaneous.
Balance conciseness and completeness.
Establish a hierarchy among frames.
Organize terms.
Provide consistency among frames.
Group information into logical categories.
Add descriptions and incorporate information when desirable.
Reveal size and scope of what is in each branch.
Insert statements directing user elsewhere or to curtail search.
Test, test, test.

Figure 10.7.   Tips for creating menus.

glance at the end of an article, even when it is spread over several pages. The entire piece can be perceived, impressions formed, and a decision made whether or not to proceed. An electronic medium is not quite so flexible; consequently it requires a short-word, short-sentence, short-paragraph style, because the frame itself must be short. The extent of its brevity is illustrated by considering the size of the common frame type mentioned earlier. The working space is a window 39 columns by 18 rows, a considerably smaller working space than that on a PC screen, which provides room for 702 characters. Allowing for space between

Style
  Mix uppercase and lowercase text according to the convention of the day.
  Sentences should be short, simple, not threatening, and structured with important points at
      beginning.
  Compliment text with *simple* line drawings.
  Use punctuation sparingly.
  Highlight text by taking advantage of the medium.
Arrangement on Screen
  Paragraphing: meaningfully segment a block of text.
  Highlighted text may need additional spacing.
  Locate labels and paragraph headings in the left margin.
  Skip a row between paragraphs.
  Start paragraphs at the left margin.
  Use column formats for words and numbers likely to be sought and for sets of numbers.
  Introduce displays logically (left to right, top to bottom).

Figure 10.8.   Recommendations for preparing text displays.

words and five characters per word, there is space for only 117 words per frame. When paragraph indentations and hyphens are included, the quantity of information that fits on the display is further reduced. Working within the constraints imposed by the amount of available space, an author must be guided by concerns for style and arrangement on the screen. The most debatable of the style issues is whether to use all uppercase text. Research on this issue has produced mixed results, partly because different equipment has slightly different fonts, letter sizes, resolution, and brightness of the display. These and other variables will impact a study of whether all upper or lower text is best. Since a reader is accustomed to uppercase letters indicating the beginning of a sentence or a label, the same convention should be used in an electronic medium unless there is conclusive reason to think that all uppercase letters are more effective.

Information should be presented in short sentences composed of familiar words. Use simple line drawings that can be displayed on character-based terminals to illustrate or supplement text if possible. Emphasize points or create labels with uppercase letters. Use punctuation sparingly, *e.g.*, US for U.S., NY for N.Y. Highlight text by taking advantage of the many variations available on electronic medium; examples include increasing the intensity of a word, displaying it in color, causing it to blink, underlining, changing the font and type style, and using an indicator, such as an arrow, that may be flashing or in color.

The arrangement of information in a display not only determines how pleasing it is to the eye, but also comprehension, the speed at which it can be read, and the time needed to search for a key point. How information is arranged to form a paragraph is one of the most important considerations. A meaningfully segmented and identifiable text has been found to improve response times on reading tasks by as much as 16%. In fact research has shown that once text has been meaningfully segmented, such cues as highlighting do not further reduce reading time unless the material is only being scanned.

A screen full of text may impair comprehension. A blank line separating paragraphs is effective for relieving text density and promoting clarity; a blank line between paragraphs also eliminates the need to indent the paragraph. Within a paragraph spacing between letters and rows of text is important, perhaps more important than changing the point size of characters to increase legibility. Bright letters (highlighting) affect spacing, so that more space is required between letters and words to compensate for the possibility of colors running together. Arrange paragraph headings on the left margin, not centered, since the eye tends to return to the left-hand side of a frame after completing each row. *Do not* change the color of successive paragraphs unless color conveys some significance; a change in color suggests significance that is not there.

Use column formats sparingly, since text will require an extensive number of hyphens due to the narrowness of multiple columns on the same screen. Numerous hyphens waste space and inhibit comprehension. Information being scanned for a single word or number should usually be arranged in columns, *e.g.*, an index of keywords for direct access of information.

The location of text on a display is also important because of the need to update or integrate nontextual information. If text is to be added or replaced at a later date, space must be allocated when the frame is created. If graphics are combined with text and both are updated at different frequencies, care must be taken to assure that updating can be easily done. Normally it is a good idea to paint or present objects including text from the top-left to the lower-right corner. It is possible in some systems to introduce objects, including text in blocks, in various parts of the screen, but a random introduction leads to rapid eye movements as the eye jumps from image to image and may be very distracting.

10.3.3.1.4. Color. We live in a world imbued with color, it stimulates, conditions, informs, warns, and entertains us. Since the eye expects to see color, and the brain is trained to respond to it, incorporating color in electronic displays is a natural and often necessary element of effective visual communication. Color perception is a response to three properties: brightness, hue, and saturation. Brightness, or luminance, is a response to the amount of light perceived; a bright object reflects a large amount of light, and a dim one absorbs light. Hue, an object's tint, is associated with the wavelength of light. Red differs from blue because of each color's wavelength and the eye's reaction to that physical condition. Saturation refers to how much a color differs from white. A pale color, such as sky blue, is low in saturation, whereas a bright blue is highly saturated; in other words, a lightly saturated color has more white.

The human eye reacts differently to the three properties of light; for example, the eye is more adept at detecting small changes in brightness than small changes in hue. The person designing a display deals with the three properties of light in the aggregate: displays are based on the color produced by the equipment, not by individually managing the three properties of light.

Print publishers have used color for centuries to help visualize information; undoubtedly color originally was added to print-based material to make it more attractive and thus increase its appeal to the buyer. Research studies on the use of color in electronic media have revealed a variety of favorable results. Color added to text increases attractiveness and if added judiciously, also improves clarity. One study evaluated increasing the detail in a display ranging from a standard text screen to video versus adding color to a monochrome display. Subjects were asked to judge how acceptable different kinds of displays would be in the home and at the office. The findings were that adding color was as important as improving the level of detail. Other studies have shown that color makes displays increasingly cheerful, exciting, interesting and colorful.

Color is employed in computer environments to improve communication and highlight key segments of a display. Color is used to isolate ideas; direct a particular response; separate information, such as the heading from the body of text; and to make subtle statements about the information being presented, *e.g.*,

expressing monetary deficits in red. Color is an important part of the communication process in a visual world. It is remarkable there are as many monochrome
screens as there are.

Although color is desirable and sometimes a necessary enhancement, it must
be incorporated carefully into a system. The types of problems that color may
introduce involve storage requirements, equipment limitations, spectral issues,
and vision handicaps. Color consumes storage in surprising quantities: A monochrome image requiring 100 Kbytes of storage will require four times that amount
if the same image is rendered in 16 colors. As more colors are added, the demand
for storage increases rapidly.

The display and image resolutions vary among equipment; consequently a
sharp and precisely colored image on one piece of equipment may be blurred on
another. Color palates are different, so that red on one machine may look pink on
another. Variation in hue among machines is further impacted by how each viewer
adjusts the machine and the level of ambient light in the room. A color display
created on one system may not look particularly good when viewed on a different
machine. A designer must certainly keep in mind that carefully planned color
coding breaks down completely when viewed on a monochrome set.

Color varies according to the wavelength of the color spectrum; short
wavelengths are at the blue end of the spectrum, and long wavelengths are at the
red end. As light passes through the lens of the eye, the wavelengths are refracted
by different amounts, which means that only one color at a time can be focused
on the retina. If the eye is focused for green, then the shorter wavelengths
representing blue will be focused in front of the retina, and the longer wavelengths
of a red image will be focused behind the retina. Consequently if the eye is focused
on the color green, adjacent objects colored red or blue will have a blurred
appearance, which is called chromatic aberration. Another problem is the legibility of objects in certain colors. Acuity is highest for green and white stimuli; the
eye's resolving power for red images is only one-third of that for white images and
only one-fifth for blue images. Consequently the colors red and blue should be
used sparingly, and they should not be adjacent when used together on the same
screen. Selecting harmonious colors avoids these problems. Harmonious colors
are near one another on the color wheel; colors on opposite sides of the color wheel
are sharply differentiated from one another, and these can induce eye fatigue.

There is a sizable minority of visually impaired people; the two most common
impairments are color weakness and color blindness. People who suffer from color
weakness are capable of seeing all colors, but they tend to confuse them, especially
when the light is dim. The color-blind tend to confuse red, green, and gray; only
about 0.003% of the population is color-blind and see everything as gray or as one
color. Color impairment differs for men and women; approximately 8% of all men
are color defective compared to less than 1% of all women. Color impairments also
vary with ethnic backgrounds; in some parts of the world, it is possible to have a
male-user population where 10% are color impaired.

Color is desirable for enhancing a display, but care must be exercised. An indiscriminate use of color for the electronic publication THE WINDOW, described earlier, could damage rather than improve public relations. There is considerable evidence that multicolored screens impair the normal reading process. A well-designed screen usually has no more than two or three colors for text, and five is probably too many. White, green, cyan, and yellow are excellent choices because of their visual distinctness and compatible wavelengths.

Designers must also take into consideration the sociological and psychological associations of some colors. Although color is the result of the eye's reaction to light waves, our response to what we see is shaped by personal and cultural experiences. Our reaction to color is well expressed in idiomatic and colloquial expressions: When we are angry, we see red; when we have a bad year financially, we end up in the red. Blue denotes melancholy. This is not to be confused with true blue friends. Green means go when we are waiting at a light or, new, if we are watching a movie with cowboys and greenhorns. Brides wear white and melodramatic plays feature blackguards.

To make matters more confusing, colors that attract attention are not necessarily ones that we prefer. Studies have shown that orange, red, blue, and black attract attention, but blue, red, green, and violet are preferred. Colors can impart subtle meaning to the content of an infobase, but reactions to color depend on age, sex, place, and a host of other human factors that makes standardization impossible. Consequently color must be carefully considered in the design process.

Figure 10.9 lists guidelines for using color; the guidelines are divided into five categories: readability, aesthetics, discriminator, highlighter, and relater. The material applies primarily to displays containing text, but many of the principles discussed are applicable to graphic and video displays. The ability to use color with text is attributable to the flexibility of the electronic medium. Color is not widely found in print-based text because it is expensive to produce and such alternatives as underlining and italics are available for less cost.

Readability is somewhat subjective, since it is a function of the physical characteristics of the medium and human perception. Too many colors confuse and distract the reader; five or less are advisable and generally two are sufficient. Colors of low inherent brightness make the best background, while colors of high inherent brightness make the best text. Blue makes a good background color, especially when used with white text. Research studies do not consistently reveal the same findings regarding what colors are best. This is probably due to variation in equipment and research designs, especially the background color chosen. The most legible colors for text are white, yellow, cyan, and green; red and blue lead to reading difficulties on a dark background. Complementary colors produce the phenomenon described earlier as chromatic aberrations, so blue and yellow, or red and green combinations should be avoided.

Aesthetics and readability are related but must be considered on their own merits. Frames can be pleasant to view and almost unreadable and unpleasant to view and very readable. The challenge is to develop a frame that is pleasing to the

Readability (text)
  Use only enough colors to satisfy the needs of the application—generally five or less.
  Colors of low inherent brightness make good background.
  Colors of high inherent brightness make good text.
  Complementary colors should be avoided.
Aesthetics
  Check colors and frames on display monitor after they have been completed.
  Avoid browns and purples for text.
  Make text and images that need to be studied sufficiently large.
  Develop a color motif.
  Incorporate surprises using color.
Discriminator
  Use colors to isolate for emphasis or comprehension.
  Break up items to facilitate readability.
  Discriminate by color change and/or luminance.
Highlighter
  Change background or foreground.
  Use brighter colors for the most important points.
  Use duller colors for less important points.
  Review the impact of highlighting on a monochrome monitor.
Relater and organizer
  Develop consistent color code.
  Relate similar ideas or elements.
  Pick distinct colors that are easily namable.
  Consider including a legend of colors if relationships extend across multiple frames.

Figure 10.9. Tips for using color.

intended audience without sacrificing readability. After a frame has been created it is important to view it on the monitor that the viewer would use. Colors tend to vary slightly among units. Avoid browns and purples for anything but graphics because they generally do not display well. Make text large enough for easy viewing. Blues, grays, and greens are unobtrusive and make nice background colors. Bright, saturated colors are fatiguing to look at and can overwhelm other colors. Develop a motif for each distinct group of frames, perhaps a common background or border; this will project an image of consistency and neatness. Incorporate small surprises periodically to enliven the presentation; this can be a small deviation in the color pattern or an interesting graphic.

Colors as a discriminator can be used in two entirely different ways: They can isolate areas of a display for emphasis or to help comprehension; they can also be used to identify items that belong together. For example, a colored line inserted periodically in a table of numbers helps the reader follow the numbers across the screen. Discrimination can be accomplished by changing the color or the luminance of the color. When using color to discriminate, it is important to remember that viewers infer meaning from a change in color. Consequently subliminal messages should not be sent unintentionally, as was discussed in the section on the psychology of color.

Highlighting is a form of color discrimination. Alternatives to highlighting with color are italics and bold type, but in some instances, these may also be part of the notation, which would create confusion. The most common form of highlighting involves changing the background color, although an electronic medium also permits changing images in the foreground. Normally colors are selected in accordance with the color dominance hierarchy. Brighter colors should be used consciously and consistently to emphasize the most important points. The brightness of colors, from most to least is white, yellow, green, blue, and red. Colors for highlighting must be chosen carefully to prevent problems with adjacent, foreground, or background colors. It is also important to recall that color cueing will be rendered relatively ineffective on monochrome displays.

Color is effective for helping the viewer understand the logical structure of information in a display and for showing relationships among its various parts. Captions, headings, hash patterns, and perimeter designs contribute to the overall sense of organization. Adding color emphasizes structural features and can make a display more aesthetically pleasing. Color coding can create a unique relationship that immediately conveys information without need for further action or explanation, such as indicating similarities. In a list of references, all items pertaining to a particular subject or items published in the current year can be related through color. A three-dimensional appearance can be simulated with a judicious use of color. Components or elements that are treated as a unified entity can be related through color. Color coding streets in a map or the flow of gasoline in an engine are examples of the effective use of color for demonstrating relationships. Since using color to establish relationships may extend over multiple frames, distinct and easily nameable colors should be selected, especially if a legend is provided somewhere in the infobase.

The best advice regarding color is to keep it simple: Color adds complexity to a basic design. Restricting the number of colors can make the design task easier and reduce the chance of creating an undesirable effect.

10.3.3.1.5. Graphics and Pictorial-Quality Displays. We live and work in visually oriented environments. Graphics, picture-quality displays, and motion video play an increasingly important role in those environments. The extent to which these various forms of imagery are needed depends on the user's expectations and the applications involved. For good or bad, American's are weaned on nontextual imagery. Communicating with the written word is important but no longer the predominant way of conveying information, as indicated by the 1990 Nielsen Report on Television, which found that the average television household can receive 30.5 channels and ninety-two million households own at least one television set. This is more than twice the number of households owning a set in

1960. The average number of hours viewed each day was 7.02 hours in 1989. Interestingly children between the ages of 2–5 years spend 27.5 hours a week in front of a television set. When they are not watching television, they are often camped in front of a VCR, which is now in more than 67% of the television homes. School books, magazines, and even some newspapers contain a heavy dose of nontextual information. Graphics and its antecedents are part of the information milieu and users expect to see them.

Some applications require graphic content to communicate ideas, and graphic illustrations can improve the appearance of a text frame and contribute visual relief and interest. Graphics can provide a conceptual understanding of what to expect from the text and contribute to the aesthetic quality of the display. Interestingly abstract, nonrepresentational pictures were not used to illustrate numerical relationships until the mid-1700s, when scatterplots, time series, and other forms of statistical graphics were developed.

While graphics and pictorial-quality images complement the written word, they are not usually a substitute for it. Language expressed in the written word is the ideal medium for communication. Words can express propositions about past, present, and future conditions. Words can form logical chains of cause and effect. They can express one's feelings and intentions. In short, images are a poor substitute for words. Graphics present information about the physical appearance of objects or illustrate the relationship among numerical quantities. The significance of what the viewer sees is however subject to the vagaries of personal taste, style, cultural heritage, education, and intention. What was hoped the viewer saw may not have been what the viewer really saw.

The ease and speed with which a graphic interpretation can be read is not necessarily in direct proportion to its degree of photographic realism or level of detail. Research shows that outline sketches, cartoons, and caricatures may be more easily grasped than photographic-quality images. This point is most evident in a map that excludes unessential details while visually clarifying only information deemed to be necessary.

Graphics and picture-quality displays entail a number of costs that must be considered before deciding to what extent they will be incorporated into the system. The amount of time required to create images must be considered carefully. Table 9.1 lists time estimates for creating frames of increasing complexity; it is not unusual for a frame containing a multicolored chart with line-art to require 10 times as long to create as a simple frame containing alphanumeric characters and no color. Not included in Table 9.1 is the time required to conceive of and develop material needed to create the frame. There is a significant amount of computer overhead associated with graphics and pictures. Generally the more sophisticated the image, the more storage it consumes. In a small computer, storage constraints and moving images to and from storage may create a system bottleneck.

The time a recipient spends waiting for an image to be transmitted is a factor that should be considered if employees are involved. Obviously this is not so critical to the company if it is the general public, customers, or suppliers who are waiting; however excessive waiting time will adversely affect the willingness of the user population to access any system as well as their satisfaction with it. Figure 9.3 illustrates the time needed to send the Corporate Report at different transmission rates and receiving equipment features that can be supported effectively at these different rates. The entire report is estimated to be 23,400 bytes. If the general public were to access the report by phone and download it at 1.2 Kbits/sec, it would take slightly less than three minutes. Whether or not anyone would wait that long or instead log in to read it on-line will be balanced by the viewer's patience and the communication costs involved.

If graphics and pictures are going to be included in the infobase, some fundamental issues should be considered. Why are graphics or pictures being considered? What will they contribute to the comprehension? What level of imagery is needed? Are detailed pictures necessary, or will line drawings be satisfactory? Who will make up the audience? How will the images be disseminated? Answers to such questions help shape the form the imagery takes. Regardless of how simple or how detailed, how conceptual or how real an image looks, there are some general guidelines that should be followed in the design and creation process. These guidelines are shown in Figure 10.10.

Images may be incorporated into the infobase to improve the communication process by making complex issues clearer, summarizing data, and so forth, or to add to the aesthetics of the display. Whichever reason prevails determines the form of the image and the cost that can be justified.

Follow the principles of good design. Communicate ideas clearly, precisely, and efficiently. Avoid distorting the image or exaggerating portions of it unfairly

```
Decide on reasons for incorporating imagery:
    Communicate information
    Aesthetic interest
Follow principles of graphical excellence:
    Communicate ideas clearly, precisely, and efficiently
    Give the viewer the greatest amount of information with the least number of bytes
    Be honest in presenting relationships and information
Create friendly displays
Minimize transmission delays and storage requirements:
    Maintain clip art in local memory
    Stage transmissions
    Overlay drawing
    Use an image for diversion
    Use images in a sensible and careful manner
```

Figure 10.10. Guidelines for planning and creating graphics and pictorial displays.

with blinking, color, and animation. Convey the greatest amount of information with the fewest number of bytes possible, since bytes use storage space, slow transmission time, and consume hardware resources at the receiving end. Be honest in presenting relationships and information. The electronic dissemination of information provides opportunities for misrepresentation that do not exist in print-based environments. Create friendly displays. In the Tufte reference at the end of this chapter, the author cites a number of things that can be done to create friendly displays, such as spelling out words, inserting explanatory messages, and choosing color schemes that make sense for both color and monochrome displays.

Minimize transmission delays and storage requirements by designing and operating intelligently. If practical maintain clip art in local memory. Insert codes in the data stream that retrieves images already stored in the receiving unit. Stage transmissions so that information is being buffered by the system while the user reviews information already received. Overlay portions of the screen as other parts are being viewed and generate complex images in layers so that feedback can begin as the image is forming on the screen. Develop an animation or electronic form of music to entertain and inform during long periods of forced idleness.

Use images in a sensible and careful manner. Images take time to design and produce; they require more time to transmit and more sophisticated hardware and software to view than text. The time needed to recreate an image on the screen, called the paint time, can be in excess of a minute if the image is being transmitted by phone and contains a lot of detail, which can be frustrating. The reappearance of the same image can be annoying and boring. Common sense and concern for the user are essential when producing graphics and picture-quality displays.

10.3.3.1.6. Audio and Video. Audio is a natural accompaniment to almost any type of display. A voice overlay can explain complex ideas or reveal subtle points. Sound can make the message more personal, increase its appeal, or redirect the listener's attention. The major difficulty when storing sound is the amount of space it can consume as pointed out in Chapter 4.

When a video is being filmed, voice information is often captured at the same time, although it may be recorded separately and later integrated with the video. When audio is being appended to what would otherwise be a silent display, moderation is in order.

Music is an appropriate complement for business presentations and training exercises. The preferred approach for generating music is through MIDI, a digital protocol for transmitting musical data. Hundreds of prerecorded musical selections are available for use at a nominal charge, and computer bulletin boards often have musical segments that can be downloaded. However, it is important to verify that material from bulletin boards is in the public domain, especially if the music will be used outside the organization.

There are other ways of incorporating music; for example, Microsoft *Windows* supports MIDI, wave form, and Red Book audio. But MIDI uses space most frugally; 30 minutes of stereo music would consume about 200 Kbytes of memory with MIDI compared to 300 Mbytes with the other two. It should also be noted that the quality of sound depends on the MIDI hardware, so that the same MIDI data may sound different when played on a synthesizer other than the one on which it was created. There are many MIDI sequencer programs priced under $500 for adding music to presentations.

Video is normally associated with video conferencing and two-way communication between people. Another dimension of video is however storing, retrieving, and presenting information in a video format. Most of the systems in which this is done are stand-alone units, such as public access terminals in shopping centers and desktop systems designed for corporate training; however advances in technology are making it practical to consider transmitting video over a switched digital network similar to the way that voice and data are transmitted today.

The drawback to this technology is the bandwidth needed. In a monochrome television picture, an image is displayed on a grid of 512 horizontal and 480 vertical picture elements, referred to as pixels. A pixel contains the brightness information for each location; it is coded in 8 bits, which enables 256 possible shades of gray. To display motion, the picture is sampled at 30 frames a second; therefore 59 Mbits/sec ($512 \times 480 \times 8 \times 30$) are transmitted each second. Adding color information may raise the data rate to 90 Mbits/sec. To satisfy data rates of this magnitude, the user must be closely coupled to the equipment.

To support a video environment and deal with the constraints imposed by the bandwidth, full-motion video may have to be reduced to a series of still frames. For example, a snapshot of a simple frame requires one-thirtieth or 1.97 Mbits/sec; a 30 to 1 compression ratio reduces this to 64 Kbits/sec, a rate available on ISDN. Compression works by eliminating redundant information and eliminating some of the detail. Most organizations introduce video by means of snapshots of video images that have been compressed. As increases in bandwidth occur, accessing and retrieving full-motion video from a remotely located infobase will become more practical. Until then public access terminals and desktop systems will predominate.

Producing video is a very expensive and complex task. A production studio, highly trained technicians, special software, expensive equipment, and a lengthy development cycle are the rule. Very few companies have the in-house experience and facilities required for the task and it is unlikely many organizations will find it cost effective to produce videos themselves. Instead they turn to specialized companies for their video images; the survey results discussed in Chapter 12 elaborate on this issue of in-house versus outside text, graphics, audio, and video production.

10.3.4. Identifying the Source of Information

Traditionally data processing has collected data, prepared it for processing, entered and maintained the data, and later distributed data as information in reports. This sequence of activities is undergoing a transformation because application and content are changing. The extent of this transformation is obvious from the composition of the Corporate Report described in the previous chapter. Recall that the Corporate Report contained text, graphics in the form of simple line and bar charts, tables, pseudoanimation, and a question-and-answer section.

Producing an electronic Corporate Report involves more of the organization than just data processing, since information sources are inside and outside the organization. In this case, the Corporate Report is patterned after the annual report, so most of the content will have already been developed, but that is not always the case. As in any decision regarding whether material should be developed in-house or acquired from external sources, lead time, in-house expertise, and costs must be considered.

Lead time involves a trade-off between how quickly material can be developed internally versus finding a suitable commercial source that already has the material or can develop it quickly. The simpler the composition (*e.g.*, numeric, text, simple graphics) of the material and the more it relates to just that organization, the more likely it can be developed and created in-house; sophisticated graphics on material that requires subject experts may have to be done by commercial sources. Cost is another factor. Many organizations simply do not have the staff and equipment for cyclical projects. Rather than staff for peak periods, it may be more cost effective to have work of this type done outside the organization.

In many instances, it is necessary to supplement corporate information with information obtained externally. The most common sources are on-line services, such as CompuServe, NewsNet, Prodigy, and Dialog. Most of these services have charges based on the time of day access occurs, the length of access, and the particular information accessed. Normally information available on-line was created during the production of a paper-based product, such as a newsletter or newspaper. One alternative to obtaining information on-line is retrieving it from a local comprehensive library; the relative cost of one particular search is shown in Table 10.3.

In this example, a person searched for and retrieved the full text of an article using four on-line services and by driving to the public library. The Dow Jones search was done when rates were the lowest; other searches were conducted during the day. Driving to the library and retrieving the article took two hours; the cost was $27.60—$5 for mileage ($0.25 per mile), $2 for parking, and $.60 for copying; time was billed at an unrealistic cost of $10/hr. In fairness to tradition and the local library, one library visit could have yielded additional information at a significantly reduced cost, but on the flip side, few people would be willing to take two hours out of their work day if the same task could be performed in two minutes.

There are also a variety of products that can be purchased and incorporated with material developed in-house. Such material includes clip art, sound tracks, pictures, and video. One of the most impressive products is the *Visual Almanac*

Table 10.3. Hidden Costs of Traditional Information Sources

Source	Cost ($)	Time spent (min)
Dow Jones News/Retrieval	6.36	3
Dialog	4.04	2
IQuest	10.28	6
Nexis	11.80	—
Driving to library and searching	27.60	120

Source: D. Coursey, The Cost of Information, *Info World* 13(31) (1991) pp. 40–41, 44.

designed by the Apple Multimedia Lab and made available to educators. The *Visual Almanac* contains 7000 multimedia objects consisting of still images, audio, and movies; it also contains compositional tools for selecting, retrieving, and manipulating these objects to create presentations.

Figure 10.6 shows menu selections for a Corporate Report; main menu items are repeated in Table 10.4 with their likely information sources. Obviously data processing is a minor source for much of the material, although it may be responsible for managing the system that enables public access. The more frequently information is updated, the greater the reliance on the sources indicated and the less on the paper form of the annual report produced once each year.

The other side of identifying source material is attributing the source. Enhanced systems incorporating video and audio are particularly susceptible to copyright infringement. The ability to grab portions of a film clip or overlay portions of a popular song create the temptation to ignore ownership rights. This is a dangerous habit to develop, even when the content is intended solely for internal use.

It is reported that commercial producers of CD-ROM and videodiscs spend as much as half of their production time securing copyright permission; sometimes it is even a major challenge to determine who the real owner is. Musical pieces, for

Table 10.4. Main Menu for the Corporate Report and Sources for This Information

Menu Selection	Source
1. Chairperson's report	Corporate headquarters
2. Markets	Marketing department and external subject experts
3. Financial reports	Accounting department
4. Financial analysis	Accounting department
5. Annual meeting	Public relations
6. Stockholder questions and viewpoints	Public relations and legal

example, may have several owners, all of whom must grant permission. Once permission is obtained and perhaps a royalty paid, legal risks may still exist. Altering song words or overlaying several media, such as video and sound, may be objectionable to one or more of the different owners.

The issue of ownership rights is a legal minefield through which the most well-intentioned producer must tiptoe. Fortunately there are catalogs and directories available to help in the search process, and there are companies that sell clips of audio, picture-quality images, and other raw material on CD-ROM and other electronic media that may be used at the buyer's discretion. However until a central clearinghouse for copyrighted material has been established along the lines of the Motion Picture Licensing Corporation or the American Society of Composers, Authors, and Publishers (ASCAP), producers will have to be careful not to cross the line separating what is legal to use from what constitutes copyright infringement.

10.3.5. Weighing the Economics

Information being prepared for resale cannot be thought of in the same way as information intended to further an organization's objectives: Commercial information providers make decisions regarding economic issues based on what they think others will pay for their information; organizations providing information to further their objectives rather than for sale view economic issues differently. An attempt should be made to determine the cost to produce a given body of information; this should include the costs from time of inception to delivery of the final product, although a commercial information provider is able to weight this cost against the expected revenue from forecasted sales, an organization furthering its objectives by providing information often cannot quantify the value of the information in dollars and cents.

The issue of tangible versus intangible benefits from using computer technology is not a new one; data processing has always had a much easier time justifying expenditures when it could show that personnel costs would decline or inventory turnover would improve. It has had a much tougher time justifying expenditures that improved the quality of data or its timeliness. The Corporate Report is intended to improve stockholder, employee, customer, and public relations; its secondary objective is to market the firm's stock to the extent allowed by law. What value should be placed on accomplishing these objectives? To firms like Exxon and Union Carbide, recovering from oil and chemical spills, the value placed on improved stockholder and public relations is high. To a firm like AT&T, whose business is communications and information services, it is high because it reinforces its image as a leader in these services. But value in this case cannot be measured in dollars and cents; therefore the decision to invest in something like the Corporate Report should be based on knowledge of the costs involved and a subjective assessment of whether or not it is worth spending that amount to accomplish the intended objectives.

10.4. SUMMARY

There is a significant effort involved in planning and producing the infobase. In addition to the work normally associated with capturing and storing data in a traditional data-processing environment, graphics, pictures, and video scripts may have to be written and produced. This added complexity requires careful planning, reviewing, record keeping, and maintenance. To facilitate these efforts, standards governing the development and creation of the infobase are often adopted.

A 10-step model was developed, and the first five steps are covered in this chapter; these steps are (1) identifying the need for the information, (2) setting goals, (3) establishing the content of the infobase and its design, (4) identifying possible sources for information, and (5) weighing the economic factors.

Two philosophies prevail regarding the creation of an infobase. The first is that if the infobase is created, applications will follow; the second is that the infobase should be a logical derivative of an identified need. Identifying a need facilitates the goal-setting process and setting goals helps sell the system to management, gives a tone and direction to the effort, and very directly influences the system's content and form.

Content is determined by need, availability of particular types of information, and the technology involved. Content is influenced by such factors as the users' profile and how the infobase will be used. Types of design issues that must be considered include formatting the displays; designing menus and other forms of access, presenting text; and incorporating color, graphics, picture-quality displays, and video. Tips and guidelines are included for many of these design issues in figures and tables.

Traditionally the source of information in computerized systems has been data processing. Enhanced systems often incorporate information acquired from other sources, sometimes outside the organization. Departments and qualified information providers in and outside the organization often provide information and respond to questions electronically, which introduces a new set of issues related to security, accountability, reliability, and timeliness of response, for example.

The economic ramifications of creating enhanced systems is always lurking in the background. Many of the benefits, such as improved communications, better customer and public relations, more meaningful information, and improved esprit de corps are difficult to evaluate in terms of dollars and cents.

REFERENCES

S. J. Brinker, Corporate Bulletin Board Systems: Customer Support and More in the 1990s, *Telecommunications* 25(11) (1991), pp. 33, 35–36.
J. Burger, Seven Steps to Color Sense, *New Media* (May 1992), p. 49.

D. Chorafas, *Interactive Message Services*, McGraw-Hill, New York (1984).

D. Coursey, The Cost of Information, *InfoWorld* **13**(31) (1991), pp. 40–41, 44.

H. J. Ehlers, The Use of Colour to Help Visualize Information, *Computers and Graphics* **9**(2) (1985), pp. 171–176.

S. Floyd, ed., *Handbook of Interactive Video*, Knowledge Industry Publications, White Plains, NY (1982).

Frame by Frame, AT&T Information Systems (1985), pp. 1–9.

W. Galitz, *Handbook of Screen Format Design*, Information Science, Wellesley, MA (1985).

M. S. Gelinne, Creating an Internal Customized News Service, *Online* **15**(4) (1991), pp. 52–57.

R. Hendall, MIDI Goes Mainstream, *PC Magazine* **11**(6) (1992), pp. 181–183, 186–189, 194, 202, 208, 217–218.

R. Johnson, Reconciling Consumer Needs with Client Wants, Designing Other People's Pages, *Videotex—Key to the Information Revolution. Proceedings*, Online Publications Ltd, Middlesex, UK (1982), pp. 295–299.

J. S. Johnson, *Structured Hypermedia Tutorial*, Software Engineering Professional Education Center, University of Houston—Clear Lake, TX (1990).

P. Karon, Electronic Publishing Faces Legal Traps Over Copyrights, *InfoWorld* **14**(10) (1992), p. s70.

C. Mink, Speaking without Words, *DEC Professional* **6**(7) (1987), pp. 32–34, 36–38, 40–42.

S. Morse, Multimedia on the Network, *Network Working* **3**(3) (1992), pp. 58–59, 62–67.

Nielsen Report on Television, Nielsen Media Research, Northbrook, IL (1990).

G. Nugent, P. J. Peters and L. Rockwell, *Designing and Producing Videotex Instructions: A Producer's Handbook*, Station KUON-TV and Division of Continuing Studies, University of Nebraska—Lincoln (1983).

B. O'Keefe, Adopting Multimedia on a Global Scale, *Instruction Delivery Systems* **5**(5) (1991), pp. 6–11.

J. Panepinto, DEC Finds Ample Use for Paperless Videotex, *Digital News* **7**(12) (1992), pp. 19–20.

L. Reynolds, *The Presentation of Bibliographic Information on Prestel*, Graphic Information Research Unit, Royal College of Art, Great Britain (March 1980).

P. Rubens, Online Information, Traditional Page Design, and Reader Expectations, *IEEE Transactions on Professional Communications* **PC 29**(4) (1986), pp. 75–80.

E. R. Tufte, *The Visual Display of Quantitative Information*, Graphics Press, Cheshire, CT (1983).

VAX VTX Information Providers Guide, Digital Equipment Corporation, Maynard, MA (1984).

L. Walker, *Hypermedia and Visual Technology*, Software Engineering Professional Education Center, University of Houston—Clear Lake, TX (December 3, 1990).

K. H. Woolsey, Multimedia Scouting, *IEEE Computer Graphics & Applications* **11**(4) (1991), pp. 26–38.

Chapter 11

Planning and Producing the Infobase: Part 2

The steps in the planning and production cycle covered in Chapter 10 set the stage for what follows. Chapter 11 explains the preproduction activities of concept development, flowcharts, and scripting, followed by the mechanized tasks of producing and editing the frames. The cycle ends with documenting what occurred and periodic audits to determine the need and appropriateness of the infobase contents. Samples of forms that can be used to help produce this infobase are explained. These forms incorporate documents from several sources. It is unlikely that one organization would use all of them; especially if the computerized authoring facility incorporates similar information. In addition many organizations will choose to maintain the forms on a computer using hypertext and other suitable technology.

11.1. PREPRODUCTION

At this stage, work can begin on creating the product. The sophistication employed in the preproduction process depends on the complexity of the message and the medium to be used. A series of frames containing text but devoid of graphics and branching does not require so much effort to plan and produce as an interactive video script. Two portions of THE WINDOW are developed to illustrate the process. Figure 10.6 showed the menu structure for THE WINDOW. The two areas selected are option 1 in the Corporate Report, called the Chairperson's Report, and option 6 in the XYZ Corporation, called Investing in the XYZ Corporation; the Investment Strategies portion of the latter was chosen. The set of forms identified in Table 11.1 are filled out for each area. These forms are intended

Table 11.1.   Preproduction Documentation

	Chairperson's Report	Investment Strategies
Concept statement	X	X
Flowchart		X
Storyboard	X	X

as guides, so they must be redesigned to fit the needs of each organization. Since the Chairperson's Report is straightforward and consists of text, it does not require a flowchart.

11.1.1. Concept Development

The first step is filling out a document called Concept Development (see Figure 11.1.) This form has two roles: First it is a management tool for estimating resource needs and creating a work schedule; second it is a guide for the scripting and production staff who are probably not the same individuals who conceived and proposed the idea. The Concept Development form can also be used as a final screening device for ideas that sounded good but which after further examination were determined to cost too much or take too long to produce.

Figure 11.1a illustrates the portion of the Corporate Report called the Chairperson's Report; note that at this point in the development cycle, decisions are being made about specific topics to include—the strategic decision to develop something called the Corporate Report has already been made. The name of the document originator appears at the top of the form followed by the name of the proposed item, where it will be located in the infobase, and the purpose for incorporating it. The approximate size in frames is estimated and the source identified for information that will be used to create the infobase; in some cases, the source may not be in the company.

As stated in Chapter 9, it is assumed that most organizations will create the magazine in a special unit to take advantage of equipment and personnel skills that are needed. After the item is created, it will be the responsibility of another unit to maintain; probably the source of the information. It is possible to incorporate multimedia features into the information, but to ensure as wide a base of users as possible, this information is limited to text with color.

Budgeting, production schedules, and marketing plans are built around the resource estimates. Very little time will be needed to create the text for the Chairperson's Report because it will be drawn from the company's annual report; this is not the case for the text for Investment Strategies; which will be a unique set of screens consisting of a series of frames with an underlying computer program.

Originator: John Doe (35)   Date: 2/01/93

Address: Administrative Services   Phone: 708-330-XXXX

          Chicago   E-Mail: JD452 FAX: 708-330-XXXX

Name: Chairperson's Report   Magazine: Corporate Report

Purpose: Communicate state of company and future directions to public.

Size/Frames: 4–8   Source of Information: Jack Armstrong

Name of Individual/Unit Responsible for

    Creating: Sandra Day

    Editing/Updating: Jack Armstrong

Features (Check all that apply.)

    Text   X   Video   _____   Sound   _____

    Graphics _____   Color   X   Other (specify) _____

    Picture   _____   Animation _____

Resource Estimates

	Flowcharts	Scripting	Production	Updating
Time (Person/Days)	N/A	0.5	0.25	0.25
Nonlabor Costs	0	0	0	0

Updating Rate (Insert a digit in only one of the choices.)

    Daily   _____   Monthly   _____

    Weekly   _____   Annually   4

Description (Check box ☐ if continued on back.)

    A narrative based on the Chairperson's Letter that appears in the front of the annual report.

Updates will occur only in the event of major developments.

Figure 11.1a. Concept development statement for the section of the Corporate Report called the Chairperson's Report.

Originator: Jack Pierce (66)   Date: 2/26/93

Address: Treasury   Phone: 303-896-XXXX

              Denver   E-Mail: JP894   FAX: 303-896-XXXX

Name: Investment Strategies   Magazine: The XYZ Corporation

Purpose: Provide information to the public about things to consider in developing an

investment strategy.

Size/Frames: 60–80   Source of Information: Treasurer's Office

Name of Individual/Unit Responsible for

    Creating: Sandra Day/Jack Pierce

    Editing/Updating: Jack Pierce

Features (Check all that apply.)

    Text   X   Video   _____   Sound   _____

    Graphics X   Color   X   Other (specify) _____

    Picture _____   Animation _____

Resource Estimates

	Flowcharts	Scripting	Production	Updating
Time (Person/Days)	2	5	0	1
Nonlabor Costs	0	0	0	0

Updating Rate (Insert a digit in only one of the choices.)

    Daily   _____   Monthly _____

    Weekly _____   Annually   1

Description (Check box □ if continued on back.)

    An interactive set of screens that will enable a person to enter personal information and

develop an investment strategy for acquiring stocks.

Figure 11.1b. Concept development statement for the section of the Corporate Report called Investment Strategies.

Planning and Producing the Infobase: Part 2287

The entry on the form entitled nonlabor costs is used to record expenses incurred purchasing information from external sources and obtaining special equipment or software. A general description of the information including sample menus appears at the end of the form.

11.1.2. Flowcharts

Information is presented as a series of sequential frames regardless of whether it is text, graphic, or video based. The planning stage of the preproduction process begins by dividing the information into blocks. When the relationship among the blocks is simple and straightforward, perhaps a series of sequential frames, scripting can begin. If on the other hand, there are a number of branching options or relationships among the frames are complex, it is advisable to develop a flowchart.

A flowchart serves two purposes: It provides a structure for content development and branching options, and it provides a guide for writing computer programs that may be needed. Flowcharts are particularly helpful when there are multiple branching options. Content for a film or a paper-based medium, such as a book, is linear: It has a beginning, middle, and end; content for an electronic medium can be prepared and delivered sequentially, but this is not usually done. One of the inherent strengths of a computer medium is its ability to respond interactively to user input. Consequently multiple paths exist to reach different end points. A flowchart helps identify these paths and provides a simple and straightforward way of developing each one.

The flowchart for Investment Strategies is shown in Figure 11.2; it provides information about what to consider when developing an investment strategy. Specific stocks, bonds, and other investment instruments for this or any other company are not mentioned by name. There are 73 frames in the application, which begins with a brief explanation followed by a disclaimer informing the viewer that information is provided for its educational value and, not for the purpose of recommending specific investments. The user provides personal information by selecting a set of menu choices. These choices are the number of years until retirement, the purpose of the investment, and the current tax bracket (Figure 11.3). Based on these inputs, an investment strategy is recommended, followed by examples of investments to consider. The model ends by reminding the user that the information is intended to educate and inform the user and actual investment decisions should be made with the advice of a professional financial counselor who will consider additional factors. The user is then directed to other areas of the THE WINDOW or invited to exit from the system.

Each symbol in the flowchart is numbered to provide easy identification, reference, and identify script segments and completed frames; numbered symbols are also the basis for identifying frames. The numbering scheme for frames is a three-part code; the first three characters reading from left to right indicate the

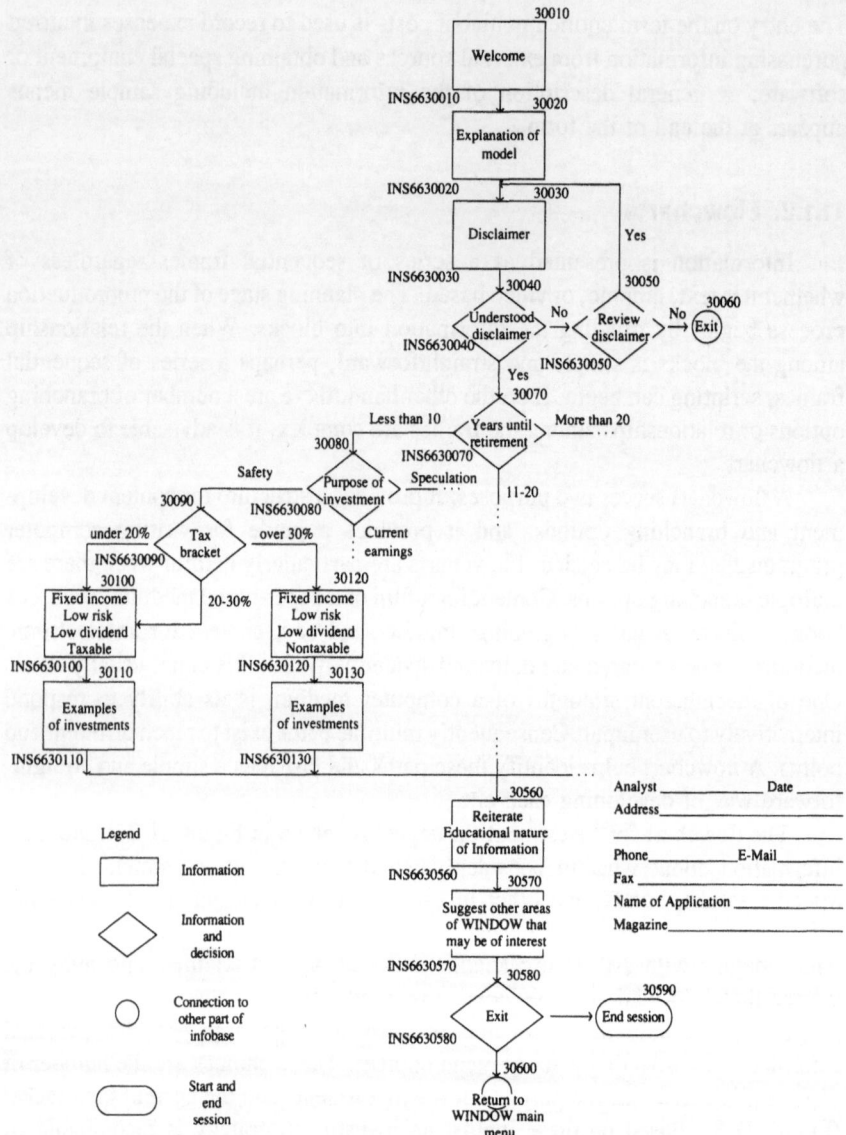

Figure 11.2. Flowchart for Investing Strategies.

location within the infobase; the next two digits identify the originator or responsible unit, which corresponds to the first entry in the Concept Development Statement in Figure 11.1; the last five digits correspond to the numbered symbol in the flowchart. This method for numbering frames is different from the treelike procedure described in Figure 10.3 because the hierarchical structure does not exist when there is extensive branching.

Inputs	Outputs
Number of years to retirement	Returns
Less than 10	Fixed income
Between 10–20	Growth
More than 20	Risk
Purpose of investment	Low risk
Safety of capital	Moderate risk
Current earnings	High risk
Speculation	Current earnings
Current tax bracket	Low dividend
Less than 20%	High dividend
Between 20–30%	Tax status
More than 30%	Taxable
	Nontaxable

Figure 11.3. Single transaction investment model for Investment Strategies.

A rectangle indicates a frame containing information, and a diamond represents a frame containing information and requiring a decision; the oval and circles represent computer functions, and since these are not displayed on the screen, they do not have frame numbers assigned to them.

In a complex series of frames or one where a lot of branching occurs, relationships among nodes should be carefully documented prior to developing a storyboard. Figure 11.4 is an example of a form to use; it is based on the Nugent and Peters reference at the end of the chapter. The form that works best will vary according to the software package used to create frames and the complexity of the application. Although the documentation described here is paper based, there is nothing to prevent it form being implemented on a computer.

The Node and Frame Relationships form contains the same basic information as the Concept Development Statement, and it is developed from the flowchart. The first two columns contain the node and frame numbers from the flowchart. The next set of columns reflect different ways that branching can occur. Unconditional branching is based on a menu selection or user-specified response, such as age or income. In a defined branch, a keyboard entry, such as m or p, routes the viewer to the main (m) menu or the previous (p) menu. Default branches are selected when the user is directed to enter a carriage return or any character to advance to the next frame.

It is occasionally desirable to freeze a frame for a specified length of time, which is done in product demonstrations where a particular pace is set and in public access terminals that deploy rolling advertisements when no one is using the terminal. The auto/advance column indicates the time in seconds that a node is displayed; the columns labeled user specified input shows the number of characters allocated and the location on the screen where the user will input such information as age or income.

Node and Frame Relationships

Address: ______________ Date: ______________ Name of Application: Investment Strategies

Phone: ______________ Magazine: The XYZ Corporation

______________ E-Mail: ______________ FAX: ______________

Node Number	Frame Number	Branching: Unconditional Entry	Branching: Unconditional Node	Branching: Defined Entry	Branching: Defined Node	Branching: Default Entry	Branching: Default Node	Auto Advance Time/Sec	Auto Advance Node	User-Specified Input Size	User-Specified Input Row	User-Specified Input Column	Comments
30010	INS6630010					CR	30020						
30020	INS6630020					CR	30030						
30030	INS6630030					CR	30040						
30040	INS6630040	1	30070	M	30010								1 = Yes
		2	30050	X	30600								2 = No
30050	INS6630050	1	30030	M	30010								1 = Review
		2	30600										2 = Exit
30060	—												WINDOW Menu
30070	INS6630070		30080	M	30010					2	13	35	Less than 10
			30240	P	30040								11 to 20
			30400	X	30060								More than 20
30080	INS2730080	1	30090	M	30010								1 = Safety
		2	30140	P	30070								2 = Current earnings
		3	30190	X	30060								3 = Speculation
30090	INS6630090	1	30100	M	30010								1 = Under 20%

Figure 11.4. Node and frame relationships for the Investment Strategies flowchart.

Node and Frame Relationships

Date: _____ Name of Application: Investment Strategies

Phone: _____ Magazine: The XYZ Corporation

E-Mail: _____ FAX: _____

Address: _____

Node Number	Frame Number	Unconditional		Branching Defined		Default		Auto Advance		User-Specified Input			Comments
		Entry	Node	Entry	Node	Entry	Node	Time/Sec	Node	Size	Row	Column	
30100	INS6630100	2	30100	P	30080	CR	30110						2 = 20 to 30%
30110	INS6630110	3	30120	X	30600	CR	30560						3 = over 30%
30120	INS6630120					CR	30130						
30130	INS6630130					CR	30560						
30560	INS6630560					CR	30570						
30570	INS6630570	1	30600			CR	30580						1 = WINDOW Menu
30580	INS6630580	2	20590										2 = End session

Figure 11.4. Continued.

The entries in Figure 11.4 correspond to the flowchart for Investment Strategies. The way that a user accesses information and proceeds through it can be managed by judiciously arranging the branching options. The reader of a paper-based magazine can browse articles, but someone viewing screens for the Investment Strategies can be forced to move along particular paths to progress to the answer, since in some cases, the only other recourse is to terminate the session by forcibly hanging up the modem. The ability to manage the information flow in this way distinguishes videotex from all other forms of media presentation.

For example, as shown in Figure 11.4, the first three frames must be viewed because the only option is a carriage return (CR). This path leads to the disclaimer where the viewer must indicate an understanding of the disclaimer before being allowed to proceed. At the bottom of node 30040 (Figure 11.2) and most of the other nodes requiring a decision (e.g., a diamond on the flowchart), there are options to return to the main menu (M) or the previous (p) decision frame. The logic of using p is to allow the viewer to reconsider earlier decisions that led to the current location. There are also numerous opportunities to exit (x) this portion of the infobase to return to the main menu for THE WINDOW. Note that node 30070 requires the viewer to enter a two-digit number indicating how many years until retirement; other inputs are made by selecting particular menu options.

The form shown in Figure 11.5 specifying media is a companion to Node and Frame Relationships. This form, Media Guide, is patterned after a document in the Bergman and Moore reference at the end of the chapter; the form specifies the type of media, e.g., the number of frames of text, graphics, or picture-quality images; the number of seconds of audio; and the number of seconds of video. There is room for a cost estimate and an identification number, similar to what appears on the Node and Frame Relationships form (see Figure 11.4).

The Media Guide documents the number of frames or amount of time for each media and provides an estimated cost; it also helps allocate storage requirements. Knowing the approximate number of seconds for audio, video, and other media forms enables the developer properly to configure the content to the physical limitations of the media. Identifying costs, storage requirements, and media permits intelligent trade-offs to be made early in the development process.

11.1.3. Storyboards

A storyboard is a script to guide the people who are producing the information. It may include the full text to be input, color choices, branching instructions, renderings, and photographs of images to be input, and any other information needed to facilitate rapid and efficient production.

There are numerous types of storyboard formats, and the amount of detail varies according to the media involved. A series of frames consisting of straightforward text that may be positioned anywhere on a frame requires very little specificity. On the opposite extreme, a series of frames containing text, video, and

	Number of Frames			Time/Secs		
Identification Number	Text	Graphics	Picture-Quality Images	Audio	Video	Cost Estimate

Figure 11.5. Media guide.

audio requires considerably more detail. The areas on each frame reserved for the different media forms have to be developed and integrated with other parts of that frame to build a compatible and informative whole.

Script development follows the flowchart and is closely aligned with the flowchart during the development process. Since by themselves, individual pages of the script bear little resemblance to the final product, the flowchart is essential for understanding branching and various elements of the whole unit. In fact the scrip sequence is often not arranged in the same order in which it will be viewed when presented electronically. Depending on the availability of particular people and necessary information, different segments of the script may be developed weeks apart and merged later.

Examples of a page for the scripts of the Chairperson's Report and Investment Strategies are shown in Figures 11.6a and b. Each page of the script is related to a node on the flowchart by the node number in the upper right corner of the form as an aid in routing the script to the correct production people later when the media features are checked. Generic specifications, such as foreground and background colors, are specified to minimize the need to make notes on the script itself. The two blocks under layout inform production personnel whether the layout is approximate or exact, which prevents unnecessary efforts and provides production personnel with latitude in producing the frame.

Space is provided on the left side of the frame for the developer to insert instructions and on the right side for the editor/reviewer to make comments or corrections; routing to other frames is shown on the bottom of the form, with a space for script approval. The area reserved for the content is broken into 20 rows and 40 columns, which conforms to the NAPLPS standard and the common frame type described earlier. The first and last lines and the last column are reserved for system usage; this provides a working area of 18 lines and 39 columns.

Figure 11.6a contains the script for the opening frame of the Chairperson's Report, which will contain text, graphics, and color. The graphics image is the corporate logo, and it will appear as a color block for those unable to receive graphic images; the characters will be black on a light blue background on a color monitor. The layout is approximate giving production personnel latitude in creating the frame; title characters are twice the default size, and two lines are compressed by skipping a space between sentences.

Figure 11.6b depicts the frame for node 70 in the flowchart. Since the user has to input the number of years before retirement at a designated location, this frame must be laid out according to instructions contained in the script. Branching for this frame is much more complex. The creation software will enable the branches to be established depending on the choice made by the user.

The storyboard for a video sequence is much more elaborate. A video sequence, like a series of text frames, must have a discrete start and finish; however video production must be much more carefully thought out. For example, outdoor scenes for multiple sequences must be shot at approximately the same time of day under similar weather conditions; similarly an actor in related sequences cannot be dressed differently if filming spans several days. The subtleties and complexities of producing video often lead organizations to hire companies specializing in this work.

The process becomes even more difficult when audio is added, because audio has to be synchronized with the other media it overlays to make sense. Naturally audio has to be produced with voices and other sounds that are consistent with the information displayed. Adding to the difficulty is the different rates at which the various media forms are retrieved from memory for display. Systems of the type

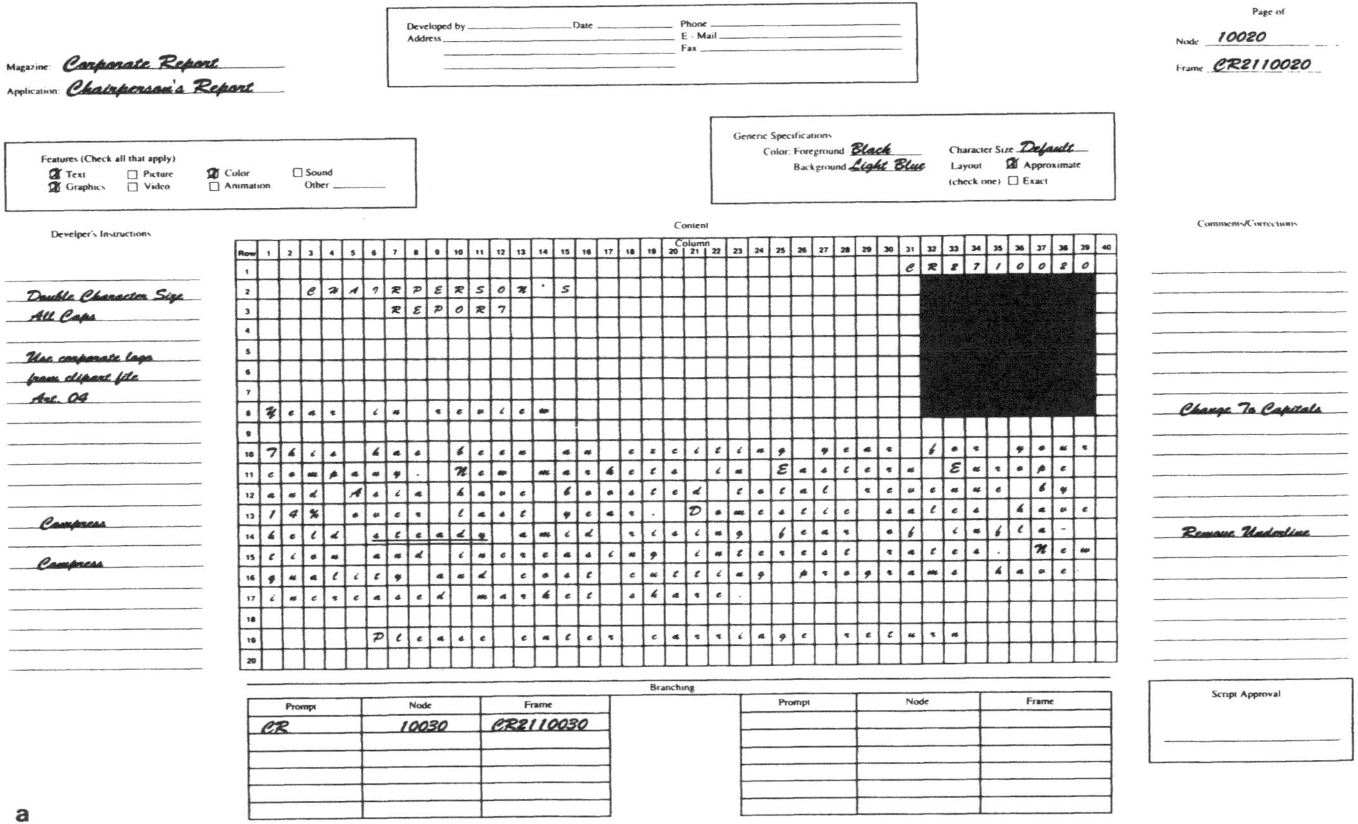

Figure 11.6a. Opening frame of the Chairperson's Report.

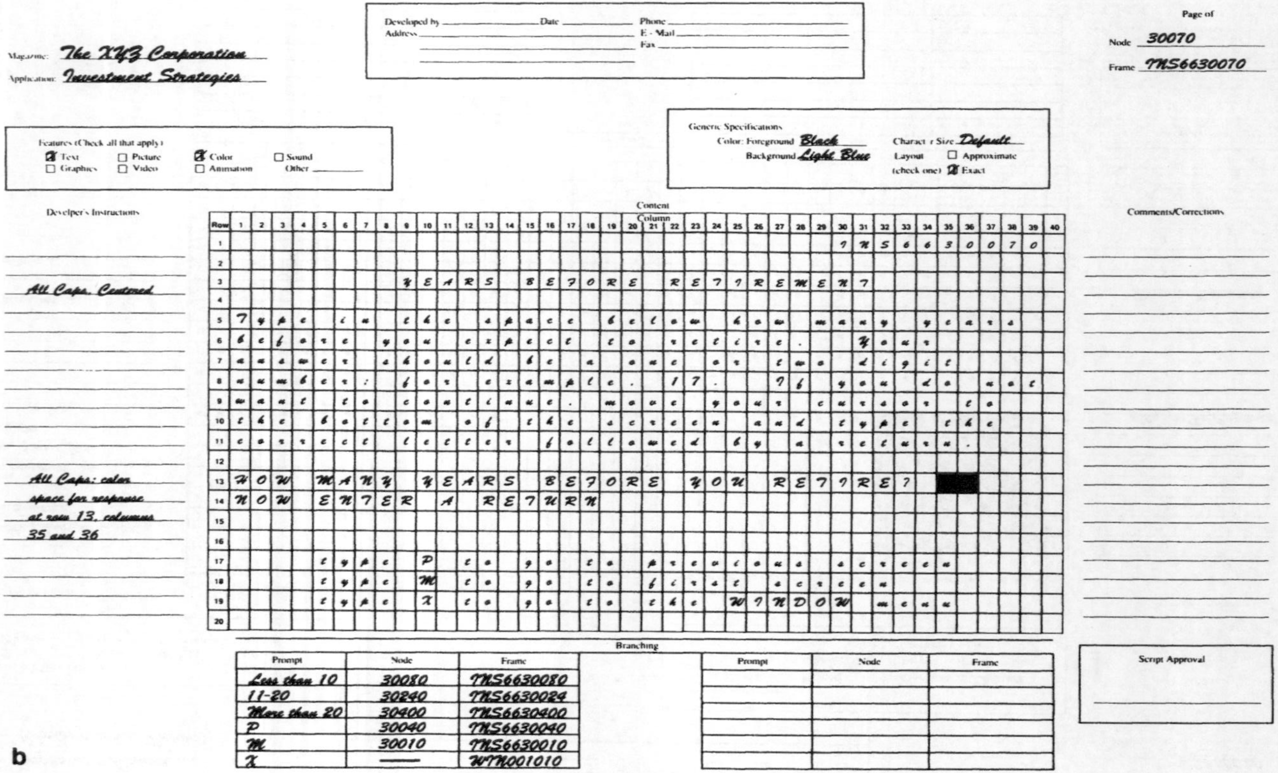

Figure 11.6b. Frame for node 70 of the flow chart.

b

described in this book have existed since the mid-1960s, albeit in a much cruder form. The IBM 1500 System was the first; it consisted of a special-purpose workstation controlled by an IBM 1800 minicomputer. Attached to the workstation were various pieces of equipment for producing audio and projecting slides; the program that controlled the equipment was a far cry from what exists today. Integrating the system and the application, regardless of the vintage of the hardware and software, can be greatly facilitated by good authoring software.

11.1.4. Authoring Programs

Authoring software can help bridge the chasm between the design and production stages. An authoring program is a special software package that aids in developing and creating the content. In Chapter 8, five types of authoring facilities were discussed, which can be condensed into two broad categories: languages and systems. The former is a specialized programming language in which a series of commands define the structure of an interactive program. An authoring system is much less flexible; it works by answering a series of questions from various menus or perhaps by building a flowchart with icons to activate displays of text, graphics, and so forth.

Authoring systems are marketed much like general business software for personal computers, such as spreadsheets and database programs. Initially buyers were graphics professionals, but the abundance of new, simple to use, and affordable packages is rapidly broadening the user base. Products like Anthology® CEIT Systems, Inc., enable audio, full-motion DVI video, still images, graphics, text, and animation to be combined to create multimedia presentations and instructional courses. In 1991 it was estimated that there were introduced 59 paint packages and image-editing tools, 51 charting packages, 39 modeling and rendering packages, and 38 desktop video packages; these new products were augmented by revisions of existing packages. The introduction of Windows 3.1 and Quick-Time is likely to increase the number of new products.

11.2. PRODUCTION

After the script has been completed, frame production and if necessary, computer programming can proceed simultaneously. This is true for the Investment Strategies example, since creating the frames does not depend on completing the computer program or vice versa.

When the script is ready, the graphic designer can complete the remaining details, and the input specialist can review the material to clear up questions and ambiguities prior to beginning production. The skills needed by the input specialist will vary according to the media involved, complexity of the theme, and the artistic abilities required. A full-motion video constructed from a segment captured by a camcorder and coupled with graphics, text, and audio requires much more equipment and a more highly skilled craftsperson than THE WINDOW, but the wide selection of available products enables any person or business to acquire suitable equipment easily and without spending a great deal of money. The Frenkel and Yager references cited at the end of this chapter outline the production process for several multimedia implementations and discuss commercial products available. These examples are considerably more complex that the examples discussed in this chapter.

The frame development process actually generates two kinds of frames: those which are seen and control frames. Creation software usually has a set of menu choices that guide the input specialist through the process. At some point, one of the menu selections will display a control frame requiring certain inputs, which include such things as an expiration date, access codes for security, branching instructions, charge codes, keywords, and so on. The control frame is coupled with the frame that is displayed, but the former is not accessible by the general user.

The input specialist must incorporate various media to produce a frame. For now and the foreseeable future the vast majority of corporate and residential users will be limited to a world of text and various forms of still imagery because of the large inventory of end-user equipment capable of handling this form of presentation and bandwidth limitations of available communication facilities.

Images that are not produced as artistic renderings or generated by graphic software from raw data are produced by scanning already available images; scanning is also a simple and inexpensive way of capturing text. Until the early 1990s, the major stumbling block to image scanning was direct access storage, since images can consume a million or more bytes of storage. Optical disc storage and data compression algorithms that reduce the byte count have made scanning a practical adjunct to frame development. Several tips that should be followed in the image-capturing process are summarized in Figure 11.7.

An important part of creating a frame is achieving balance among its various parts. Graphic illustrators expend considerable time and effort achieving balance in print-based media, but it is often overlooked in electronically presented information. Part of the problem lies in the ease with which an image can be

---

Maintain consistency among frames.

Avoid excessive detail that is not germane to the information value of the image.

Make images as large as practical.

Balance text and images and maintain their proximity.

Adapt colors of digitized images to display clearly on monochrome receivers.

Figure 11.7. Image capture using scanners and video cameras.

manipulated after it has been digitized. Images should be treated consistently throughout a segment of related frames. Borders, background, captions, placement, and other elements of good design should be followed.

Excessive and extraneous detail wastes precious storage space and communication bandwidth. Images scanned or captured by video are often accompanied by background detail that does not contribute materially to the message conveyed by the image. Such unnecessary detail should be removed.

A balance must be established between the size of the image and the amount of text that accompanies it. Although we live in a visual world, there is nothing more precise than a few words to convey information. Make images as large as practical, but do not sacrifice the precision and clarity of the written word to do so.

Keep text and images in close proximity to each other. Since captions are often not included with electronic images, the text that relates to a particular image should be on the same or immediately adjacent frames.

Color can create problems. In photographs and pictures scanned from a print-based source, a number of colors may be present, which when displayed on a low-resolution monochrome monitor, may not be clear. The input specialist must view images on the same type of equipment as the user to adjust hues and levels of brightness so that frames will be legible to everyone.

11.3. REVIEWING AND EDITING

After a frame or a group of frames have been produced, each should be reviewed and edited prior to insertion in the infobase. In an age when information can be duplicated and transmitted around the world almost instantaneously, it is essential that only grammatically correct, aesthetically pleasing, and accurate frames are released.

Accuracy extends beyond simply checking factual information. Branching options, too, must be checked to verify that they lead to the intended destination. Programming routines must be examined to assure that they do not contain logic or arithmetic errors. Utilities that graph statistical information and report the percentage of correct responses to questions must be tested.

The task of editing frames involves reviewing frames and correcting mistakes, noting deficiencies, making recommendations, and updating, which can be done either off-line or on-line, interactively or in batch mode. Updating is a time-consuming and costly activity, as listed in Tables 9.4 and 9.5. It should be carefully considered and planned for when designing the information base.

The infobase ages unevenly: Portions may be relatively static and have a long life while other parts may be of little or no value after several minutes, *e.g.*, ticker prices of stocks. Several things that may be done from a design standpoint to accommodate this uneven aging and facilitate the editorial process are listed in Figure 11.8.

```
Avoid mixing short- and long-term information.
Mix text, static images, and video judiciously.
Add calendar dates to information.
Incorporate expiration dates in control frames.
Develop templates for updating.
Create bulk update techniques to minimize manual intervention.
```

Figure 11.8.   Suggestions for aiding the editorial process.

Avoid mixing short- and long-term information if possible, since the need to change the former may impact the integrity of the latter. Whenever a frame is updated, it is necessary to review its entire contents for accuracy and aesthetics, and reviewing information that has not changed is wasteful. For the same reason, mixing text with other media forms must be done judiciously. This is especially important since producing graphics and video sequences is very costly. If changes to text render graphic or video material obsolete, unanticipated and significant expenses to update the material may be necessary.

It is a good practice to date frames, particularly those containing time-sensitive information. It may even be wise to incorporate an expiration date for information that could be misleading or incorrect after a certain time. When common features appear across a set of frames, templates are a cost-effective way of providing consistency and simplifying updating.

Whenever possible bulk-updating techniques should be developed to simplify converting information from other sources. Such techniques may convert data in the corporate mainframe or incorporate text prepared by a secretary with a word-processing package. Whatever the information is, attempts should be made to avoid manually recreating it.

11.4. RECORD KEEPING

Among the tasks involved in any job, record keeping is usually the least desirable. In data-processing environments, programmers and systems analysts are all too ready to have someone else develop and maintain documentation that is essential to operating the systems; it is unlikely that the existence of enhanced systems will change this attitude about record keeping. Documentation is normally necessary for operating, performing maintenance, and recovering from system crashes in any computer system; in enhanced systems, additional considerations exist: The orientation of the entire system is user focused; information is delivered directly to the end-user; and multiple media forms exist. The contents of the system are created by individuals whose skills are beyond the scope of traditional data-processing departments (*e.g.*, graphic illustrators, writers, and animators). Responsibility for maintaining the contents may be distributed to originating information providers. Some information may be acquired externally and integrated into the systems; some applications may be transaction based and oriented toward the public.

In addition to the traditional forms of documentation maintained by the data-processing unit, the items listed in Figure 11.9 are often helpful. Many of these items can be kept in electronic form to simplify retrieval and updating. The snapshot, *i.e.*, copies, of the first and last several frames of each menu selection is a common method of establishing copyright protection. Since the infobase is changing constantly, it is often impractical to maintain a current copy of everything; instead a snapshot is taken periodically, *e.g.*, monthly. This would be necessary only for frames accessed by the public for which proof of ownership could be needed, such as the Investment Strategies exercise.

The keyword list and frame directory are necessary to ensure that unique keywords are assigned and to update frames, respectively. The frame map is a diagram of what points to what; some software packages maintain such a map automatically, which facilitates changing pointers when menus and frames are added or removed from an existing infobase. The issue of standards and policies is particularly important because of the distributed control and the multiple information providers. Everyone involved in using, creating, and updating the system benefits from carefully worded instructions about what should and should not be done and how the system can be most effectively used.

11.5. AUDITING AND REJUVENATING THE SYSTEM

The 10-step model for planning and producing the infobase that was introduced in Chapter 10 started by identifying a need and setting a goal. Periodically a

---

Concept development statement

Flowcharts

Storyboard

Snapshot of first and last set of frames for each menu selection

Keyword list

Frame directory

Frame/routing map

Standards and policy statements regarding updating, security, access and so forth

---

Figure 11.9. Documentation for supporting an enhanced system.

system must be audited to assess how well it is doing and to add new vigor. The purpose of the audit is to compare intentions with outcomes. If the need has been identified by surveying potential users or assessing the competition, then it is necessary to reassess the outcome by resurveying users or reexamining the organization's competitive position. If particular goals were the reason for launching the project, then it must be determined whether or not the goals were attained. Techniques for assessing primary and secondary goals were shown in Table 10.1.

There are several ways of reinvigorating a system: Introductory frames can be changed periodically; colors can be altered, icons changed, and new sections added; bulletins can be posted about items of interest and systems alerts deposited in mailboxes. The most important thing that can be done however is to make sure all information is accurate and timely.

Statistical analysis is an integral part of auditing the system and determining what requires rejuvenation. Statistics may be generated through telephone and mail surveys, interviews, or other forms of active development; they may also be collected electronically by the system. A thorough audit requires a combination of both collection techniques. Figure 11.10 illustrates types of statistics that may be collected by the system. By employing statistical analysis, other measurement, such as averages and correlations can be developed.

System statistics are divided into two categories here for purposes of explanation. The traffic measures include the obvious information relating to date, time of day, and duration of the session. The distinction between registered users

```
Traffic
    Date
    Time of day
    Session duration
    Registered user or guest
    Blocked calls
    Form of access
    Terminal type
    Reason for session end (normal disconnect, line drop, exit to other service)
Application (by session)
    Frequency of access (last access and aggregation)
    Frame accessed
    Preceding frame accessed
    Succeeding frame accessed
    Time per frame
    Frame charge (if applicable)
    User input (bytes)
    User output (bytes)
```

Figure 11.10. Examples of statistics that may be collected by the system for audit and rejuvenation.

and guests is important for systems open to the public, such as THE WINDOW infobase. Portions of THE WINDOW, such as the Corporate Report, are intended for the public at large, while other portions, such as Employee Opinions and Suggestions, are only for employees. Consequently access is restricted for the latter. Some areas of the infobase are of more interest to the public or employees than others, so when collecting statistics, it is important to take into consideration who is using what.

To configure the system properly, information about the number of unanswered calls is helpful. If there are multiple forms of access, such as through a PBX and a LAN, each form must be incorporated into the analysis. Some software packages designed for public access on different terminals require the terminal to be identified as part of the login process, which is useful for determining the technical capability (*e.g.*, downloading, graphics, *etc.*) of the audience, and this information should be collected. The last item under traffic lists how the session ended; this information is useful for assessing the need for exit points on screens and revealing problems in the communications links.

Statistics for assessing how the system is being used involve recording access at the frame level. Information is captured detailing the frequency of access, what frame is accessed, the path taken to and from a particular frame, and how much time is spent viewing a frame. The path information is helpful for establishing routing options and analyzing behavior patterns. In systems that charge a fee for particular frames, information is needed about which frames are selling and which are not. And information regarding the amount of user input and output measured in bytes is useful for determining communication and storage parameters and determining how many users can be supported by a particular system.

11.6. SUMMARY

Two segments of THE WINDOW are developed to illustrate the process involved in producing information for an electronic medium. The preproduction steps are outlined beginning with development of the concept. This is an important step because it helps establish costs and schedules. This is followed by flowcharts which provide a structure for development and a guide for writing whatever computer programs may be needed. Once these two steps have been completed, the script can be developed; it will serve as the guide for producing the content. The script includes the text, branching instructions, color choices, and other attributes that will affect what the viewer sees. Special software called an authoring program can help bridge the process of moving from script development to production.

Once all of these preproduction steps have been completed, the work of producing the contents and if necessary, the computer programs can start. The frame development process generates two types of frames: frames that are seen and those that control frames. A number of tips were suggested for developing the content.

The process does not end when the frames have been produced. It is important to hold a formal review and edit the contents to ensure a product of high quality. Among the more mundane follow-up chores is documenting what has occurred. The process of planning and producing the infobase never ends. Auditing and rejuvenating the contents are essential to keep the user's interest.

REFERENCES

A. F. Alber, *Videotex/Teletext: Principles and Practices*, McGraw-Hill, New York (1985).

N. J. Bamberg, *Videotex Production: A Case Study*, report no. 2 of the Electronic Text Report Series, an Annenberg/CPB Report, San Diego State University, CA (1984).

R. E. Bergman and T. V. Moore, *Managing Interactive Video/Multimedia Projects*, Educational Technology Publications, Englewood Cliffs, NJ (1990).

M. Eagle, On Opening the Number One Bottleneck in Online Service—Data Entry, *Proceedings of the Eighth National Online Meeting*, Learned Information, Medford, NJ (1987).

Frame by Frame, AT&T Information Systems, (1985), pp. 1–9.

K. A. Frenkel, Peeking behind the Interface, *Publish* **6**(7) (1991), pp. 58–64, 66.

G. Gery, *Making CBT Happen*, Weingarten Publications, Boston, MA (1987).

E. Holsinger, A Show with Character, *Publish* **5**(9) (1990), pp. 97–102.

M. Mann, Authoring Software Is Making Strides toward Maturity, *PC Week* **9**(16) (1992), pp. 95, 98.

G. Nugent, P. J. Peters and L. Rockwell, *Designing and Producing Videotex Instruction: A Producer's Handbook*, Station KUON-TV and Division of Continuing Studies, University of Nebraska–Lincoln (1983).

J. Phillipo, An Educator's Guide to Interfaces and Authoring Systems, *Electronic Learning* **8**(4) (1989), pp. 42, 44–45.

S. Reisman, Developing Multimedia Applications, *IEEE Computer Graphics & Applications* **11**(4) (1991), pp. 52–57.

T. Yager, Practical Desktop Video, *Byte* **17**(4) (1992), pp. 106–110, 112, 114.

T. Yager, Raw Material, *Byte* **17**(5) (1992), pp. 129–130, 132, 134, 136, 138.

Chapter 12

Survey of Enhanced Information Systems

A survey of companies thought to be using enhanced systems was undertaken in late 1990. A rich body of information was collected about a variety of projects. It is evident from the survey results that many companies are employing these systems for operational and strategic purposes, and they are planning even more elaborate projects for the future. Most of the information collected from the survey is reported in Chapter 12.

The three chapters that follow describe in detail 10 projects that were surveyed. Site visits were made to the unit responsible for each of these 10 projects, and the individuals responsible for the project were interviewed. The 10 cases were updated in June 1992. These visits added considerable depth and breath to the information collected from the survey.

12.1. MAIL SURVEY

A set of research questions was developed that the survey was designed to answer; a copy of the survey used appears in the appendices. The purpose of the survey was fivefold. First and foremost, the survey was to provide information about how firms are using enhanced systems. Much of what is written about videotex and multimedia systems originates from hardware and software vendors of these systems. Consequently the information is often self-serving and seldom reflects the using organization's views. Second the survey was to discover why organizations invest the time and effort in developing enhanced systems since the technology for these systems is complicated and expensive, what are the outcomes expected? Third the survey was to uncover what technology was being used. There are many ways of mixing and matching software and hardware options. Some firms are just beginning to explore the options available while others have advanced along the learning curve.

Little is ever heard about the process followed or the experiences encountered in designing, implementing, and operating enhanced systems, yet these topics are central to the existence of a successful system. Organizations are usually reluctant to reveal difficulties they have encountered doing something. The fourth purpose was to discuss some of the trials and tribulations of using enhanced systems. Hopefully information provided on the topic will prove invaluable to the reader. The final purpose was to gaze into a crystal ball to discover from the perspective of individuals involved how enhanced systems are expected to evolve and their likely impact on the organization over the next five years.

12.1.1. Survey Instrument

The survey contained the following definitions:

Enhanced system: An easy-to-use, interactive computer system that *possesses some or all* of the following features: text, graphics, color, picture-quality displays, audio, and video
Project: A work effort that you can think of as an entity
Public access kiosk: An enhanced system in a public location, such as a shopping mall, hotel lobby, or conference facility, to be used by any passerby

Cover letters accompanying each survey provided additional details about the nature of the study, so no one would confuse a traditional data-processing system with an enhanced information system. There were three different letters that varied in content slightly depending on the target. Respondents were asked to follow these three guidelines:

Please select a specific project from anywhere in your organization. While computers may be used throughout your organization, *select just one project* that contains portions or all of the elements of an enhanced system as defined above.
The project may reside on a computer system operated by your organization or operated by another party, *e.g.*, an on-line commercial service provider. Users may or may not be employees.
If a question applies and you are not absolutely sure of the answer, please make an educated guess.

There were 21 questions, many of which had multiple parts or were open ended. Several questions were designed to check for internal consistency among answers, and some respondents were contacted to clarify points made in the survey. Participants were asked to include any narrative material, such as internal documents or brochures, that described the project in greater detail than was possible on the survey instrument. Many of the respondents supplied such material.

12.1.2. Survey Targets

Survey targets were identified from articles appearing in the print media, from referrals by vendors of equipment, and from the author's personal knowledge. Many of the targets were contacted by telephone, and an individual familiar with the project was asked to participate. On several occasions, people declined to participate because they felt that the success of the project provided a competitive advantage and they did not wish to publicize the outcome.

Individuals in 85 organizations were mailed a survey, and 50 usable surveys were received for a response rate of 59%. In addition three companies offered to distribute several copies to clients, and one usable survey was received from this effort, yielding a total of 51 usable surveys.

Because of the nonrandom manner in which organizations were contacted, it is important not to draw erroneous conclusions about the prevalence of enhanced systems in business today. For the most part, participating organizations represent forward-thinking and often venturesome enterprises. These organizations provide ample evidence of how enhanced systems are used, but they do not necessarily represent a balanced cross section of corporate America.

12.1.3. Respondents

Table 12.1 illustrates the Standard Industrial Classification (SIC) divisions to which the respondents belonged. The heaviest concentration of activity was in manufacturing and services. In each of these two divisions, enhanced systems were used in a variety of ways. In the manufacturing division, applications ranged from networked kiosks used to improve internal communications and thus favorably affect the work environment to systems that promoted the purchase of goods. In the service sector, systems were used for such purposes as establishing a closer bond with customers and disseminating information more rapidly and accurately to customers.

The concentration of projects in a handful of SIC codes may be partly due to the nonrandom way of selecting target companies; however there are probably many more opportunities to employ enhanced systems in industries where computer technology is pervasive or there is extensive contact with the public. Agriculture, mining, and construction are probably not represented for a combination of these reasons, and public administration is probably not represented because of financial considerations; that is, the systems under study often represent a significant investment in financial and human resources, which are always in short supply in the public sector. It should be pointed out, however, that there are

Table 12.1. Respondents Grouped According
to Standard Industrial Classification Codes

Division	Industry	Number
A	Agriculture	0
B	Mining	0
C	Construction	0
D	Manufacturing	17
E	Transportation, communications, utilities	7
F	Wholesale trade	1
G	Retail trade	4
H	Finance, insurance, real estate	5
I	Services	17
J	Public administration	0
K	Nonclassifiable	0
Total		51

governmental units experimenting with enhanced systems, but those who were sent a survey chose not to participate.

Forty-four of the projects were located in the United States, six in Canada, and one in Australia. Canadian organizations have extensive experience with enhanced systems due to government sponsorship of the Telidon program from the late 1970s to the mid-1980s. During this period, many of the technologies being studied here were financially and morally supported by the government to strengthen the Canadian communications industry and develop new employment opportunities. The survey from Australia was from a client of a company that had mailed several surveys for the author.

12.1.4. Intended Audience

Thirty-eight of the projects were considered permanent, and 13 were described as trials; however only 7 of the 13 trials had scheduled termination dates, which indicates that the life span for some of these trials may depend on their perceived success.

Intended users were almost evenly scattered among employees, customers, and the general public, as shown in Table 12.2. The responses add up to more than 51 because many of the projects have more than one audience. In the other category were groups composed of retirees, students, the U.S. government, art specialists, travelers, people in agriculture, and teachers. In all but one case, projects including these others had as their principle audience either employees, customers, or the general public and were targeting a subset of their principle audience. In a sense, the systems were being used as the electronic equivalent of direct mail.

Table 12.2. Intended Users of the System

Intended Users	Number of Projects
Employees	30
Customers	30
General public	25
Other	9

12.1.5. Applications

A rich variety of applications was revealed by the survey. For purposes of analysis, the applications were divided into the following four categories: information distribution, selling information, selling a product other than information, and training and education (see Table 12.3). The most frequently occurring application was information distribution; selling goods and services electronically closely followed by training and education were second and third, and selling information lagged far behind.

12.1.5.1. Distributing Information

This medium clearly supports information distribution both inside and outside the company. This application occurred almost as often as all the others combined because it is easier to accomplish than the others. Systems for information distribution are either passive or active; 15 were passive.

A passive information system informs the viewer but does not elicit action. Interactivity enables more information to be obtained, but it does not enable an action to be initiated; the user will have to resort to the telephone or some other channel for action. One example of a passive system was a companywide information system used to access statements of company policies and practices, training catalogs, newsletters, and even credit union rates. Another example was a university information system providing 24-hour access to need-to-know information for educational enrichment, life in the campus community, and help with certain courses. A third passive system was operated by an electrical utility for

Table 12.3. Applications

Application	Frequency
Information distribution	25
Selling information	5
Selling a product other than information	11
Training and education	10

displaying and retrieving energy efficiency and energy management information. A fourth system was used to promote in-store products and hopefully increase sales.

There were 10 information distribution systems classified as active. Two supported management decision making. In each case, managers were able to access information and use integrated decision-making tools for analysis and action. Two other systems provided benefits information to employees. Employees were able to read about particular benefits, also ask questions through the system, and modify their benefit package.

12.1.5.2. Selling Information

Only five organizations responded that were selling information; three of these were publishers, two of which were selling an electronic version of a paper-based product. For these two publishers, only a nominal cost had been involved modifying the paper-based product for electronic distribution. The third publisher added value to the paper-based product by selling software programs and databases; surveying readers on behalf of advertisers; and by establishing communication (E-mail) among readers, advertisers, and the editorial staff of the paper-based product.

Another organization was preparing to help researchers who visited its facility access pictorial information on-line. The information was also going to be sold on laser disc for off-site researchers. The last organization collected information from users that was then used to make a legal will.

12.1.5.3. Selling a Product Other Than Information

These systems were serving as the conduit for making a sale. In most instances, they provided complimentary information to facilitate the selling activity. The item was delivered by mail, the customer picked it up, or it was available where the user accessed the system. In most cases, the seller eliminated a middleman, or if the seller were the middleman, he/she extended the market area by using an enhanced system. In every case, the ease and convenience to the buyer were increased.

The projects can be divided into three categories: reservation systems, order taking where the customer is an industrial buyer, and order taking where the customer is a consumer. Reservation systems consisted of companies in three different segments of the travel industry. The first is a large company that owns and operates a customer reservation system. The system is PC based and has been operating for nearly six years. The second company is a travel agency that maintains its own mainframe. The agency allows overseas customers to log onto the host to make travel reservations. One feature of the service is immediate confirmation of travel plans. The third project is a reservation system for renting cars. The company offers its rentals through a commercial videotex network operated by another company.

Order taking for industrial buying involved two organizations that supported dealer networks. The first operates internationally and provides parts and consumables to manufacturer's dealerships. The second company has been operating a regional parts and supplies distributorship using videotex since 1983. Its primary objectives for offering the service are to sell products, improve customer relations, and increase market share. It has had considerable success over the years and believes the system has helped retain customers. Both of these companies permit their clients to connect directly with their system to place orders and check order status, among other things.

Order taking for consumer buying involved six companies in retailing, brokerage services, and groceries. One of the companies in retailing operated a 500-plus terminal network that provided customers with a video catalog of the entire product line. Terminals were located in kiosks at each store and allowed easy ordering via touch screen, next-day shipment, free delivery, and satisfaction guaranteed. In addition to serving as a selling vehicle, the system was used to train store employees. The elaborate network of terminals is supported by two help desks. One monitors terminal status, and a second tracks and troubleshoots orders. The second company is one of the largest retailers in the country; it is conducting a trial using kiosks for gift merchandise. Units are placed in such high-traffic locations as train stations and lobbies of large office buildings for customers to order gifts and flowers.

Two companies were operating brokerage services that targeted the home market. Each of these firms offers a broad range of financial management services that enable customers to do such things as buy and sell stock, check quotes, manage portfolios, monitor company and other news developments, engage in various forms of research, and check account information. One of the firms offers nationwide customer service consisting of hot-line telephone support and user groups.

The last two companies serving the home market sold groceries on-line. Both operate in a similar fashion, although the level of service varies significantly. The customer accesses the store via a PC or a special terminal. Menus, random selection, name, generic categories, product number, personal order selections, and life-style-based ordering options lead the customer to desired items. Nutritional information and ingredient statements offer an added value to those on special diets and the health conscious. One of the systems even accepts coupons. At the end of the session, the customer can comment or provide delivery instructions. After the order is received it is sequenced according to the warehouse. The order is then picked, the actual weights of meat and produce items are entered, the invoice is printed and the order is delivered to the customer. Although a service like this may seem frivolous, it is a tremendous boon to the handicapped, the place bound, the busy, and those who simply hate to go grocery shopping.

12.1.5.4. Training and Education

Ten organizations that responded had systems whose primary purpose was training and education. Three were located in museums (natural science, science and industry, and sports), three were directed specifically at the educational market, and four were in companies.

Two of the museum systems incorporated text, color, audio, graphics, picture-quality images, and video; and the third included everything except audio and video. All the museum systems were located near exhibits and were intended to provide additional information about exhibits. The interface to all three was a touch screen and two were networked. Two of the systems reported usage figures, and the numbers were very impressive.

Two of the three educational systems were created for grade schoolers. One was a prototype using digital video interactive technology, intended for educational programs in museums. The system incorporated a number of innovative features, including elements of virtual reality, so that when viewers moved a joy stick to indicate the direction they wished to walk, the video scene changed accordingly. The second system is marketed to public schools by a publisher. It consists of special software and a network that enables students to send and receive data, receive maps and charts based on data submitted, and send electronic mail to other students on the network. The third system, located in a university, is essentially a self-pacing tutorial for students. Student machines are connected through a LAN, and the system instructs and assesses learning. Text, color, picture-quality images, and video are used.

All four corporate systems contained text, color, audio, graphics, picture-quality images, and video. Three of the systems were PC based, and the interface was by touch screen. The fourth was housed on a mainframe with access gained via a PC and usually a WAN. Two of the systems were in electronics companies; one was in a large insurance firm; and the fourth was in the manufacturer's facility, where it was used to train employees and demonstrate products to potential buyers.

12.2. ORIENTATION AND OBJECTIVES

The technology is well developed and components are readily available to create the types of systems just described. The knowledge needed to design and build enhanced systems properly, and the management farsighted enough to see that it gets done are less easily found.

12.2.1. Form and Focus

In Chapter 1 the information investment was classified according to its form and its focus, where form refers to whether the investment is used for operational or strategic purposes. Operational investments enable an organization to carry on the day-to-day activities incidental to running the business. Strategic investments shape or support the organization's competitive strategy, and they have a long term impact on the organization's competitive health and vitality. Focus refers to whether the information investment is directed inwardly or outwardly. An inwardly focused investment impacts the organization and its employees and might result in improved communications or a positive change in the organizational climate. An externally directed focus might lead to improved customer service or product differentiation and new business opportunities.

The 51 projects are classified in Table 12.4 according to the nature of the information investment. Sometimes the task of classifying the application is not easy for two reasons. Some applications produce unexpected results; the SABRE system described in Chapter 1 is a perfect illustration. It was originally focused on automating American Airlines' reservation system, and it was operational in nature; but within a short time, its strategic value was realized, and it became a competitive weapon in the marketplace. Since information systems are dynamic and evolve over time, an organization may design a system to benefit employees but discover how to serve customers in new and innovative ways as well. EAASY SABRE, the PC world's incarnation of the SABRE system, is just one example. Likewise a system that is innovative and strategic today may become operational tomorrow as it is emulated and becomes the *modus operandi* of the industry.

Most of the projects studied were strategic and external. Not too much should be read into the imbalance among categories in Table 12.4, since the sampling was not random, and organizations surveyed were often those that were written about in the literature and easy to identify as potential targets. The discussion of objectives and their attainment in the next section reveals the merits of investing in information regardless of the investment's form or focus.

12.2.2. Objectives

A great deal of information was collected about the objectives behind the information investment. Each respondent was asked to check on a list of possible objectives all those that applied to the project being described. Then the respondent was asked to rank from highest to lowest all the objectives checked and finally

Table 12.4.   Form and Focus of Projects

Form	Focus	
	Internal	External
Operational	6	6
Strategic	9	30

to check each objective that was accomplished. There was ample space to enter objectives not contained on the list.

Organizations have a variety of reasons for initiating these systems, and most projects are implemented with multiple objectives in mind. The average number of objectives checked for the 51 projects was 5.7. Only one objective was checked for two projects, and one project had 12 objectives checked; Table 12.5 lists the objectives and their frequency. The two most frequently checked objectives were to distribute information and improve accuracy of information. Clearly the electronic medium is viewed by many organizations as a way of improving traditional forms of communication, such as the print media. Many of the most frequently checked objectives are geared toward saving time and money or improving the process or quality of communication. Such objectives are often operational in nature.

As a group, strategic objectives such as increasing market share, improving the work climate, differentiating products or services, and entering new businesses, are not cited so often as operational objectives. This may be because it is easier to conceptualize operational objectives, such as cost savings, since they can be quantified, or perhaps those answering the survey think more in terms of operational than strategic issues. In any event, operational and strategic objectives were commingled by the respondents, indicating that systems of this type are viewed as having both operational and strategic value.

Space was provided to write objectives not included in the check list; Figure 12.1 shows an edited list of some of the responses. These objectives and those in Table 12.5 show the variety of expectations that exist and can be used to help justify

Table 12.5. Frequency with Which
Each Objective Was Checked

Objective	Frequency
Improve communications	31
Reduce cost of communication	15
Improve accuracy of information	33
Distribute information	35
Distribute information sooner	31
Sell products or services	24
Improve public, customer, or stockholder relations	20
Increase market share	21
Reduce operating costs	16
Improve the organizations's work climate	12
Differentiate products or services from competitor's	22
Enter new business opportunities	16
Other	19

Education and motivation

Improve productivity

Reduce paper

MIS awareness of multimedia capabilities

Provide quality, consistent training nationwide while decreasing training time

Develop new channels of distribution

Provide an alternative format for student evaluation

Children's science education

Provide product information

Position company for future communications

Figure 12.1.   A partial list of the respondents' objectives.

videotex and multimedia projects. The extent to which these expectations were realized is discussed later.

Looking at just how often an objective is checked belies its importance; conceivably everyone could have the same objective without considering it very important. To understand how important each objective was, respondents were asked to rank the objectives they checked. The order of the first seven rankings is shown in Table 12.6.

The desire to improve communications was ranked first most often among those projects for which it was an objective. In contrast distribute information

Table 12.6.   Ranking Order for Each Objective*

Objective	Rank						
	1	2	3	4	5	6	7
Improve communications	1	5	3	2	4	6	0
Reduce cost of communication	0	1	2	3	3	3	3
Improve accuracy of information	1	1	1	4	2	3	3
Distribute information	1	3	3	2	4	4	5
Distribute information sooner	2	2	4	3	1	5	5
Sell products or services	1	5	4	2	5	3	0
Improve public, customer, or stockholder relations	2	2	3	5	4	1	0
Increase market share	3	2	1	4	3	4	3
Reduce operating costs	2	0	2	1	2	3	3
Improve the work climate of the organization	2	1	1	0	0	1	1
Differentiate products or services from competitor	2	3	1	2	2	4	4
Enter new business opportunities	1	2	3	3	3	0	0

*Calculated by counting how many times each rank occurred for an objective and then assigning 1 to the rank with the largest count, 2 to the second largest, and so on; 0 indicates nothing was recorded.

sooner, which has a 1 in the rank fifth column was ranked fifth among those projects for which it was an objective. Using this approach, objectives considered first in importance include improving communications, improving accuracy of information, distributing information, selling products or services, and entering new business.

12.2.3. Attaining Objectives and Benefits

Respondents were asked to report whether each objective checked was or is being accomplished; overall 73% of the objectives checked were or are being accomplished. Table 12.7 shows the degree to which each objective was being satisfied. Improving communications, distributing information sooner, and improving the accuracy of information were the three objectives most frequently satisfied. These are all fairly standard outcomes of computerizing an application, so there is no surprise that they were achieved.

The objectives least frequently satisfied were improving the work climate, reducing operating costs, and increasing market share. Since changing the work climate and increasing market share are much more complex issues than, for example, improving communication, it is not surprising that they are more difficult to achieve. The impact on operating costs may reflect the expense of implementing these systems.

In general videotex and multimedia systems were successful when measured against their objectives. Recall that 13 of the projects were reported to be trials rather than permanent systems. Since some of the trials had just gotten under way, it is possibly too early to determine their outcome. When these 13 trial projects are removed from the database (not shown), all but one of the objectives

Table 12.7.   Degree to Which Objectives Were Accomplished

Objective	Frequency Checked*	Frequency Accomplished	Percentage
Improve communications	31	26	84
Reduce cost of communication	15	10	67
Improve accuracy of information	32	26	81
Distribute information	34	26	76
Distribute information sooner	31	26	84
Sell products or services	24	17	71
Improve public, customer, or stockholder relations	20	15	75
Increase market share	21	9	43
Reduce operating costs	15	8	53
Improve the work climate of the organization	12	7	58
Differentiate products or services from competitor	23	17	74
Enter new business opportunities	16	12	75

*The slight difference between this column and the statistics in Table 12.5 is due to several cases that were discarded because accomplishment was not reported.

has a much higher level of accomplishment and the overall level of accomplishment for all the objectives jumps to 80%.

In addition to the objectives that were satisfied, respondents were asked to specifically indicate benefits that resulted from implementing the project. Over three dozen specific benefits were cited; an abbreviated list is shown in Figure 12.2.

Since respondents stated the benefits in their own words, it was not practical to try to quantify their responses, but there were several recurring themes. Several organizations were involved in these projects to gain experience with the medium and to experiment. Others were using enhanced systems to test ideas and concepts about new ways of conducting business. Several were aggressively pursuing new approaches to promotion and marketing, and still others were using enhanced systems to reach customers and employees more effectively. It is often difficult to separate illusion from reality when new applications of technology are being implemented, but in this case, it is clear that many organizations reaped tangible benefits and were excited about the potential for doing more with the technology.

12.3. TECHNOLOGY

There is a temptation to become enraptured with the technology involved in enhanced systems. Hype over reason and sizzle over substance are the consequences when this occurs. While technology is the enabling factor, it is not the

---

Increased understanding of the role of multimedia in education and learning

Showcase particular products to the consumer

Gained leadership position in videotex technology

Increased confidence in computer usage and on-line information by user

Improved public relations and more efficient operations

Educated salespeople at point of sale about our products, so they were more knowledgeable and had more confidence selling our products

Increased loyalty and retention of current customers

Viable test of alternative channel of distribution

Retained existing accounts

Revenue stream

Electronic marketing is considered to be the wave of the future, gives us experience in the field

We hope to deepen the cognitive impact of the museum visit and provide interpretive material appropriate for a wide range of museum visitors.

Figure 12.2. An abbreviated list of benefits cited.

raison d' être: Without technology, the systems cannot exist, but with it, there is no assurance that the systems are useful. The clearest manifestation of the technology is how the material is presented by the system.

12.3.1. Presentation

Table 12.8 illustrates the forms of presentation found in the systems surveyed. Clearly the dawn of multimedia computers has not obviated the need for text: All 51 projects depended on it; however this text is unlike that found in a traditional data-processing application. It is not simply names and numbers, and the style is usually journalistic; the text is often accompanied by other presentation forms. In fact there were only six projects in which text was not combined with something else. All but one of the text-only projects were old and well established, and several are in the process of adding complimentary forms of presentation in the near future. At the time of the survey, the average age of the six projects was 4 years. In the life of videotex and multimedia systems, that is old!

Color was nearly as pervasive as text and occurred in 42 projects. It was not present in nine projects, six of which were the text-only systems previously discussed. The three remaining projects were a home grocery shopping service and two systems implemented on CompuServe, whose basic service at the time was not color oriented.

Audio was incorporated into 19 projects; 16 of these also included video. On the site trips, it was discovered that audio is the bugaboo of multiple media presentations. Perhaps even more than video, it is the last media form incorporated, although there was no particular reason uncovered. It is probably a combination of the difficulty of dealing with audio and the absence of radio-quality voices among the content producers.

Capturing and synchronizing audio with screen displays are a companion rather than singular activity. Most of the time when other media forms are visible on the screen, they occur by themselves as a frame of text, a graphic, or a short video sequence. Even when different media forms appear together, such as text

Table 12.8. Forms of Presentation

Presentation	Frequency
Text	51
Color	42
Audio	19
Graphics	41
Picture-quality images	25
Video	19
Other	6

with a graph in a window, each media form is able to stand by itself as an entity conveying information; however audio never appears by itself—if it did, a tape recorder rather than a computer would suffice. Audio is superimposed over, rather than used in conjunction with, the other media forms. Audio is also beyond the technical expertise of most systems personnel. Although graphics, images, and even video are much different than pure text, most systems people can work with them in their digitized form. Audio is a less familiar medium and requires special expertise, especially if it involves music and other forms of nonvocal communication.

Finally recorded audio still retains the flaws of its human creator. Although practically everyone can create the spoken word, only a few people have voices with the resonate qualities that we like to hear. Consequently when audio exists, it is often created as part of the video production—either because technical expertise resides with the video creation staff or because audio is captured live while shooting a video.

Graphics is almost as common as color (41 times) and in two instances occurred in projects that did not incorporate color. Graphics runs the gamut from simple line drawings to fairly detailed renderings. The proliferation of business software incorporating such graphics as spreadsheets, databases, and GUIs similar to *Windows* indicates graphics is almost a *sine qua non* for computer-based applications.

Picture-quality images were found in 25 projects; however what constitutes picture quality is in the mind of the beholder. The gray area between graphics and picture-quality images is difficult to resolve; from the author's viewpoint, the bottom end of picture-quality imagery is a display created with NAPLPS, the videotex presentation standard defined in the glossary. Because of bandwidth requirements for transmitting high-quality imagery, there appeared to be no systems encountered that maintained high-quality imagery through a computer network, although such systems exist. Most of the systems were either stand-alones and workstations or kiosks. And even though the kiosks might be networked, they did not necessarily transmit the image portion of the content on the network.

Nineteen projects encapsulated video. Video and audio constitute the most sophisticated computer media form. Video requires trained personnel to produce studio-quality content and special hardware and software to display it. The spreading popularity of CD-ROMs, and to a much lesser extent, laser discs, alleviates the problem of hardware and software availability, but the personnel issue remains. The DVI technology was cited in three surveys, but it may have been used in others. The DVI sites were developing applications for a museum, a training work station, and a trade show kiosk.

Six organizations used a media form other than those previously described. Three provide paper output for the user to carry away. One project distributes software electronically; another has a text-to-speech option, which is slightly different than what is thought of as stored audio, and so this form was recorded as other; in the sixth project, a credit card access unit is attached to the system.

12.3.2. System Host and User Interface

There are a variety of ways of segmenting hardware for analysis. In Figure 12.3 host systems are separated into public access kiosks and computer systems. A kiosk was defined in the survey as an enhanced system in a public location used by anyone. Kiosk systems can be further divided into those that are networked and those that stand alone.

Computer systems were divided into those operated in-house and public systems. In-house systems were divided into those dedicated to supporting just the enhanced system and those that shared the host with other applications. Public systems may be subdivided into commercial systems, such as CompuServe and Prodigy, and community systems. Community systems, such as Heartland Free-net, are noncommercial and free; they contain community-oriented information and services.

Although the surveys provided considerable detailed information about the host, in a few instances an educated guess was made about how the system fit into this classification scheme. The number of projects in each category are shown in

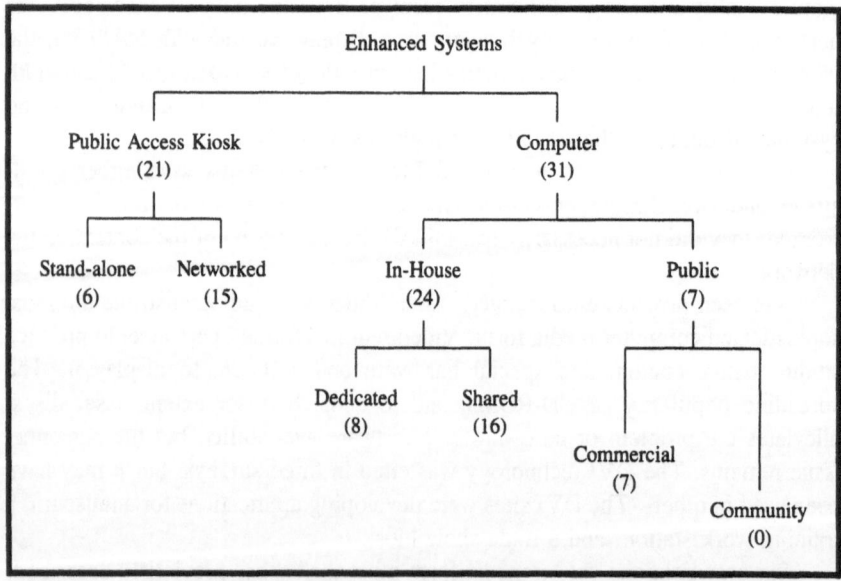

Figure 12.3. Classifying the system host. Total adds up to 52 because one organization offered two versions of its service.

parenthesis in Figure 12.3. This classification scheme illustrates the diversity in system implementation, which was one of the goals of the study.

Twenty-five of the host platforms were microcomputers. Twelve systems were built on mini and mid-sized computers, and 12 systems were implemented on mainframes; two surveys did not indicate the host system. The unusually large number of microcomputers was due to the number of kiosks: Seventeen kiosks were built on a microcomputer, three used a minicomputer, and one a mainframe. The mini and mainframe systems were essentially terminals of varying capability that depended extensively on support from the larger host.

A keyboard was the primary means of accessing the contents followed by a touch screen; 30 systems depended primarily on a keyboard, 15 on a touch screen, and one a joy stick. There was no response on five surveys. Twelve of the touch screens were installed on kiosks, which typically have several presentation forms; 10 of the 16 projects that incorporate all six forms of presentation also depended on touch-screen technology. Clearly touch screens rather than keyboards are the preferred access instrument when multiple media forms are used, at least in the minds' of the designers of these systems.

12.3.3. Software

Twenty-seven projects used software obtained from a vendor, 15 developed software in-house, and six projects used a combination of these two sources. Digital Equipment Corporation's VAX VTX(R) software package was cited most frequently and was used on three projects. This software supports text, color, and some graphics capability.

Software is a major hurdle for many organizations. The unique hardware that is assembled for videotex and multimedia projects, especially those projects containing sophisticated multimedia displays, defy off-the-shelf software solutions. Microsoft's release of Windows Extensions in July 1991 was a first step in establishing an industry standard that users and other software developers could build on. However for the foreseeable future, organizations will be struggling with the choice of creating their own software or forcing their application to fit into a package from a growing selection of candidates.

12.3.4. Networking and Communications

Thirty-nine of the systems were networked to other computers, but most of the links were low-speed telephone lines. Telephony was the only form of communications in 24 projects, and it was coupled with other forms of communications in six projects. Other links consisted of LANs (nine), WANs (six), and unspecified (three).

The presentation form requiring the most bandwidth is video. An examination of the 10 projects with video that were networked revealed only one had the bandwidth necessary to transport video, and this was in a company whose business is selling communications services. Three others used low-speed networks, and the balance of the networked projects with video used telephone lines. It is clear that with one exception all networked projects with video were using CD-ROMS or videodiscs. Although companies are rapidly moving into the more bit-intensive forms of presentation, such as video, they are doing it by integrating equipment with the work station rather than by transmitting the content. Given the relatively low cost of one or two dollars to copy a CD-ROM disc versus the major expense of installing a high-speed network, it is likely that this approach will continue in the foreseeable future.

Table 12.9 shows the communication speeds, network configurations, and protocols reported. Many of the respondents were unsure of the technical specifications of the communications systems, so these figures suggest what is being done and are not exact.

12.4. DESIGN AND IMPLEMENTATION

Approximately 20% of the survey consisted of questions about designing and maintaining the contents of the project being described, which was assumed to be a significant impediment for some organizations. Furthermore it was assumed that as the presentation form became more sophisticated—that is, moved from text to video—impediments would be greater. This was borne out by survey results, although a number of the organizations studied are developing expertise in the area of creation and maintenance.

12.4.1. Initial System Design

The initial design was more than twice as likely to be done by company employees as by external sources. In 29 projects, company employees were the sole design agent, while in 14 projects, external parties whose business was

Table 12.9. Number of Reported Network Speeds, Configurations, and Protocols

Speeds		Configurations		Protocols	
0.300 Kbits/sec	7	Star	15	CSMA/CD	4
1.200 Kbits/sec	17	Ring	1	Token	0
2.400 Kbits/sec	17	Bus	3	Other	13
9.600 Kbits/sec	8	Mesh	2		
Other	7	Other	8		

Person responsible for the initial system design	
Company employees	29
External sources	14
Company and external	7
No response	1
Total	51

Department responsible for the initial design*	
Business unit	4
Corporate communications	3
Data processing	12
Marketing	1
Personnel	1
Training and development	3
Other	8
No response	4

*Includes organizations where employees and external sources both worked on the design.

Figure 12.4. Person and department responsible for the initial system design.

designing systems of this kind performed the function (see Figure 12.4). In seven projects company and external sources worked side by side.

Four of the organizations set up special business units to do the initial design. This occurred when the project produced a product for sale or was so unique that it did not fit into the normal operations of a unit in the company. Three projects were under the jurisdiction of the corporate communications unit, and 12 were developed under the auspices of some unit that traditionally falls within data processing's dominion. These units functioned under such names as advanced office systems, analytical support, information center, or information networks. The functional areas of marketing and personnel each had one project, and training and development three; eight projects were designed in units whose names defied categorization, and four respondents did not provide an answer. There was a tendency for the more advanced forms of presentation, such as video and audio, to be developed outside the organizations; however some organizations had sufficient in-house expertise to design the system.

12.4.2. Creating the Initial Contents

The most expensive and time-consuming part of developing these systems is acquiring or developing the source information. As discussed in Chapters 10 and 11, this involves considerable work prior to creating the displays themselves. No effort was made to collect information to compare how much labor was involved in such preproduction activities as identifying the need for the information, locating it, and designing how it would be presented and the actual production of the content. Some estimates in the literature show that 80% of the cost of operating these types of systems is tied to content creation, and a sizable portion of that is for preproduction activities.

Table 12.10 reports how the initial content is created. The column labeled Both lists the number of projects whose contents had been done in-house and externally. The last column, which is the sum of the three previous columns, shows the number of projects reporting how the initial contents had been created. The total varies slightly from Table 12.8 because several organizations reported that a particular form of presentation was being used but did not report how the initial contents for that form were prepared.

Information in the form of text was prepared in-house more often than any other type of presentation. A distinction was made between text presented as information in a journalistic sense and text produced as part of a report created by data processing; text or any other output produced as an ancillary part of a data-processing report were not included in the findings reported in this book.

The level of comfort organizations have with text is reflected in Table 12.10, which shows that text is more than three times as likely to be produced in-house. Three factors contribute to in-house text production: (1) Some of the text is already available in print publications and requires limited or no editing to be distributed electronically; in fact in some instances the print production sequence involved producing the information in an electronic medium on, for example, a word processor. (2) The means of transforming the text is readily available, it may involve rekeying or scanning with optical character recognition equipment. Either of these approaches involves very little training and expense. (3) The expertise for producing written text usually resides within the organization. Even in organizations where text was produced externally, research indicated that the basis for the content was collected internally and often proofread and approved in-house early in the preproduction process.

On the continuum of difficulty, text is the simplest type of content to produce, followed by graphics and picture-quality images. Audio and video are difficult to separate because audio is often captured as part of the video process and seldom

Table 12.10. Creating the Initial Contents Displayed on the Screen

Presentation	In-House	External	Both	Total (n)
Text	36	10	5	51
Audio	8	9	1	18
Graphics	19	17	4	40
Picture-quality images	10	14	1	25
Video	5	11	2	18

created in isolation. If audio involves a separate process consisting of special sound effects, computer-modified music, or special voice overlays, it requires technical expertise. The statistics in Table 12.10 reflect this continuum: Graphics are as likely to be initially created internally as produced externally. Graphic images in the organizations studied included simple line drawings, scanned images, and detailed drawings and maps.

Picture-quality images are normally produced with a video-capture board from a photograph, digitally captured or enhanced image, or a video still frame. Photographic-quality images are available on CD-ROMS much as clip art is available on floppy disks for desktop publishing. Most likely some of the organizations that reported creating picture-quality image content have actually been using images purchased on CD-ROMs. Creating images requires highly specialized equipment and the operators require more training than is necessary for simply scanning an image. Some organizations reported using firms that specialize in creating images.

Audio and video are often created outside the firm. The statistics for audio production are slanted for already stated reasons. It is extremely unlikely that any of the organizations surveyed were equipped to create audio tracks, although many were capable of recording voice and music. Several of the firms that reported producing their own video were very large and had personnel and physical facilities for making training and promotional videos. These in-house resources were used to create a video tape that was then sent to a company that created the videodisc or CD-ROM used in the system.

12.4.3. Maintaining the Contents

The Achilles' heel of videotex and multimedia systems is maintaining the contents. Some information ages quickly: Directories, financial databases, and news often have a life measured in minutes or at most hours; on the other hand, maps, the text of articles, and historical information may have a life span measured in years. Systems are often regarded as a way of delivering the latest information but operated as if the contents never had to be updated.

Creating content is usually the single largest expense encountered; however maintaining the content once created will often surpass the creation costs over time. A hypothetical illustration of this fact was presented in Chapter 9. The survey contained questions about who updated the contents and how frequently each presentation form was revised; Table 12.11 compares in-house and external maintenance.

A few more organizations updated the content in-house than created it in-house; however the shift from external to internal production was surprisingly small. It was hypothesized that once organizations had some experience working with the various presentation forms, they would do the work themselves. The fact

Table 12.11. Maintaining the Contents Displayed on the User's Screen

Presentation	In-House	External	Both	Total (n)
Text	34	9	8	51
Audio	9	10	0	19*
Graphics	20	16	4	40
Picture-quality images	12	12	1	25
Video	9	10	0	19*

*This figure differs form Table 12.10 because a respondent knew how the contents were maintained but not how they were created.

that the shift had not occurred in the period studied may indicate that a fairly steep learning curve is involved or resources for internal maintenance do not exist.

A predictable shift occurs in the updating frequency as the content becomes more complex. Text is updated most frequently and audio and video least frequently (see Table 12.12). The availability of equipment to produce text, the large number of individuals to operate the equipment, and the relative cost *vis-à-vis* mastering a new CD-ROM or videodisc make it practical at times to update text frequently. There is a widely held view that once something has been digitized, it can be maintained cheaply; this naive assumption is debunked by the facts. One of the most important steps in planning enhanced systems is developing an updating strategy consistent with the requirements of the application that is economically acceptable.

On the opposite end of the spectrum from frequent change is no change; as shown in Table 12.12, some infobases changed very little or not at all. Two projects that used all five forms of presentation were classified as trials, and so there was no plan to update the contents during the trial. A third project was a kiosk system in a museum whose contents were static unless the exhibits changed. No informa-

Table 12.12. Updating Frequency

	Text	Audio	Graphics	Picture-Quality Images	Video
Daily	14	1	3	1	—
Twice each week	1	—	—	—	—
Weekly	9	1	2	1	1
Twice each month	1	1	1	1	1
Monthly	6	1	5	3	1
Quaterly	1	1	1	1	—
Semiannually	3	2	4	2	2
Annually	8	5	13	6	3
Never	3	4	5	6	7

tion was available about the rationale for not changing the contents or changing infrequently.

When updating occurs, it is done selectively. Respondents were asked to estimate the percentage of the contents changed each time an update occurred. Changes for the different forms of presentation ranged from 1–100% for text, graphics, and images; for audio and video, it ranged from 2–100%. In each project where several forms of presentation were used, the percentage that changed varied; text was usually changed the most and the more complex forms of presentation the least within the same project. The average change each time an update occurred for the different forms of presentation is shown in Table 12.13. The amount of change is amazingly consistent for text, audio, graphics, and picture-quality images. The reason for the much larger percentage of change for video is probably related to the nature of video: When video is produced, it is edited, and the result is a sequence of frames that run as a unit for a few seconds or several minutes. Consequently, when any of the unit is modified it affects the continuity of the whole unit necessitating more extensive change than is necessary for other presentation forms that are produced and accessed in one-screen units.

12.5. PROBLEMS, PITFALLS, AND COMPLAINTS

The cliche is, "Experience is a hard teacher. She gives the test before she gives the lesson". There is little known and much to learn about designing and implementing enhanced systems. There is no body of literature to be researched nor people with work experience to be recruited. The one thing the organizations surveyed had in common was several individuals who dared to try something new and different. Most of these people learned as they went and acquired experience as they were tested. One segment of the survey dealt with these experiences. The question asked was, "What was the single biggest problem that had to be overcome at any point in the system's design, implementation, or operation?" Fifty out of 51 respondents had no difficulty coming up with an answer; their responses are grouped under the headings design, corporate politics, technical, and user centered (see Figure 12.5).

Table 12.13. Average Percentage of Contents Updated*

	Text	Audio	Graphics	Picture-Quality Images	Video
Average percentage	22	28	20	23	39
n =	39	9	23	12	6

*This average was calculated by adding the percentage updated regardless of whether the updates were made daily, weekly, and so forth, then dividing by n (number of cases).

Design
- "Major task was conceptual—how to use the technology to take advantage of the special capabilities to teach the subject."
- "Integrating the ——— concept into the entire company. It was so successful, it left many departments in our company behind. . . ."
- "The single biggest problem was nontechnical—providing the right information to the right locations (from the perspective of the user)."

Corporate Politics
- "Lack of understanding about this technology. Terminology got in the way so there was no administrative support in the end. . . ."
- "Overcoming exec's reluctance to use computer technology."
- ". . . not invented here theory."

Technical
- "Hardware integration and performance."
- "Existing software had to be rewritten and debugged to accommodate new system."
- "Speed . . ."

User Centered
- "Downtime equals negative employee reaction."
- ". . . Timing of changes and user notification is difficult."
- "Understanding the needs of the consumer and clarifying mystifying industry data."

Figure 12.5. A partial list of the single greatest problem cited.

Design issues ranged from system conceptualization to figuring out the proper content. The respondent describing a project in a museum stated that the major task was conceptual—how to take advantage of the special capabilities of the technology to communicate with patrons. At the other end of the spectrum were five organizations that toiled with content issues. Figuring out what questions the user would pose to the system was a major hurdle in two firms. In a slightly different context, several organizations encountered difficulty in pinning down the right product. One of the largest retailers in the country grappled with incorporating something unique in its electronic kiosk beyond the presentation of information about products. Another, a large manufacturer, had the doubly difficult task faced by any business that wishes to promote its products through an on-line videotex service, such as CompuServe or Prodigy. The message had to be designed to appeal to the subscriber profile of the electronic service being used as well as to the traditional customer, since failure to attract the former meant that the latter would never access the information.

Between these two ends of the spectrum were organizations faced with the daunting task of integrating the project with the way the company does business. Representative of this group is a retailer with one of the largest and most successful kiosk systems in the country; in fact the kiosk system was so successful that it left many departments in the company behind. The respondent explained that the entire company must be aware of the system, how it affects each area of the business, and what each area must do to support it successfully.

Internal politics are an inevitable part of every organization; in some cases, the opposition came from predictable quarters. One respondent in a training unit encountered resistance from the data-processing department. Hidden agendas and a not-invented here mentality delayed the project considerably. In other organizations, problems stemmed from a lack of understanding; in one case, terminology "got in the way," which eventually eroded administrative support. Compounding this was a lack of computer literacy among the information providers. In another project, it was difficult to obtain legal approval at the corporate level.

Selling the idea proved to be difficult in some firms. One respondent reported that convincing management that the company should be in the business of providing information electronically was a major hurdle, and this was a firm whose business was print publishing. In yet another case, the executive's reluctance to use computer technology had to be overcome.

The technical problems were almost equally divided among hardware, software, and communications issues. Due to the nature of the applications, some developers were working with prototype hardware that did not always function as it was supposed to. Even more difficult was the problem many organizations had integrating equipment from different vendors. In some of the more sophisticated systems, it was necessary to interconnect equipment that was not designed to work together. Existing software had to be rewritten and debugged to accommodate new applications. Software for special applications may be difficult to find and expensive, while applications running on commercial on-line services, such as Prodigy and CompuServe, are constrained by on-line service limitations.

Speed of and access to communications networks haunted several projects. These twin issues combined with the location of some systems and the rigorous use they received only compounded the problem.

Whether users caused some problems or were the unfortunate victims depends on perspective. One person felt that the greatest challenge faced in his project was "keeping the system so simple that anyone, computer literate or not, could use it." This lament was repeated in different forms; in another project, it became "serving customers with little or no experience with personal computers" and in another, the continuing challenge of coming up with an interface that ensured ease of use; and again, "developing an interface that communicates a wide range of user options while remaining accessible to naive users." The lesson is clear: Never underestimate the naiveté of the user and plan for it.

12.6. Evolution and Impact

The survey concluded by asking the following questions: What do you think will be the evolution of systems of the type that you described during the next five years in your organization? What will be the impact on your company, if your prediction in question 20 is accurate? There were few Luddites. There was a firm belief that information technology would be increasingly used and it would significantly impact the organization.

Some of the people involved in these projects have unwittingly become change agents in their organizations. They have embarked on a path that actively uses information technology to realize the goals of the firm in new ways. In this role, many of the respondents foresaw the projects evolving naturally as technology advanced as comments such as the following indicate:

"Additional applications/ use of CD-ROM technology/ faster data transmission/ more error control"

". . . the use of graphics and images on the system; more applications that integrate the use of electronic mail, electronic forms submittal, and approval and interfaces to databases"

"-enhanced graphics, touch-screen applications, access to on-line databases, *e.g.*, access to medical claims databases as well as payroll databases and savings plan databases; ability to access information from any PC in organization's office or employee's home"

"Expand and improve; quality will improve; enhancements to capture user information, provide transactional services, allow multiple platforms"

"Much more data will be placed on CD-ROM, with the addition of graphics and sound."

"Enhanced capabilities (*e.g.*, multimedia database, full-motion video, voice-annotated E-mail) will extend depth into organization."

"More services meeting more targeted needs."

"More detailed information will be maintained with better presentation methods and increased download and interactive functions; also more use of direct-response marketing techniques to enhance cost effectiveness of system."

And then there was the realist: "It all depends on money we receive!"

What do these predictions portend for the organization and its future? A few respondents prophesied very little: "In my opinion, the impact will be minimal"; "No impact at all." But these were minority opinions addressing special situations. Most people expected the applications made possible by these technologies to have a significant impact (see Figure 12.6).

The way the organization conducts its business will change. A museum director anticipated an increase in attendance and more enjoyable presentations. A retailer expected the quality of the work done by people on the floor to improve, resulting in increased sales. The network communications manager for a manufacturer foresaw a more powerful and productive environment for knowledge workers. And the manager for field training in a large insurance company envisioned a day when training becomes transparent and learning becomes part of the daily work routine. These respondents along with many others viewed

---

Operational
- "Shifting emphasis to electronic delivery of information and greater emphasis on inter-activity and customizing of information."
- "Improved quality and effectiveness. Better customer satisfaction, reduced time to market with new products."
- "Positive impact—faster, better communications as company expands globally."

Economic
- "Big bucks!"
- "Greater revenues, better competitive position."
- "Continued reduction of operating overhead through elimination of paper systems for information dissemination and information storage and retrieval."

Marketplace
- "Keep what we have—hopefully increase; preserve our company from chain competition."
- "We will have increased market share plus a stronger position in those markets with an increase in profits. In some respects, a more important result will be a much stronger awareness and acceptance of electronic delivery."
- "Acquisition of more customers."
- "Electronic marketing will begin to carve out its segment of media options available to marketers.

Figure 12.6. A partial list of what respondents felt the impact of these systems would be on their organization.

videotex and multimedia computer systems as facilitators enabling work to be carried out more efficiently and effectively.

Others felt these systems would have an economic impact either by increasing revenues or reducing costs. One respondent succinctly described the impact on the positive side of the ledger as, "big bucks!" and another as, "more profit". The manager of training for a large manufacturer of electrical products anticipated that the multimedia training system would reduce fixed and variable costs for a variety of reasons, presumably stemming from the impact of better trained employees. Still others felt that these systems would affect their competitive position in the marketplace: "preserve our company from chain competition" and "gain competitive advantage". In a nation and a world being reshaped by such factors as deregulation, global competition, instant information, and new geo-political alliances, information technology is recognized as another tool for competing successfully.

12.7. Summary

Many organizations are implementing videotex and multimedia computer systems. A survey of 51 of these organizations revealed that the intended users are relatively evenly divided among employees, customers, and the general public; the systems are most often used to distribute information; the focus is most likely to be external and designed for strategic purposes.

The most frequently cited objectives are to distribute information and improve the accuracy of information. The objectives most frequently accomplished are improving communications and distributing information sooner. Many benefits were reported in addition to the objectives, that were satisfied such as developing alternative channels of distribution and retaining existing customers.

Sixteen projects used all six forms of presentation; there were 21 public access kiosks, 15 of which were networked. The initial design was done by company employees on 29 projects, and creation of the initial contents was most likely to be done internally when the contents were exclusively text. The more sophisticated the form of presentation, the more likely it was done outside the organization; however once the contents had been created, it was just as likely for the more sophisticated contents to be updated internally as externally This probably reflects an increasing level of comfort experienced by the units responsible for the projects. Each time updating occurs, about a quarter of the contents are revised except for video, where 39% of the contents are updated each time changes are made.

Implementing videotex and multimedia systems is not without its difficulties. These problems may be related to design, internal politics, technical problems, and user-centered issues. Despite these difficulties, respondents were optimistic about the evolution of the systems and their impact on the organization.

Chapter 13

Cases:
Implementing Operational Systems

Visits were made to 10 sites to study in greater detail the projects described in the mail survey. Chapter 13 describes two projects that illustrate operational uses of the technology. Chapter 14 describes five projects that have been implemented for strategic purposes. Since so many organizations are implementing kiosks, the last chapter discusses three very successful kiosk systems. Figure 13.1 lists the organizations and projects involved.

Chapter 1 introduced the difference between operational and strategic systems. An operational system enables an organization to carry on its day-to-day activities more efficiently and thus more economically by providing better information and cost effective communications. These systems may also generate additional forms of revenue. Twelve of the projects surveyed by mail were categorized as operational, six of which had an internal focus and six an external focus. The applications involved were information distribution (five), selling information (one), selling a product other than information (two), and training and education (four); none of the 12 were considered trials. The site visits described in Chapter 13 are Clark Distribution Services, Inc., in Chicago and the National Library of Medicine in Bethesda, Maryland. Both projects have an external focus.

13.1. CLARK DISTRIBUTION SERVICES, INC.: CLARKNET II

Clark Distribution Services, Inc., is a business unit of the Clark Equipment Company. The parent company had revenues of $1.19 billion in 1991. Its primary business is designing, manufacturing, and selling forklift trucks and other material-handling equipment, construction and agricultural machinery, such as

```
┌─────────────────────────────────────────────────────────────────────────────┐
│  Operational Systems (Chapter 13)                                             │
│     Clark Distribution Services, Inc.        ClarkNet II                      │
│     National Library of Medicine             HARPP                            │
│                                                                               │
│  Strategic Systems (Chapter 14)                                               │
│     E. I. DuPont De Nemours                  Corporate Homes for Sale         │
│     Fort Worth Star Telegram                 STARTEXT                         │
│     Ford Motor Credit Company                Ford Credit—Prodigy             │
│     Ford: North American Automotive Sales    Ford Electronic Showroom—CompuServe │
│        Operations                                                             │
│     AMR Corporation                          EAASY SABRE                      │
│                                                                               │
│  Kiosk Systems (Chapter 15)                                                   │
│     Franklin Institute                       Unisystem                        │
│     General Motors                           The Town Crier                   │
│     Merck & Company                          Heartfelt Advice                 │
└─────────────────────────────────────────────────────────────────────────────┘
```

Figure 13.1.   Organizations and projects visited.

the Melroe series of front-end loaders sold under the Bobcat name, and power train components for industrial vehicles and trucks.

The distribution services company's mission is to provide a commercially competitive after-market parts distribution service to manufacturing companies. At the time this case was written, it served two corporate business units and a joint venture with Volvo AB, VME Americas, Inc., using an on-line order and information service, ClarkNet II.

13.1.1. Background

The precursor of the present system was ClarkNet I, begun in 1980. In the earlier system, the dealer called a host computer and was on-line for the duration of the transaction. The dealer's terminal was uniquely designed for the application by Texas Instruments, and the service was available only when the host was operating.

ClarkNet II was begun in 1985. A group of six dealers operating on ClarkNet I were asked to serve as guinea pigs for a streamlined and updated version of the service. Following a test that lasted almost nine months, the service went live. From its inception, it has been widely used in the company. In the period from January 1 to December 31, 1991, the system processed 300,000 orders and 2 million part numbers.

13.1.2. ClarkNet II

When the dealer logs onto the service, a main menu presents the following options: 1. Order Entry/Cancellation; 2. Inquiries & Order Status; 3. Administrative Messages; 4. Classifieds; 5. Trader Joe Update; 6. Billboard. Option 4, Classifieds, was not implemented at the time this case was written. The system incorporates a full range of order entry and cancellation options available 24 hours a day. There are four sets of order classes based on such factors as the urgency of the request and the warranty status of the product. Accompanying the ability to order on-line is also the option to cancel some orders on-line. The process begins off-line. The dealer accesses the ClarkNet II software resident on the PC and enters part numbers up to a maximum of 99. The dealer enters the entire order, which he/ she can either view or obtain hard copy. Other options include shipping instructions, routing, and even the ability to direct ship to the customer. When the dealer is satisfied that the order is complete, option 1 is selected, and the PC automatically dials the access number. A minicomputer answers the call, accepts the order, and transmits any appropriate messages to the PC before disconnecting. The dealer can then view the results of the transaction or print them later.

The Inquiries and Order Status option tells the dealer if the part is in stock, its price, what it is, whether multiple units must be ordered, and so on. The dealer can determine whether the order has been shipped and what parts were actually shipped. It is also possible for a dealer to check accounts payable status and a history of orders placed. The latter includes a history of open orders, back orders, and stock orders, among other things.

The Administrative Messages option is a full-featured electronic mail system for contacting the dealers and for the dealers to contact one another. Whenever there is a mail message from another party or a notification regarding an order, the dealer automatically receives it after logging onto ClarkNet II. The message or notification may be viewed immediately or stored on the PC for retrieval later.

Trader Joe Update is a listing of participating dealers, and either portions or all of their inventory on hand. Trader Joe Update provides an alternative source for parts; it is normally used by dealers if Clark Distribution Services does not have the part or if they order from another dealer, Trader Joe Update expedites delivery of the part.

Billboard is a graphic presentation of information used for promoting products to customers and for educational and training sessions for dealer employees. In a typical promotion, an item, such as a front-end loader, is pictured in color with its specifications on a PC monitor (see Figure 13.2). The monitor is placed in a high-traffic area, such as a counter top, and turned toward the customer side of the counter. The monitor is programmed to scroll through a series of screens about the front-end loader or whatever products the dealer is interested in promoting. Content for the Billboard is created and updated by an outside company and then mailed to the dealer on a disk; this feature has been well received by dealers.

Approximately 800 authorized dealers subscribe to ClarkNet II. These dealers account for 80% of the company's business. They are scattered throughout North America, Mexico, and overseas in Australia and Sweden. Dealers may choose not to subscribe for a number of reasons. The volume of activity may not justify purchasing a PC, fees levied may be a deterrent in some cases; inertia

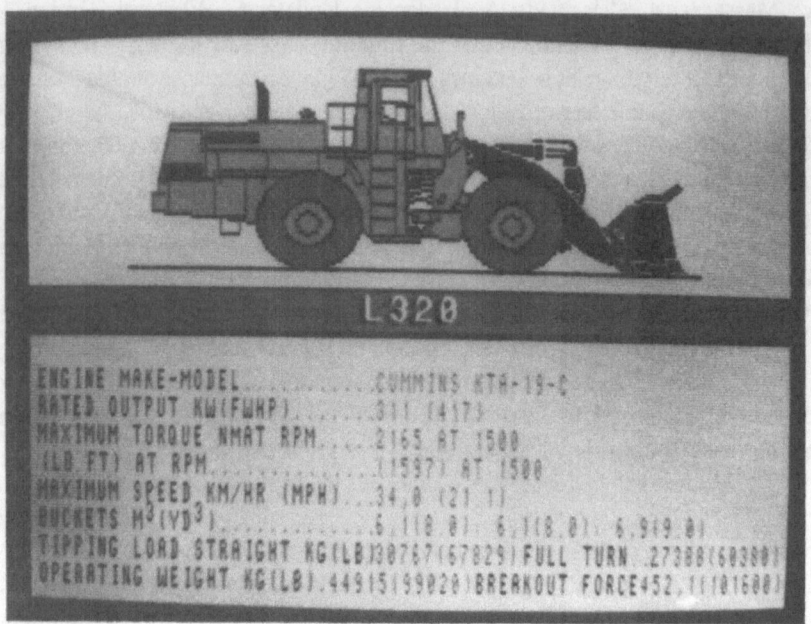

Figure 13.2. An image of a front-end loader from the ClarkNet II Billboard. Courtesy of Clark Distribution Services, Inc.

sometimes plays a part. Since there is a phone-in service with order takers, some dealers may be reluctant to change from a procedure they are accustomed to and understand. ClarkNet II is promoted among dealers in several ways: Meetings of user groups and dealer seminars provide forums for the company, but there is also a reliance on word of mouth by dealers who use the service.

In addition to PC equipment, there are several costs that must be borne by the dealer: There is a monthly subscription fee of $30 whether or not the service is used; the software and user manual cost $300, but updates for both are currently provided free of charge. There are two usage fees—one is a session fee determined by the average number of lines transacted per month; this fee can range from $.50–$1.00 per session. The other is a line item fee, which varies according to the application used; for example, inquiries are less expensive than order entries. A typical inquiry session covering several part numbers may cost $1.50.

13.1.3. Objectives

The company is justifiably proud of ClarkNet II and enthusiastic about the impact it is having on various parts of the business. The respondent felt that essentially all of the objectives listed on the survey applied and most of them were being attained to some degree. Three specific benefits of the new system were cited. The original ClarkNet I operated on a mainframe, and service was not available over night when mainframe on-line applications were shut down. ClarkNet II resides on its own minicomputer, an IBM Series I, which interfaces with an IBM mainframe. With the exception of a few minutes of scheduled maintenance performed late each evening, service is now available 24 hours a day. The second benefit is a reduction in labor: It is estimated that approximately 16 positions are no longer required because of the dealer on-line access through CharkNet II. The third benefit the service provides is a competitive edge over other parts wholesalers because of the applications it supports and its ease of use. One dealer commented that a competitor's system, which is also used, requires approximately 20 minutes to initiate a session to perform similar functions.

13.1.4. Technology

Each ClarkNet II user needs an IBM compatible PC with DOS Version 3.1 or higher, a minimum of 640K RAM, a color monitor, EGA/VGA board, 20-MB hard drive, and a 1200/2400 Hayes-compatible modem. Most dealers access the system using 800-number telephone service; the exception are dealers in countries that do not have 800 service. The system supports up to 24 simultaneous users, although the largest number encountered at one time has been 16. Communications speeds supported are 1.2–2.4 Kbits/sec.

At this point in time, graphics are not supported on-line. The Billboard application is updated by mailing disks to the dealers. However the company is studying the possibility of downloading graphics to the network at some time, perhaps through the classified ad application listed on the main menu but not yet activated.

13.1.5. Design, Implementation, and Operation

The initial system design was done by company employees; the initial creation as well as updating text was done both internally and externally. Graphics material for the Billboard was initially created and is currently maintained entirely by the same firm that produces a segment of the text. The text and graphics contents do not change much over time, and when infrequent updates do occur, approximately 5% of the respective contents is changed.

Dealer training is handled almost exclusively by telephone and tutorial. The typical training time via phone is 2 hours. Training time continues to be reduced as familiarity with PCs increases among dealer organizations; this is in contrast to the 8 or more hours of training formerly required with ClarkNet 1. The tutorial provided is quite extensive; its orientation is nontechnical, and it goes through each of the screens and available options. The tutorial is built on Story Board®, a software presentation program from IBM. A well-written and relatively complete user's manual provided to each dealer has tended to reduce many of the difficulties associated with new and unfamiliar software. In addition to the manual and tutorial, a software program called Anywhere is installed on each PC, which enables headquarters staff to troubleshoot on-line problems for the user.

13.1.6. Problems and Predictions

ClarkNet II has evolved over a 7-year span, and many of the tribulations of starting the service have been faced and forgotten. However one problem that never quite disappears is training and serving customers who have little or no computer experience. Extensive user documentation and on-line training have been a partial antidote but the problem never completely goes away.

Some very unusual and sometimes amusing incidents occur; for example one dealer was experiencing trouble communicating with the host in Chicago. After extensive troubleshooting the telephone company was called and asked to inspect the telephone line. Upon investigation the repair person discovered the telephone line lying in a pool of water and spliced together after apparently having been cut. Another recent incident occurred during a telephone call when an operator was asked to turn off her PC. She turned off the dealer's main computer instead. These are the sorts of problems that can not be anticipated and make going to work in the morning interesting.

As ClarkNet II evolves, it is anticipated that additional applications will be added. These applications may incorporate such things as CD-ROMs, faster data transmission, and improved error detection and control features. The eventual goal is to have 80% of the volume of order entries, inquiries, and technical support carried out through the system and to use ClarkNet II to entice other companies to use Clark Distribution Services to supply their dealer networks.

13.2. NATIONAL LIBRARY OF MEDICINE

The National Library of Medicine compiles, verifies, and organizes citations from over 350,000 new articles a year. To accomplish this Herculean effort, it relies extensively on computers for storing, updating, and accessing information. A principal means of access is a terminal, modem, and the public telephone network. The on-line facilities support over 50,000 registered biomedical users around the globe by providing immediate access to a vast pool of information. The extensive use of information technology creates a positive climate for developing new and innovative ways of using computers.

The Prints and Photographs Collection is located in the History of Medicine Division of the Library. This unit contains prints and photographs documenting the social and historical aspects of medicine from the Renaissance to the present. An example of the type of images is illustrated in Figure 13.3. The History of

Figure 13.3. An example of the images contained in the History of Medicine Division, Archival/ Retrieval Picture Project. Courtesy of National Library of Medicine.

Medicine Division Archival/Retrieval Picture Project (HARPP) is a system for electronically retrieving images; it becomes operational in December 1992.

13.2.1. There Must Be a Picture Somewhere!

The Prints and Photographs Collection began in 1879 and has grown to contain approximately 60,000 items. Material comes from two main sources: individual items that have been donated or collected and may consist of caricatures, photos, fine prints, ephemeral, and portraits; and illustrations from books and journals in the library's collection from which patrons requested a picture.

The subjects are medically related and generally in the public domain. The focus of the collection is not clinical. Pictures of cells, surgical procedures, and other clinically oriented imagery is generally not collected. The latest major addition is a group of 5000 contemporary posters collected since 1990 concerning present-day public health issues, such as aids, breast feeding, and drug abuse. The collection is being increased continually, but relative to the current size, annual growth is modest with the exception of the poster collection; typically between 200–500 items a year are added.

Patrons of the collection are usually doctors and others in health fields who are writing articles and books and staff from publishing companies searching for illustrations. However it is not unusual to find individuals with commercial interests ranging from someone creating wall hangings for doctors' offices to the writer of a television documentary.

The traditional approach typical of many still picture collections requires anyone searching for a specific print or photograph to journey to Bethesda, Maryland. Once there the researcher goes through a card catalog containing a brief description of the item, card by card, looking for what is needed. When an item in the card file is selected, the researcher asks a librarian to retrieve the original. Once the item has been viewed, the researcher may request a photograph. If the item has been requested before, a photograph is made from a negative; otherwise the original is set aside, and once a week, a photographer under contract comes to the library to make several negatives of the item; afterward prints are made. The whole process takes a minimum of two weeks.

There are several obvious disadvantages to this system. First it requires a researcher to travel to Bethesda without knowing beforehand whether or not the required image even exits. Then it is necessary to look through a card catalog to select an image based on a description of a few words. A librarian must spend time retrieving the item requested; then the item must be handled by several people, leading to wear and tear on the artifact, some of which date from the fifteenth century. If the item is selected, a researcher must then wait for a photograph to be made from a negative or perhaps from the original; the latter generally necessitates mailing the copy. In the early 1980s, someone at the library suggested putting the collection on videodiscs to reduce handling. In 1982 a pilot project was undertaken in which 1000 pictures were stored on a disc. The disc and a catalog were sent to approximately 100 libraries, and the response was universally favorable.

13.2.2. HARPP Is Born

In 1988 when the decision was made to put the entire collection on videodisc, HARPP was born. Fifty-seven thousand images are stored on one videodisc. The control program is currently run on a LAN from a server. The control program allows keyword search for images and catalog information. The program drives the disc player and retrieves the image as the catalog entry is selected. Images may also be retrieved by the images bar code number, the library's call number, and the computer system's frame number. Initially the system is being used only on site, but it is hoped that discs will be made available to other libraries, resource centers, and the general public. Discs will be packaged in a 13 by 13 inch container less than an inch thick.

Not only is the collection going to be available to anyone with the necessary equipment, there will now be a way for anyone located anywhere to use the collection. When an item is desired, it will be ordered from an organization contracted to store the negatives and positives and make prints from the 35-mm negative or positive used to create the videodisc. This will allow the library to concentrate on its primary mission of obtaining and maintaining items for the collection.

Those who obtain the package of discs will not be precluded from copying noncopyrighted material; however the quality of the copied image will be not that good, and most users will lack the sophisticated equipment to enhance the copy. Also each of the stored images has been reduced in size, but the true proportions are preserved. There are no close-ups; instead the full image is shown with a mat around it.

13.2.3. Objectives

Objectives cited for the project were in their order of importance: distribute information, improve accuracy of information, and reduce operating cost. The most important objective was to make the information more portable so researchers would not have to visit the library. By reformatting the collection and putting it on videodiscs, it becomes available to anyone who has access to the equipment. The price of the videodisc is expected to be very affordable and in the price range of even those consumers with the appropriate equipment for playing the videodisc.

Ready access to the images is expected to increase the accuracy of the search process, since a user can literally scroll through the entire collection in a few hours or search for keywords and subject terms. It will no longer be necessary to read index cards and try to visualize an item.

In the process of capturing the collection on discs, the library has succeeded in reducing the amount of staff time needed to serve patrons, since it is no longer necessary to search for and retrieve items from the archives or to assist patrons in formulating a search strategy. Because actual handling of items in the collection has been reduced, there is less opportunity for misfiling and less time spent searching for misfiled items. Wear and tear on the collection caused by handling has been almost entirely eliminated, and the library will shortly be freed of the laborious task of providing copies of images to patrons. Furthermore the project will offset its original cost by reducing overhead. It should not be forgotten that this has all been accomplished using an approach that will produce some revenue, albeit small due to the expected low price of the discs.

13.2.4. Technology

The HARPP can be installed on any IBM compatible PC capable of supporting videodisc and CD-ROM players simultaneously. Although it is possible to put images and text on the same videodisc, a CD-ROM will be used to allow simultaneous access to both the pictures and the catalog listings. The original image base was developed using dBase III and an enhancement program. The final version of the product is dBase oriented.

The first step in creating the image base was bar coding each item in the collection and transporting it to another location in the building where it was photographed and then returned to the archives. Photographing the collection was a laborious job that required 14 months and two full-time workers who did nothing but photograph the collection using 35-mm film; this was done by a firm specializing in this type of work. Another four months were spent transferring photographs to videotape from which the videodiscs were made.

The process of photographing the images was done in the following way: An item was placed on a camera stand and a bar code reader run by the operator recorded the number. The bar code was then automatically integrated with the picture by using a special computer. Next the operator used a laser pen to mark the edges of the item. The computer forwarded the dimensions of the picture to the camera, which was then automatically focused and the picture taken. Originally it was hoped that 500 items could be photographed daily, but it turned out that 300 images were deemed a good day's work.

A third person was employed full-time for an equal period to create the catalog records. The source for this information was the cut line (*i.e.*, caption) for the image and other information contained in the card catalog. The library staff provided whatever assistance was needed. The videodisc was then mastered by an outside firm using the 35-mm film, and up to 54,000 images were stored on each side of the disc. The control program that integrates the videodisc and CD-ROM was written by library personnel.

13.2.5. Design and Operation

The system includes text, color, graphics, and picture-quality images. The design for the system was generated in-house with extensive support from the library's audiovisual unit, the group that developed the kiosks and computer-driven displays in the lobby and reference areas of the library. The text and graphics development were done by library personnel. As explained earlier, a contractor was hired to photograph the entire collection. When the discs are updated, a similar procedure is expected to be followed.

A four year old could learn to operate the system, which is as easy as changing channels on a television set. For example, an image having the cut line U.S. Veterans Administration Hospital, Muskeegee, Oklahoma, could be located and retrieved by entering Oklahoma, Muskeegee, Veterans U.S., Veterans Administration, and so on. There are plans to enhance each catalog record with complete information, including standard subject words and terms.

The ability to perform compound searches enables the researcher to isolate topics quickly; for example, a search of only the first 13,000 images using the term nurse produced 542 possibilities. Searches based on the terms nurse and Russia yielded two images. With help from the collection's curator, who is familiar with keyterms, the search process can be refined considerably to locate images that are particularly difficult to uncover.

13.2.6. Problems and Future Developments

No significant difficulties were encountered in developing the system, although the job of photographing the collection was more work than originally anticipated, but that was a one time effort. Periodically updating the images and the catalog will be much less laborious.

For the foreseeable future, information will be distributed to users via the package of discs. However it is anticipated that some day cataloging information and perhaps even images will be available on-line through Medlars or a similar dial-up access. Due to the bandwidth requirements for transmitting images, this is unlikely to occur soon, but it would be possible for organizations with suitable networks in-house to allow users to access contents from those locations served by the network. But even without on-line access, HARPP is a tremendous boon to researchers and an excellent example of how an organization can extend the scope of its services through technology.

Chapter 14

Cases:
Implementing Strategic Systems

Strategic systems shape or support the competitive strategy of the organization for the intermediate and long term. These systems are intended to produce a strategic thrust of some type, which occurs when an edge is gained or maintained over a competitor or a competitor's advantage is decreased. A strategic system may be manifested internally by changing the organizational climate or effecting a new or better set of business opportunities. Externally a strategic system may be implemented for improving client relations, differentiating products and services, and engaging in new business opportunities.

Thirty-nine of the projects surveyed by mail were classified as strategic. Nine of the projects were internally focused, and 30 were externally focused. By their nature, strategic systems are more risky then operational systems. Of the 39 strategic projects surveyed, 13 were trials, although a few of these were beyond what most would consider a trial period.

The primary application of 20 projects was distributing information. Four projects involved selling information, and nine projects sold a product other than information; six were developed for training and education purposes.

The systems described in Chapter 14 were implemented at E. I. DuPont De Nemours, the *Fort Worth Star-Telegram*, two units of the Ford Motor Company, and AMR Corporation. The project at DuPont was focused internally and the others externally. Two of the projects were considered trails, although they were more than two years old and both still existed when this book went to press. Three projects were considered a permanent part of the system repertory in each organization or at least as permanent as anything can be in the dynamic environment in which a business functions today.

14.1. E. I. Du Pont De Nemours and Company

Du Pont is a diversified international company with operations in more than 48 countries. In 1991 it had approximately 145,000 employees worldwide and sales of $35 billion. Its product lines are varied, but it is best known for developing, producing, and marketing agricultural and industrial chemicals, biomedical products, pharmaceuticals, and polymer products. Its domestic plants are concentrated in the Eastern-half of the country, and it has sales offices located in principal cities throughout the United States.

Du Pont has a well-developed information technology infrastructure and a corporate culture that supports innovation. This environment has been conducive to forming an extensive series of videotex applications that permeate the whole organization. One example of the many hundreds of applications is Corporate Homes for Sale; a companywide service for employees who are relocating within the United States.

14.1.1. You Have Been Transferred! What about the House?

Internal transfers are a fact of life in any organization, but when the organization has sites throughout the country and around the world, the impact on every member of the family can be more significant. One of the major issues to be faced is selling your house. Du Pont tries to lessen the severity of the problems a home owner faces with its Corporate Homes for Sale application. The system enables owners, buyers, and the curious throughout the world to post and read information about employee properties in 46 states, it has become a central and recognized source for this type of information within the company. This activity was not handled nearly so efficiently or expeditiously using paper and the postal service; especially when the inquiring party was overseas.

Corporate Homes for Sale is just one of approximately 1000 videotex applications that are presently implemented within Du Pont. Another much more widely accessed service is a newsletter published three days a week, which is approximately two typewritten pages. It is estimated that perhaps as many as 50,000 people look at each issue. The basis for such widespread acceptance of videotex is tied to the corporate culture that prevails. Du Pont is generally regarded as an innovative company that attempts to stay on the leading edge of technology, whether in engineering, chemistry, or the information systems arena, and videotex is consistent with that philosophy.

Videotex and the traditional data-processing services within the company compliment one another but are handled by different units. Whenever possible videotex and data-processing units work in tandem. For example, in one application a cost and financial accounting system for the largest department in the corporation has all of its cost sheets presented through videotex, but the information is provided from the database maintained by data processing.

The videotex infrastructure is well organized, which contributes to the success of videotex in the corporation. Each site has a videotex coordinator responsible for promoting videotex, training information providers, assisting in the implementation of new applications, and serving as a liaison to the corporate videotex support unit, which consists of eight people. The unit provides overall support and is charged with monitoring advancements in technology, enhancing corporate videotex software, and troubleshooting problems that may occur at the sites. The corporate support group also develops specialized applications that a site may request but does not have the technical resources to do.

14.1.2. Corporate Homes for Sale

Videotex applications are organized by site. A person accessing the system at a site supporting videotex sees the main menu listing information indigenous to that location. One of the menu options leads the viewer to a main menu for the corporation. A facsimile of the corporate main menu at the time the case was written is shown in Figure 14.1.

Item number 4, Employee Relations, leads to the department that provides information about Corporate Homes for Sale. A facsimile of the menu of applications maintained by the Employee Relations Department is illustrated in Figure 14.2. A wealth of information is provided to employees, ranging from organization charts to compensation, safety, and corporate homes for sale.

Choice 8 leads to a list of homes for sale organized by state (see Figures 14.3 and 14.4). By choosing a state, the viewer is presented with a list of homes by (1) address, (2) the broker's name and address, (3) the name of the real estate agent, (4) telephone number, and (5) listed price.

14.1.3. Objectives behind the Application

Reasons for implementing the application are given in Table 14.1. The principal objective was to improve communication but closely allied with this was a desire to reduce the cost of communication while improving the accuracy and process of distributing information. It is hoped that through this process, the work climate will be improved.

Objectives related to impacting the communication process and the accuracy and timeliness of information have demonstratively been attained. The influence on the work climate is affected by too many variables to determine the impact of any single application, but it is thought that this application and many other videotex applications are having a positive influence.

The videotex service provides a central source of information and a familiar vehicle for retrieving it. Even when employees do not know if what they are seeking is in the system, they frequently begin their search using videotex. This creates a self-fulfilling process as more and more units recognize the importance

```
--------------------------------------------------------------------
                    Du  Pont   Corporate  Videotex   Services
--------------------------------------------------------------------

              0  What's New in Du Pont Corporate Videotex???

     1 Agricultural Products              8 Finance
     2 Central Research and Development   9 Information Systems
     3 Chemicals and Pigments            10 Legal
     4 Employee Relations                11 Materials & Logistics
     5 Engineering                       12 Petrochemicals
     6 Automotive Products               13 Polymer Products
     7 Fibers                            14 International Services
--------------------------------------------------------------------

        For Du Pont Internal Use Only          Last Update: 25-Jan-199X
--------------------------------------------------------------------

        98  FEEDBACK          99  HELP            More --

     Choice:
```

```
--------------------------------------------------------------------
                    Du  Pont   Corporate  Videotex   Services
--------------------------------------------------------------------

                    Other  Du Pont Information Services

     15 Corporate DEC OIS Technology      22 News Around Du Pont
     16 Electronic Information Security   23 BASISplus TIMs
     17 Standards and Guidelines          24 Technical Report Database
     18 Weather Reports                   25 Southeast Reg. Res. Sharing
     19 Du Pont Training                  26 Separations Process Guide
     20 Du Pont Consultants               27 General Services Operations
     21 Corporate Surplus Materials       28 Computer & Systems Function
--------------------------------------------------------------------

        For Du Pont Internal Use Only          Last Update: 7-Aug-199X
--------------------------------------------------------------------

        Left Arrow Key to return to first menu page

     Choice:
```

Figure 14.1. Facsimile of the Du Pont Corporate main menu. The main menu is periodically revised.

EMPLOYEE RELATIONS DEPARTMENT VIDEOTEX SERVICES

What's New ... 0

1 (Future Option)	9 Pers. Dev. Courses (Future)
2 How to Use Videotex	10 Records Retention (Future)
3 Index	11 SIP Unit Val. & Loan Int. Rate
4 Contracts/Organization Charts	12 Wilmington Site Information
5 Compensation & Benefits	13 Du Pont Network News
6 Saftey & Health	14 Finance Department (EX&B)
7 Mail Users' Handbook	15 Other Du Pont Infobases
8 Corporate Homes for Sale	16 Du Pont Country Club

Table of Contents ... 97   (Future)   Feed Back ... 98

Help ... 99

** FOR DU PONT USE ONLY ***
VTX Site Coordinator: John Doe 777-7777

Choice:

Figure 14.2.   Main menu for the employee relations department.

Corporate Relocation Section
Corporate Homes for Sale

For Info: Contact the broker listed with the property

1 ALABAMA - AL	10 GEORGIA - GA	19 MAINE - ME
2 ALASKA - AK	11 HAWAII - HI	20 MARYLAND - MD
3 ARIZONA - AZ	12 IDAHO - ID	21 MASSACHUSETTS - MA
4 ARKANSAS - AR	13 ILLINOIS - IL	22 MICHIGAN - MI
5 CALIFORNIA - CA	14 INDIANA - IN	23 MINNESOTA - MA
6 COLORADO - CO	15 IOWA - IA	24 MISSISSIPPI - MS
7 CONNECTICUT - CT	16 KANSAS - KS	25 MISSOURI - MO
8 DELAWARE - DE	17 KENTUCKY - KY	26 MONTANA - MT
9 FLORIDA - FL	18 LOUISIANA - LA	27 NEBRASKA - NE

Note: Two digit state abbreviations may be used in keyword searches.
Press the HELP key to view keystrokes for your terminal.
HELP: IBM/GATEWAY - PF1 DEC ALL-IN-ONE - GOLD/H DEC VMS - PF2
Choice: Page Text truncated to screen size   More ---

Figure 14.3.   The first screen of corporate homes for sale, organized by state.

ILLINOIS- PROPERTY ADDRESS	BROKER NAME AND ADDRESS	AGENT NAME/PHONE LISTING PRICE
17620 W. WINDSLOW DR. GRAYSLAKE IL 60030	COLDWELL BANKER THORSON REALTORS 1225 W 22ND ST OAK BROOK IL 60521	ROCKI DE VOY 708-573-5940 $169,900
1920 ASHBURY LANE ASHBURY COUNTRY HOME INVERNESS IL 60067	COLDWELL BANKER THORSON REALTORS 1225 W 22ND ST OAK BROOK IL 60521	ROCKI DE VOY 708-573-5940 $264,900
1932 ASHBURY LANE INVERNESS IL 60067	COLDWELL BANKER THORSON REALTORS 1225 W 22ND ST OAK BROOK IL 60521	ROCKI DE VOY 708-573-5940 $264,900
Press HELP for options.		More ---

Figure 14.4. A selection from Corporate Homes for Sale.

of videotex and ask to put information into the system. In doing so, a common and easily understood interface consisting primarily of menu selections has evolved; this eliminates the need to train people how to use different interfaces and function keys.

An obvious benefit of services of this kind is the reduction in paper. Du Pont has a corporate goal of being an environmentally safe company. An internal study has shown that videotex services have been directly responsible for significant

Table 14.1. Principal Objectives
for Implementing the Application

Rank	Objectives
1.	Improve communication
2.	Reduce cost of communication
3.	Improve accuracy of information
4.	Distribute information
5.	Distribute information sooner
6.	Improve the organization's work climate

reductions in the quantity of paper consumed. This has an impact on the environment beyond the company, while saving money and producing the benefits already discussed.

14.1.4. Technology and Networking

The system is based on Digital Equipment Corporation's (DEC) VAX VTX, which Du Pont has enhanced with a variety of features, including several that facilitate remotely creating and updating contents by information providers. The infobase is distributed across a number of computers and a user selecting a particular option may be accessing information from anywhere in the system.

Since the contents are limited at this time to text, the text-only interface and the bandwidth available are not constraints. Most of the terminals are either DEC or IBM products, but virtually any type of equipment can gain access using terminal emulation software. Most of the employees access the system at work through premise networks but, interfaces for dial-up off-site are provided, and employees can enter the system from home and other remote locations via the telephone network and a modem. A variety of communication speeds are supported.

14.1.5. Design and Operating Issues

Information in the system is created and maintained by individuals throughout the company. This is possible because the contents are strictly text. The videotex support unit provides technical assistance and training, but it does not usually play a part in creating and maintaining information. If graphics or other forms of presentation existed, special support would probably be necessary.

It is estimated that about 5% of the content changes daily. To ensure that information is kept current, an effort is made to give the unit responsible for the contents a sense of ownership. Part of this sense of ownership is a natural outcome of the unit's request to publish its information through the videotex system. The sense of ownership is further heighten by usually identifying the name of the responsible unit with the information being presented. Each menu often identifies a contact person also. The best approach is having the source of the information responsible for maintaining it. The result of these efforts is a relatively well-maintained infobase.

Inputting information is straightforward. The user creates a file containing new or updated information and sends it to a designated account. The system then automatically posts the file to the infobase for public viewing without intervention by the videotex support group or any other unit within the company. Such a practice requires a great deal of trust and confidence in the participants, but is the only practical method for creating and maintaining a large and current infobase without establishing a large and centralized support group.

14.1.6. Problems and Future Developments

Implementing such a system as this throughout the company requires a great deal of planning and nurturing. Perhaps the single greatest problem was the myriad of equipment that exists in a multinational company like Du Pont. Connecting computers not designed to communicate with one another creates its own set of difficulties that are then compounded by different types of terminals that access the system; this problem has been largely overcome.

Another issue that has to be addressed continuously is how to make information that presently resides in a data-processing environment available for access by technically unsophisticated users employing a broad range of equipment. This problem may never completely disappear, but it will be alleviated as data processing incorporates the distribution of information via videotex within the design of its own applications.

Typically the system has grown by exposing employees at the lower levels to videotex and what it can do. These users in turn sold it to their management who were receptive because so many others were using it. Management in turn found the system helpful and exposed those above them to the system. Originally management was asked to use videotex for particular applications; however management felt there was no need for videotex.

Tremendous growth of the videotex system is anticipated for the remainder of this decade. One estimate is that the number of applications could increase tenfold from 1991 to 1994. The basis for this level of growth is videotex's acceptance in the company. The system appears to have reached a critical mass: There is sufficient information in the system to justify its use, and a large enough cadre of users exist to influence the rest of the company.

The support staff's goal is that by 1995, employees will have adopted the attitude that the place to start looking for answers is the videotex system. If this behavior occurs, further enhancements of the system will be necessary. Two of the most important enhancements will be developing alphabetical and subject indices and creating a keyword feature to facilitate the automatic location and retrieval of information. The vision is to create an automated process for information providers that will aid in the indexing process and keyword management.

In a paper-based environment or with electronic mail, the intended recipient of information can be sent a memo that will wind up on top of the desk or in an electronic mailbox. Once there the recipient must make a conscious decision what to do with it. In a videotex system, the contents can change, and the people who need to know of the change may be oblivious to it. A feature that is being discussed is the ability of an information provider to notify a group of people that new information has been added or an update to the infobase has occurred. In the words of one manager discussing the dissemination of information: "Our challenge is to make it available easily and quickly"; in doing so Du Pont is unlocking the puzzle of how to keep employees informed and talking with one another when they are spread across the globe.

14.1.7. Addendum

Two significant changes took place after the site visit had been completed. The first one was the removal of the Corporate Homes for Sale information from the system. At one point, it is estimated that this particular menu item received as many as 13,000 accesses a week. Its removal was the result of a change in management philosophy regarding the type of information that should be maintained on the system.

The second event was the result of corporate downsizing. The number of employees in the company videotex unit was slashed from eight to four, which has essentially curtailed the development of any major new applications, so the focus has shifted to one largely of maintenance. However the personnel reduction has not slowed the expansion and use of videotex as a communications medium within the company. Information is continuing to be added by various departments and usage remains high. Several major additions to the infobase include a listing of jobs both in and outside the company and the library catalog. A feature called the Corporate Index has been implemented along with new menu structures, especially for the main menu, designed to make the system easier to use and to locate specific information.

14.2. *FORT WORTH STAR-TELEGRAM*—STARTEXT

The *Fort Worth Star-Telegram* serves the city of Fort Worth, Texas. The newspaper is published daily and has a circulation of 256,824 each weekday, which climbs to 329,815 on Sunday. It is a subsidiary of Capital Cities\ABC, Inc., a $5.4 billion media conglomerate with publishing and broadcast holdings throughout the United States.

StarText is an on-line information service begun in 1982 by the newspaper and Tandy Corporation. The newspaper assumed full responsibility for the service in 1983 and has watched it grow about 30% a year through the first six years. Today StarText offers news, classified ads, electronic mail, stock quotes, EAASY SABRE, Online Academic American Encyclopedia, subscriber-maintained forums, and a variety of other text-only on-line services to approximately 4000 users, most of whom live in the greater Fort Worth/Dallas area.

14.2.1. Great Expectations

At the beginning of the eighties, many videotex services were being launched around the world, and it was prophesied that shortly we would all be reading newspapers by PC or a television set modified with a special adapter. The British videotex service, Prestel, was launched, and the French were in the midst of major trials. In the United States, a host of companies were planning videotex trials, and newspapers were leading the thrust in partnership with other companies.

The first commercial service catering to the local market to come on-line was Viewtron, a business unit of Knight-Ridder Newspapers. Located in South Florida, the service promised "a world of information at your fingertips." In Chicago the *Sun-Times* launched Keyfax, and on the West Coast, the *Los Angeles Times* started Gateway. Serving the national scene were CompuServe and TheSource. There were very few survivors among the dozens of startups; by the close of the decade, only CompuServe remained of those just listed.

While some companies were investing tens of millions of dollars trying and failing to launch news-based videotex services, a small entrepreneurial effort was being launched in Fort Worth. Beginning in the fall of 1981, the *Fort Worth Star-Telegram* began a series of discussions with the Tandy Corporation, parent of Radio Shack, which is headquartered in Fort Worth. At the time, Tandy was experimenting with a videotex system that it wanted to market to newspapers. Tandy was looking for some test sites and approached the newspaper about being a partner in a trial. A partnership was formed in which the newspaper was to provide the content.

The system was launched on May 3, 1982. It was run on a Tandy TRS Model II using Tandy software and could support 16 telephone lines. Unfortunately to access it, you needed a Tandy PC or a TRS-80 Videotex Information Retrieval terminal costing approximately $400. Describing the service as on-line is a bit of a misnomer. A subscriber logged onto the computer and requested whatever was sought with a single request. The computer then delivered the information, and the session ended. If additional information were needed, the subscriber had to re-dial the computer to submit a batch request.

At the end of the first six months, there were approximately 30 customers paying $5 a month for unlimited access to news and classified advertising. At the 6-month point, new software was installed that supported a variety of PCs and operated in a true on-line environment. By the end of the year, the subscriber base had grown to 260 customers, and by the end of the following month it had climbed to 400 as home computer sales began to increase. As the newspaper tried to expand the system to handle this rapid growth, it became evident that the Tandy system was not going to be adequate, so the newspaper struck out on its own.

14.2.2. Instant News That Does not Rub off on Your Hands

In June 1983, the newspaper amicably parted company with Tandy and began operating the service using a DEC VAX/750 and software developed in-house. Since then there has been slow, steady growth. The hardware and software have been continually upgraded, and new features have been added to the infobase. Figure 14.5 indicates the range of information available.

As StarText has grown and evolved, a few newspapers around the country began once again to show interest in operating on-line versions of their paper. Newspapers in large cities, such as Washington D.C. and St. Louis, have offered a

Updated news	News from the Associated Press, the *New York Times*, and other major services, and the *Fort Worth Star-Telegram*
Stock markets	Complete business news, reports from Wall Street, commodity trading, mutual funds, options, and the closing quotes from major financial markets for the last five business days, and much more
Classified ads	New classified text ads are available the evening before they appear in the newspaper.
Grolier's encyclopedia	Full use of the 30,000-topic Academic American Encyclopedia, which is updated every 90 days.
Electronic mail	Every subscriber is issued a mailbox.
Special interest news	Extra coverage in such areas as science, medicine, computers, and technology
Subscriber columns	Subscriber-produced information on topics ranging from computers to cooking and auto repair
EAASY SABRE	Free use of EAASY SABRE, the personal travel reservations system from American Airlines

Figure 14.5. Information available through STARTEXT for subscribers to the basic service.

limited amount of information through national electronic services, such as CompuServe. Newspapers in New York City, Spokane, Atlanta, St. Louis, and Kansas City are offering limited versions of the newspaper edition, but StarText is the most successful. According to management, there are several reasons for this success. First and foremost was the decision to focus on text rather than graphics early in the life of the system. When other newspapers in the early eighties were selling expensive and dedicated terminals along with a hefty subscription fee, StarText was enabling anyone with an ASCII unit to subscribe. Second the newspaper concentrated on local content. It was believed that customers would want electronic information relevant to where they lived for the same reason they buy local newspapers.

Another factor contributing to the success of StarText was its flat subscription charge. Most if not all other newspapers in the initial flurry of services charged a considerable subscription fee and tacked on to that a per minute connect charge. StarText began with a $5 a month fee, raised that to $7.95, and in November 1984, increased the fee to $9.95, where it has remained. Following a strategy of making the system easy to access, easy to use, low in price, and local in nature, the service has grown to 4000 subscribers. Interestingly this philosophy and the decision to limit the infobase to text have led to a phenomenal low break even point of 2000 subscribers, a point reached at the beginning of the fourth year.

A text-only content and heavy reliance on automation have enabled the staff to remain very small. The service began with four people and has grown to six full-time and three part-time staff members. The full-time staff consists of a marketing director, who has been with the organization since it started, and an office manager for handling billing and customer relations. There are four editors, one of whom is managing editor of operations. The editors have typically been recruited from among students majoring in journalism at nearby colleges. The experience of the company has been that it is easier to find recent graduates with a professional interest in journalism and train them then to try to retrain individuals who have extensive experience in print journalism. The total size of the staff is quite small compared to the number of employees who often numbered in the many hundreds for some of the early videotex services operated by newspapers. By limiting most of the content to what is available in the print version of the newspaper, StarText avoids the expense of creating original information. In a sense it verifies the adage that an organization can't duplicate the local newspaper without becoming a newspaper. Complementing local content with information that can be obtained quickly and easily by an editor from the wire services and posted directly to the infobase further limits expenses.

14.2.3. Objectives

The newspaper's objectives in delivering information electronically are straightforward. First and most important is to enter new business opportunities; this extends to other markets and other products. StarText is a model for other newspapers to follow and has recently become a demonstration site for software and expertise developed by the company. In January 1992, a separate company was formed—StarText Technology and Information Services (STIS)—to market Star-Text hardware and software, primarily to newspapers interested in starting their own operations. To date licensees include the *St. Louis Post Dispatch* (offering a service called PostLink), and the *Kansas City Star*, which plans to be operational in late 1992.

Two additional objectives are to increase the newspaper's share of the local market and to distribute information sooner. Management feels both objectives are being met.

14.2.4. Technology

Until recently the system was housed on a two-machine cluster of VAX/750s. It could accommodate 64 simultaneous calls on local dial-up lines at a bit rate of 300/1200 bits per second. In September 1990, the system was moved to a PC-based Novell LAN with file servers. There are 80 incoming telephone lines in the initial configuration. The system currently supports speeds of 0.3, 1.2, and 2.4 Kbits/sec and MNP-5 and Xmodem protocols. The-PC based network, which is modular, will enable the system to be expanded incrementally to accommodate a slow and steady rate of growth. This is in contrast to the former VAX cluster, which would have required a third machine to be added at some point to accommodate that next customer.

14.2.5. Operating and Promoting the System

The initial system design for StarText was done by company employees, and the service is maintained by the relatively small staff described earlier. Approximately 70% of the *Star-Telegram* winds up on StarText each day. Normally all the major news stories, some of the feature material, and the classified ads are transported to StarText. Interestingly about half of StarText's subscribers also subscribe to the *Star-Telegram*. Apparently there are enough dissimilarities so that people want both sources of information. Classified advertisements placed in the newspaper are also placed on StarText one day in advance; this is one of the most popular features. Retail and display ads and the comics are not transferred, primarily because that would require a graphic interface.

A great deal of information available through StarText does not appear in the newspaper. StarText subscribes to several wire services and databases maintained by other newspapers. These sources are monitored and information routinely selected and routed to specific areas for access. For example, articles about science news may be grabbed and put in one location called Science; this enables special areas to be set up dealing with such topics as medical news, computer news, and business news, which permits much more thorough coverage of topics than a newspaper can provide.

Contents for StarText are built from information written by the newspaper staff or taken from news wires and stored in the mainframe. Editors from the *Star-Telegram* automatically send locally written content to a special work queue where StarText editors select what they want. StarText editors continually scroll through leads on articles from the wire services to locate material from outside the region. When editors encounter something in the work queue or on the news wires, they transfer the information to the correct area in the StarText infobase electronically and instantaneously. On any given newsday, between two and three hundred stories are transferred this way. In addition several days' worth of stock quotations and the classified ads are inserted. At this time, the only element of the service that does not reside in the StarText computer is EAASY SABRE. In this case StarText serves as a gateway, transferring a subscriber through an X-25 connection to American Airlines' computers.

Recent changes to the system provide several ways of accessing information in addition to keywords and menus. A new tracking system allows up to 10 key topics to be stored so that each time the subscriber logs on, these 10 items are displayed. Subscribers can also type WHATSNEW and receive a list of recent updates to different areas in the infobase. An initial keyword can be selected so that whenever a subscriber logs on, the topic related to that keyword is the first thing seen. Even the E-mail functions may be accessed by menu or a set of commands.

14.2.5.1. Supporting Documentation

The one-hundred-plus page *StarText User Guide* provides complete instructions on how to obtain the most from the system. It is distributed in a 3-ring binder and contains a variety of information, beginning with how to log on to the system, how to define screen size, use keywords, and perform other rudimentary activities. A list of keywords is included to facilitate directly accessing information. There is also a complete directory of the approximately 320 categories under which classified ads are grouped, and each of the categories also lists the reference number for directly retrieving items in that category. Information about the type of customer service available and technical information explaining how to troubleshoot problems that may arise are also included.

StarText Ink is a free monthly newsletter designed to inform subscribers about new features on the system and to encourage greater use. A recent issue was devoted to stories, chain novels, and soap operas contributed by subscribers. Several of the literary pieces were replicated, and winners of various literary contests were listed. Contributions and comments from subscribers are also included.

14.2.5.2. Promotion

StarText is promoted in a variety of ways. A special relationship has been established with over 400 local school libraries that developed from a contact with a single school in 1984. Lesson plans have been built around StarText, and some students and teachers have become pen pals with people at other schools through the electronic mail feature. A number of special-interest products are under development or have been introduced recently to fill small niche markets, and in the aggregate these are expected to attract new subscribers to the service. One example is the Business Edition. This is a premium service priced at about 33% of the base price, and in many ways the additional charge is analogous to charging for another entertainment channel on a cable television system. Business Edition consists of 15-minute-delayed stock quotes, an executive business summary, special columns and news stories along with the regular StarText features. In 1992, StarText launched the Sports Edition. Priced at $19.95 a month, it offers all the regular StarText features plus expanded sports coverage, statistics, spectator sports (baseball, horse racing, football), and horse-racing information for five major regional tracks.

StarText is promoted in the newspaper, and ads and supplements are included periodically. On a more limited scale, a reciprocal agreement was arranged with WRR, a classical music station when it was discovered that their audiences were similar. StarText carries the station's music list, and in exchange, the station includes a reference to StarText in its newsletter and 30-second commercials. Two other commercial electronic information services are aggressively promoting on-line information in the Fort Worth area: U.S. Videotel, which began in Houston, is spreading to different locations throughout North America. Prodigy, a firm formed by IBM and Sears had been operating in several dozen large cities and in late 1990 expanded nationwide. Both of these services have advertised heavily in the Dallas—Fort Worth area, and the promotion has actually made people more aware of electronic information services, thereby increasing demand for StarText.

14.2.6. Challenges and the Future

Maintaining an acceptable response time as the number of subscribers grows is a major problem. But overshadowing this is the challenge of trying to sell an electronic service. The hardware, software, and content issues can be overcome through time and with money, but persuading people to abandon old habits and develop new ones is difficult. Management has found that the task of promoting the service on its limited budget involves two steps. First it is necessary to explain what videotex is and how it can assist the subscriber. This educational process is relatively easy to accomplish with the computer literate, but it is more difficult with the uninitiated. When the educational process has been completed, the task of selling StarText rather than one of the competing services remains. This involves explaining the convenience and immediacy of obtaining information electronically and by focusing on different niche markets where the service is particularly applicable. The school system is one such niche; another is the handicapped community. Shortly after StarText was started, a subscriber called to say, "Thank God for StarText"! This particular caller was a quadriplegic as the result of an accident 10 years earlier and StarText was the first newspaper he had read unaided since his accident. Others suffering from crippling injuries and illness have reported that the service has had a profound impact on their lives.

For the immediate future, the system will remain as text only, since management believes that customers do not require graphics for the type of information being provided. Another factor supporting this decision is the relatively low transmission rates of 0.3, 1.2, and 2.4 Kbits/sec used by the service. Many graphic images would require an intolerable length of time to transmit at this rate and thus frustrate subscribers. But perhaps the most important reason for avoiding graphics at this time is the special hardware or software that the subscriber would need. This additional expense is antithetical to the idea of offering a low-cost electronic service that practically any PC can access. To sum up management's opinion, "Our feeling is that this battle is going to be won or lost on content, not on graphics."

14.3. FORD MOTOR COMPANY

Ford is the second largest industrial and the second largest car and truck corporation in the world; it has about 381,000 employees and operates around the globe. Its two major units are the Automotive Group and the Financial Services Group. Ford is an innovator in electronically marketing its products and services. Ford has been using Prodigy almost since its inception and has advertised on CompuServe for four years. The following cases describe Ford's activities with these two commercial services. The first case traces Ford Motor Credit Company's experience on Prodigy, and the second describes how Ford North American Automotive Operations uses CompuServe. In addition to illustrating how companies pursue their business interests on commercial videotex services, the cases reveal how units within a large corporation implement information technology.

14.3.1. Ford Motor Credit Company

Ford Credit exists to increase the sales of Ford products. It carries out this function by indirectly providing consumers with funding to purchase or lease vehicles, buying contracts at a discount from dealers. Its role in the company is unique because of what it does and its relationships with customers, who are primarily manufacturing units and the dealer network.

A dealer of Ford, Mercury, and Lincoln products must obtain the new products it sells from the Ford Motor Company. However that dealer does not have to do business with the credit arm of Ford. It can arrange financing for its own needs and its retail customers from a variety of other sources. Consequently some dealers do not do business with Ford credit, which adds complexity to the business of financing vehicle loans. To work within this type of arrangement, the credit company's marketing activities are a little different than most other financing companies. It markets its services to consumers who are really the dealer's customers rather than its own in the hope of generating some business. Truly an unusual situation. But it has the overall goal of encouraging consumers to buy a Ford whether or not the dealer who makes the sale winds up selling the contract to Ford Credit. Within this obtuse environment the credit company must develop its promotional strategy.

The decision to turn to electronic marketing grew out of a strategic-planning activity that examined ways of marketing the unit's services. The strategic plan led the unit to Prodigy, which had just begun operations and had 15,000 subscribers. Prodigy was chosen for three reasons: It had a flat rate monthly subscription regardless of how many times it was accessed; it was the creation of Sears and IBM, which added to its credibility; and it seemed more innovative than some of the other services examined. Flat-rate pricing was considered important because it allowed the user to explore the contents thoroughly without worrying about cost. The presence of Sears and IBM provided the economic clout and resources to expand rapidly into major markets as well as staying power as the system evolved.

Ford Credit launched its service in January 1987 as a trial. During the intervening years, the service has been steadily upgraded in conjunction with other electronic services maintained by Ford Motor Company on Prodigy. Figure 14.6 illustrates the type of information that Ford presently (June 1992) has on Prodigy and breaks out in more detail the credit services content.

14.3.1.1. Strategy

The Ford Credit section of Prodigy is designed both to stand by itself and to be perceived as a segment of the much larger Ford Motor Company segment of the infobase. Considerable information about Ford Motor Company is contained in the system: Graphic images of Ford, Mercury, and Lincoln cars with their specifications may be viewed. A user can move between the automotive section and credit information or go directly to either one without accessing the other. How a user views the relationship is considered important, and it is treated in a very subtle way. The intent is to move a viewer from thinking about buying a car to actually calculating how much it will cost to lease or buy a particular car, then offer an opportunity to send for brochures electronically and even leave an E-mail message for a Ford representative. This is a nonintrusive way of leading a viewer through the process of inquiry, shopping, and deciding on a course of action while thinking Ford.

At this point in the evolution of electronic marketing in Ford, the dealers are relatively oblivious to efforts being made on their behalf. Most dealers are much more aware of passive forms of promotion, such as newspaper and magazine ads. Unless a dealer subscribes to an on-line service or a customer states why he/she is interested in a Ford product, this form of promotion is largely unnoticed by the dealer network, although dealers have been informed by the company that these marketing activities are being undertaken on their behalf.

14.3.1.2. Objectives and Outcomes

The purpose of Ford's electronic-marketing effort is to provide financial information to complement products and persuade potential financing customers to purchase Ford products.

The specific objectives checked on the survey are listed in Table 14.2 with their rank. It has not yet been determined if this form of electronic marketing sells more products or services, and thus its impact on market share is uncertain because there are so many other forms of promotion and factors that influence a car buyer; however the other objectives listed are being satisfied. Perhaps more important than some of these objectives is the experience it provides. Such experience will help shape further electronic-marketing endeavors while it strengthens current advertising and promotional activities.

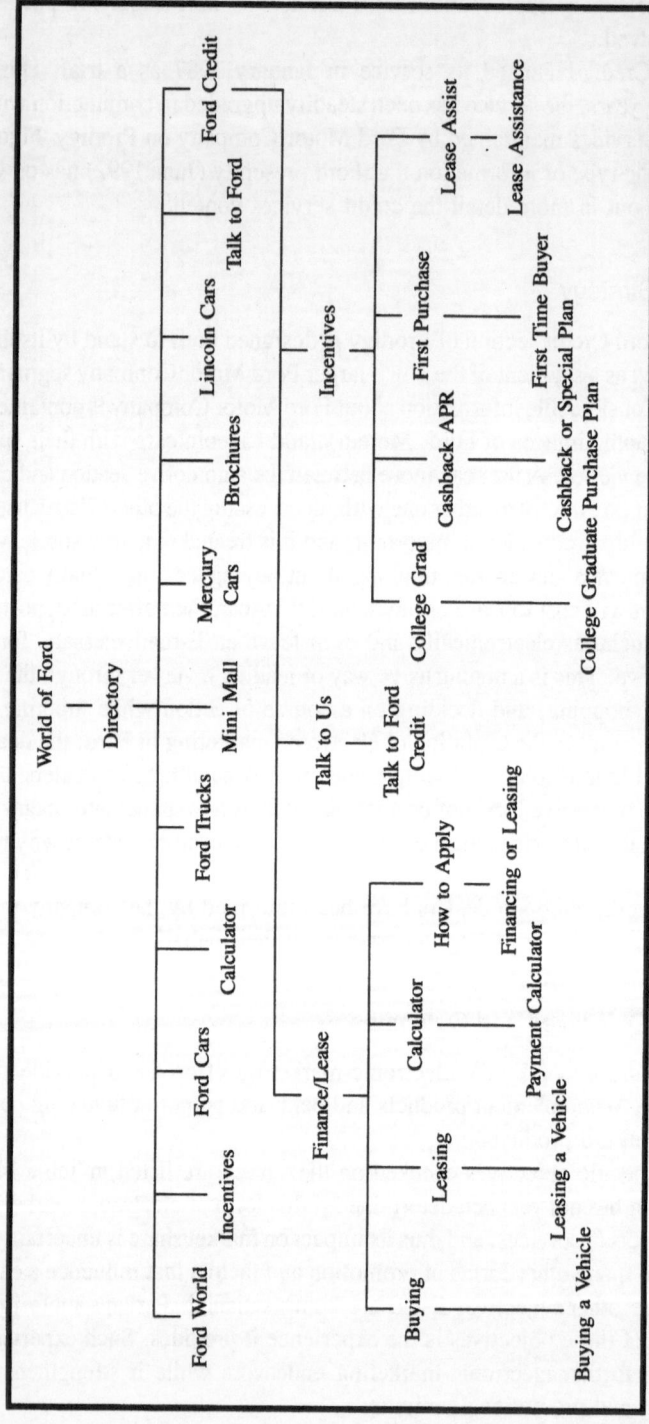

Figure 14.6. Several levels of Ford Credit information in the Prodigy infobase.

Table 14.2. Objectives of Electronic-
Marketing Efforts on Prodigy

Rank	Objectives
1.	Sell products or services
2.	Increase market share
3.	Distribute information
4.	Differentiate products or services from competitors'
5.	Improve public, customer, or stockholder relations
6.	Improve accuracy of information

14.3.1.3. Presentation and Communications

The project is limited to forms of presentation supported by Prodigy; these are text, color, and graphics. Limited animation is also possible on Prodigy, but it is not included in the content for Ford Credit. Access to Prodigy is made through a combination of its own network and Tymnet, a value-added carrier. Communications speeds of 1.2 and 2.4 Kbits/sec are supported between the customer's location and the local Prodigy or Tymnet node. Figure 14.7 illustrates the Prodigy Services Company system architecture, which is designed to serve a national clientele; approximately 90% of the U.S. population can reach Prodigy with a local telephone call.

The point of access with the network and its contents is the user's PC. The software installed on the PC is called the Prodigy Application Language or PAL; it is an interpretative system for processing chunks of information known as objects. An object may be data, a graphic, text, a program, or even a template of the screen that the subscriber sees. In Figure 14.7, objects flow downward from the host to the PC.

When the PC is logged onto the network, it constantly asks for and receives objects or sends and receives messages. When a user calls the service, he/she logs onto either a local node, called a PLS (for Prodigy Local Service), or a Tynmet node, which is connected to a regional PLS node. In mid-1991, there were 110 PLS nodes distributed around the country. Each node consists of one or more small computers capable of handling 256 simultaneous calls. The Tymnet network is used in small cities and sparsely populated areas where it is not economically practical to set up a PLS.

Logging onto the PLS establishes a link that lets the program interact with the PC. Anytime the PC needs an object to satisfy a subscriber's request, it looks on its own RAM and then its hard disk. If the PC does not find what it is seeking, it goes to the PLS; 90% of the time that the information is not on the PC, it will be found at the PLS. The remaining 10% of the time, information must be retrieved from the Delivery System, which is an IBM mainframe. The Delivery System acts

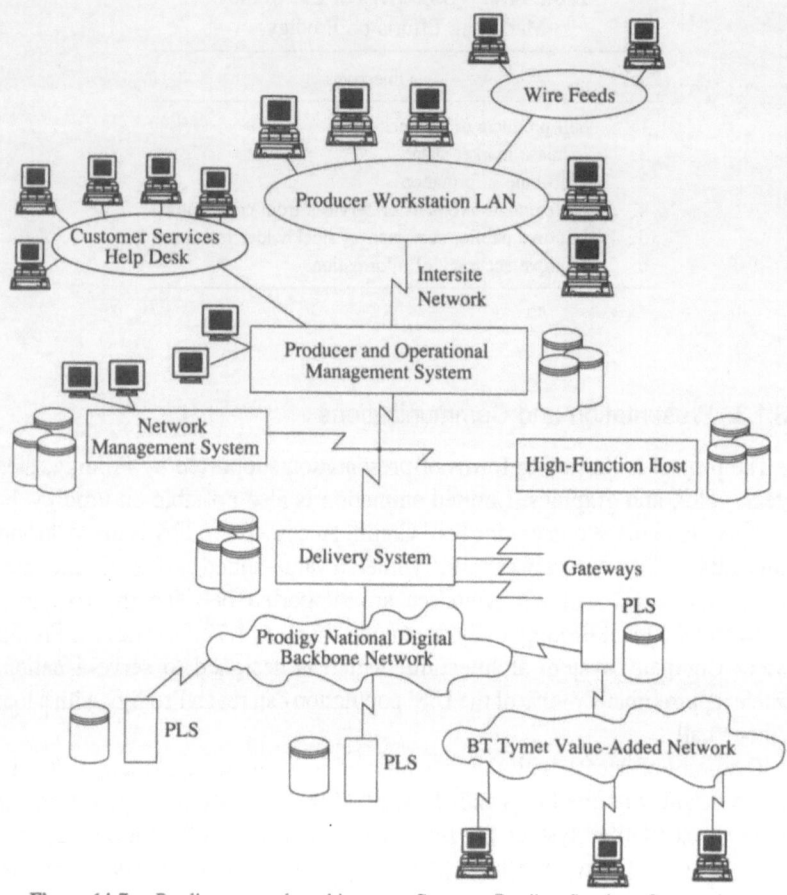

Figure 14.7.   Prodigy network architecture. Courtesy Prodigy Services Corporation.

as a central archive for the network of PLS nodes, and it is also the gateway to information maintained on nonProdigy computers; EAASY SABRE, Dow Jones/ News Retrieval, and on-line banking services are three examples of systems that use a gateway to access information on nonProdigy computers.

The Prodigy National Digital Backbone Network, which connects the PLS and the Delivery System, consists of T-1 and T-3 lines. Once information is located or accessed via a gateway at the Delivery System, it is sent over the backbone network to the PLS. On receipt at the PLS, it may be stored and is forwarded to the subscriber who initiated the request. Located at the next level of the hierarchy is the Producer and Operational Management System, which is the depository for content under development. It is connected to the Customer Services Help Desk unit and the Producer Workstation LAN.

When information in the form of an object is being updated or replaced, the object is sent by the Producer and Operational Management System to the Delivery System. On receipt the Delivery System notifies local services to delete that particular object. The next time a PLS needs that object, the latest version is obtained from the Delivery System. The Producer Workstation LAN is located at the top of Figure 14.7. This is where the contents of the service are created; it is also the point of interface with that part of the system where content is assembled from various wire feeds. One of the wire feeds is the AP Wire Service. Information received over the AP wire is routed to an editor who reads the contents, does some editing, then places the story in a message, which is inserted into an appropriate area in the Producer and Operational Management System.

There are three forms of support shown in Figure 14.7: (1) The Network Management System monitors the network to detect problems, then takes appropriate action. For example, if a modem goes down somewhere in the network, it detaches the unit and records the failure. (2) The Customer Services Help Desk enables employees to check on customer account information maintained in the Producer and Operational Management System and respond to subscriber queries. (3) The High-Function Host supports special applications.

14.3.1.4. Design, Implementation, and Operation

The initial 1987 design and content selection were a joint effort of personnel from Ford Credit and Prodigy, with production work done by Prodigy. In 1990, J. Walter Thompson/OnLine (JWT OnLine) of Detroit, which produces and manages other Ford Motor Company infobases on Prodigy and CompuServe, was given the responsibility of redesigning the Ford Credit infobase and updating its contents as required. The contract with JWT OnLine includes creative, production, management, and fulfillment responsibilities. Some organizations that advertise or sell goods and services through a commercial on-line service like Prodigy have allowed the service provider to design and create their content for them. For many organizations, this is not good policy for a variety of reasons. (1) The service provider does not and cannot be expected to know the business of each and every client as thoroughly as the creative services firm that normally handles advertising and marketing. (2) The service provider is not aware of ad campaigns and other promotional activities that are planned. (3) The service provider does not have a history of involvement with the organization and has not established the level of rapport that the creative services firm has normally developed. The primary business of the service provider, who in this instance is Prodigy, is developing, marketing, and operating the on-line service, and not creating ad campaigns and promotion strategies for the various organizations whose message is carried on the service. The client whose information is carried often does not have personnel who know and understand the complexities of the electronic media, thus they are not able to advocate positions and approve decisions made by the service provider.

For these and other reasons, organizations that have their information carried on a commercial on-line service often work through such third parties as JWT On-line.

Another third party that Ford Credit uses is called a fulfillment house. When someone orders a brochure or buys a copy of the Ford Simulator, a computer-driving game offered as a promotional item, the fulfillment house electronically retrieves the name, address, and other necessary information, fills the order, and ships it.

The text portion of the Ford Credit infobase is updated weekly and the graphics portion annually. Approximately 15% of the text and 25% of the graphics material are changed with each update.

Management considers graphics an important part of merchandising automobiles and the accompanying services, not just for the information conveyed but also because graphics help hold a user's attention. However the Ford Credit portion of the infobase is largely devoid of graphical content; therefore a way had to be developed to present information to hold a viewer's attention long enough to learn about Ford Credit. One way this has been done is with a calculator for computing leases and purchases, which provides a reason for proceeding through the automotive section gathering information and then enables the user to interact with the system in a meaningful way.

14.3.1.5. Problems and Plans

The greatest problem faced was designing the infobase so that it would appeal to the typical Prodigy subscriber while incorporating the calculator for analyzing lease versus buy decisions. Developing a calculator is much more difficult than it appears to be due to the usury laws in various states and federal laws pertaining to truth in lending and truth in leasing. A further complication resulted from using the same calculator for financing and leasing calculations. The calculator must also consider lease-end values and produce appropriate estimates that are sufficiently accurate to be of value to the user. Thus the same calculator accessed by people all over the country must not violate any laws anywhere or mislead the user.

By the mid-1990s, management hopes to make it possible for consumers to apply for preapproved credit on-line. The calculator and information in the infobase are forerunners in teaching the customer what information must be provided to obtain preapproved credit.

The long-term impact of electronic marketing on the company as a whole is expected to be significant. The average earning power of the buying pool should increase because Prodigy and other on-line commercial services have users who are considerably wealthier and better educated than the average buyer. Also preapproved credit, the dissemination of information, and tools to assist in decision making are expected to impact favorably on market share as new customers are attracted.

14.3.2. Ford: North American Automotive Sales Operations

The marketing department of Ford's North American Automotive Sales Operations is continually exploring new media to market Ford products. In 1988 a trial was initiated on CompuServe to assess the potential for direct electronic marketing to consumers. The trial was viewed as a way of exploring new opportunities for transferring information and communicating with a group of people who were not accessible through traditional media, such as print advertising.

14.3.2.1. Ford's Electronic Showroom

Information incorporated in the system enables potential customers to learn about specifications, equipment, and prices without visiting a dealer. Ancillary activities on the system are designed to build on user curiosity and also to enable the company to form a contact list of potential buyers from a very focused market segment of the buying public. The contents of the infobase are continually changing; a description of information taken in June 1992 follows. The opening screen welcomes the visitor and makes a brief statement about financing and incentive programs to pique the viewer's interest. There is also an invitation to examine a particular vehicle or vehicles. This screen is followed by a menu listing options shown in Figure 14.8.

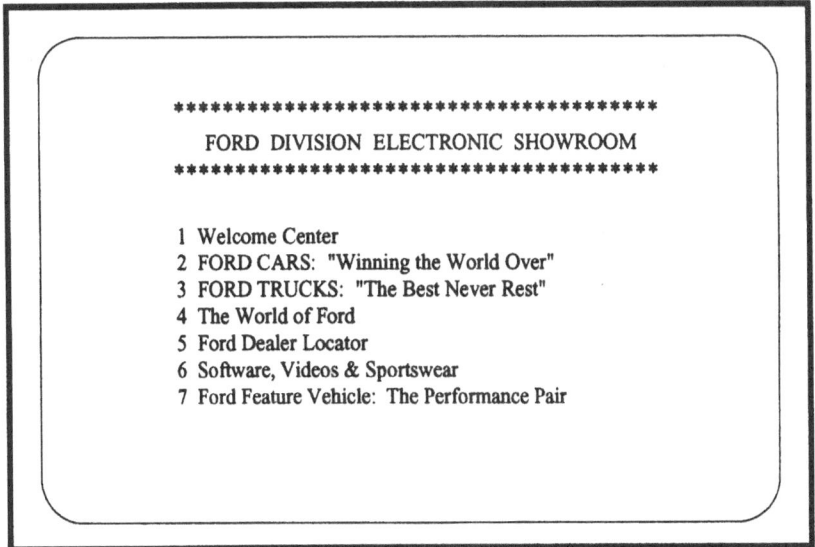

Figure 14.8. Main menu for the World of Ford Electronic Showroom on CompuServe.

The Welcome Center screen leads to a screen explaining how to send a message, how to become better acquainted with what the World of Ford has to offer, how to obtain additional information about current incentive programs, and how to obtain an overview of Ford's warranties and customer satisfaction programs. The screen explaining how to send a message carefully points out that this feature is for commenting or asking questions about information on CompuServe, the Ford Motor Company, or Ford products. The viewer is explicitly warned that this feature is not for solving automotive problems or contacting a dealer.

Options 2 and 3 enable the viewer to access information about cars and trucks, respectively. This information consists of vehicle specifications and the suggested manufacturer's retail price. Each vehicle also has a menu option for ordering a brochure on-line. Option 4 accesses a screen promoting new features of the service and options that refer to some of the same frames accessed by options on the menu in Figure 14.8.

Option 5, Ford Dealer Locator, enables a viewer to locate the nearest dealer by selecting any of the following search methods: city, state, or province; zip code (U.S.); postal code (Canada); and telephone area code. Option 6 accesses a specialty department for ordering software, video tapes, sportswear, and a variety of accessories. One of the software packages combines vehicle information with driving games; a description of one example, the Ford Simulator follows. Option 7 features a particular vehicle or vehicles, which in June 1992, included the Ford Taurus SHO and the Ford Thunderbird Super Coupe.

The text-based content of CompuServe is considered by management to have some advantages over a graphics-oriented service because of the economics involved, since the incremental cost of providing additional information as text is generally less expensive. This cost savings permits an in-depth treatment of the subject. The infobase on CompuServe historically generates about 1000 requests a month for additional information. The nonscientific checking that has been done suggests that those who request additional information are seriously considering buying a car.

One of the promotional items that can be ordered on-line or through magazine advertisements is the Ford Simulator III, which is a clever combination of information about new cars and the simulation of a drive to Lake Wakatonka with graphics of cars and you as the driver. The simulator is packaged on two 3.5 disks to be installed on a harddrive. The opening screen invites the viewer to examine the model line, explore a model, drive a car of the user's choice to the lake, or visit the customer center. Especially innovative features are the ability to choose a particular model and see it in various colors; create a worksheet for a particular model with various costs, including tax; and the ability to compare two different vehicles on the screen at the same time, feature for feature. The customer center contains a questionnaire that is filled out electronically, then printed and mailed to Ford. There is also an offer for real aficionados to buy a T shirt featuring exciting graphics from the driving game. The most fascinating aspect of Ford Simulator

III, however, is not its contents but the idea that people are willing to pay $6.95 for it. Ford has been able to sell a new version of the disk each year by modifying the games and updating the contents to reflect the new model year. Over 200,000 versions of the simulator have been sold.

14.3.2.2. Objectives

Several very specific objectives for pursuing electronic marketing on Compu-Serve are shown in Table 14.3. Management feels that the three highest ranked objectives are being met and progress is being made toward the others. Two benefits were cited in addition to the objectives listed, both related to the learning that has occurred in working with on-line media services. The first benefit is learning how to make user participation more effective in an electronic media; the second is knowledge gained about this particular market segment.

14.3.2.3. Technology

In comparison to the other cases, the technical sophistication incorporated in CompuServe is low, but it is a tried and proven form of electronic communication, and its text-based content is the lowest common denominator for effecting interaction with any user. Most PCs with a modem and communications software can be used, and the communication speeds supported are 0.3, 1.2, and 2.4 Kbits/ sec so the cost of equipment is minimal.

14.3.2.4. Design, Implementation, and Operation

Ford provided the initial main menu and indicated what it wanted to incorporate into its infobase. Working with this information, CompuServe then established the structure and the links. The actual contents were produced by JWT OnLine using information assembled and forwarded by the Sales Operations unit.

Table 14.3. Objectives of Electronic-
Marketing Efforts on CompuServe

Rank	Objectives
1.	Provide product information
2.	Position company for future communications
3.	Distribute information
4.	Improve communication
5.	Distribute information sooner
6.	Improve public, customer, or stockholder relations
7.	Increase market share

The same procedure is followed for maintaining the contents. Part of the process of developing the information involves a lengthy legal review and approval cycle, which significantly lengthen the lead time for updating the contents.

Requests for brochures, the Ford Simulator III, and other items are automatically routed to several mailboxes depending on what the item is. Then a fulfillment house that accesses the mailbox on a regular basis downloads those requests to a PC, which generates letters and labels.

About 5% of the text is changed weekly, 15% monthly, and 90% yearly. The updates are produced at the JWT OnLine agency in Detroit and then uploaded by modem to a staging area in CompuServe's computer. CompuServe is then notified and makes the changes at the proper location in the service. Approximately 33,000 accesses were made to the Ford contents in one year, and the average length of the session was 17.5 minutes.

14.3.2.5. Challenges and Plans

The greatest problems have been continuously creating copy to keep the system current and developing techniques to upload input easily and dependably. Over the next several years, the system is expected to incorporate better presentation methods and increased downloading and interactive functions. More use of direct response marketing techniques to enhance the cost effectiveness of the system from Ford's perspective are anticipated.

One aspect of the change in presentation is expected to be graphics. CompuServe is presently experimenting with graphics, and some graphic content unrelated to Ford's products is available. One feature involving graphics recently introduced by Ford is downloading VGA-quality images and vehicle information. Downloading in the future may include copies of the Ford Simulator. Part of the justification for investigating downloading is the price sensitivity of subscribers when the service or telephone charges are billed according to the time used. Downloading could be free, or it could be done during periods when rates are reduced.

Where Ford's involvement with electronic marketing will lead is anyone's guess at this point. Perhaps someday Ford will be able to justify piggybacking a consumer-based service on its own networks modeled after its Prodigy and CompuServe efforts. But one thing is sure: Electronic marketing has begun to carve out its segment of the media options available to marketers, and Ford Motor Company is right in the thick of things!

14.4. AMR CORPORATION—EAASY SABRE

The AMR is a holding company with annual revenues of nearly $13 billion. Its major subsidiary is American Airlines, the largest passenger carrier in terms of revenue in the Western world. One of American's largest divisions is SABRE Travel Information Network (S.T.I.N.) which markets SABRE, the world's largest privately-owned real-time computer system. Beginning with its Sabre reservation system in the 1970s, AMR has been a leader in using computer technology for strategic gains in the airline industry. EAASY SABRE continues this tradition and will contribute significantly to AMR's success in the 1990s.

EAASY SABRE is a travel information and reservation service that can be accessed by travelers rather than travel agents. Access is made using a PC or terminal equipped with a modem and the telephone network. The service is available through more than twenty on-line information services, such as Compu-Serve, GEnie, and Prodigy.

14.4.1. A Strategic Thrust

A strategic thrust can be defined as an action designed to gain or maintain a competitive edge or lessen a competitor's advantage; AMR made such a thrust in 1975 with the introduction of its Semi-Automated Business Research Environ-ment, which is popularly called the Sabre System. The heart of the system is a massive computer complex in Tulsa, Oklahoma, that stores over 45 million fares and has created as many as 1.2 million reservations per day. The Sabre can handle over 3100 messages a second and serves 24,000 travel agencies in 57 countries on six continents.

Sabre was originally conceived as an internal system to automate the seat reservation process, but by the late 1970s, it had emerged as an important strategic element in AMR's quest to protect its business and expand its market share. During the late seventies, Sabre was first leased to travel agents as was Apollo, United Airlines answer to Sabre. These systems fulfilled their purpose by mak-ing it easier and faster to reserve a seat, but they also did much more.

Early reservation systems were programmed with a bias to facilitate bookings on the airline of the company operating the reservation system. At this point, automated systems moved beyond the point of being computerized reservation systems and became a strategic weapon to draw business that might otherwise have gone elsewhere. Sabre and Apollo became so effective at capturing busi-ness that by the early 1980s, together they accounted for 80% of the computerized travel agency market.

By 1985 Sabre was contributing more to AMR's profit before taxes then the company was making flying passengers. Revenue came not only from seats sold on American Airlines, but from fees paid by other carriers whose seats were sold through Sabre and from fees earned from reservations for car rentals, hotel reservations, and even from travel agents who leased Sabre terminals. In addi-tion to the money earned, there were other benefits in the form of information obtained about individuals making reservations, the volume of traffic on compet-ing routes, the booking activity of particular travel agencies, and even the pricing plans of competitors who had to submit fare changes to Sabre in advance. The Sabre provided these and other competitive advantages to AMR, but nothing this good could last forever. The Civil Aeronautics Board prohibited bias on screen displays, which was upheld by the courts in 1985. Competing airlines, seeing the writing on the wall, aggressively began carving a niche for themselves by joining forces and starting rival systems. However before these forces were fully in place, AMR began planning countermeasures.

14.4.2. EAASY SABRE—A System for Lookers and Bookers

The genesis for EAASY SABRE was a study done at the Center for Futures Research at the University of Southern California. The study was funded by nine corporate sponsors, one of which was AMR. One of the products of the research was a book entitled *The Emerging Network Marketplace,* which prophesied a number of developments, such as the emergence of the home and office PC.

At the completion of the study, AMR set up a small unit in 1979 to monitor developments in this emerging market and begin laying groundwork for a consumer-oriented version of the Sabre system. One of the important developments monitored was the spread of PCs among consumers, particularly those who flew frequently. A principal source of this information were on-board surveys conducted twice a year for four years by American Airlines. Some interesting trends were revealed: Between 1983 and 1985, the percentage of the flying public that owned PC's increased from 18 to 23%; the percentage having access to them at work increased from 26 to 45% during the same period. Even more interesting, the percentage using on-line services increased from 12 to 30%. These percentages were greater for the flying public then for the population as a whole throughout this period and probably remain so today.

By 1983 it was becoming apparent that the time was approaching when a sufficient demand for a consumer-oriented on-line travel and reservation system would exist. At that point, the planning unit was moved from data processing to a group responsible for market automation; today the unit is known as STIN, the Sabre Travel Information Network. After a lengthy incubation period, the service was launched on Halloween, October 31, 1985.

A convenient way of thinking of EAASY SABRE is as a gateway to the Sabre system. Through it the consumer or business traveler can access 360 airlines to make reservations and 651 airlines for flight schedule information. In addition reservations can be made at 27,000 hotels and with 57 car rental agencies throughout the world. Augmenting this service are weather information from the National Weather Bureau; access to the user's frequent flyer account status; information about discounted travel opportunities; and a computerized directory of leisure travel information, products, and services, called the Official Recreation Guide.

Enhancements and new features are being added regularly. By August 20, 1992 EAASY SABRE had enrolled 883,509 subscribers, and plus there were many other users who used the system but had not subscribed. EAASY SABRE is available through more than twenty on-line services. To an observant user, menus on many of these systems vary slightly; a generic main menu is shown in Figure 14.9.

Revenue is generated in several ways. The content of EAASY SABRE is divided into basic and premium services, with the larger portion falling into the former category. Users normally pay the operator of the on-line service a subscription fee and some of that revenue may be shared with EAASY SABRE, but access to the basic services on most systems is free. However a surcharge is levied for premium services, such as Travelers Access and the Official Recreation Guide.

A major source of revenue is the booking fee generated by the Sabre system. EAASY SABRE extends the market of the Sabre system to every home or office in the world equipped with a PC or dumb terminal. This market is expanding rapidly and its potential size is huge. Every time a subscriber books through EAASY SABRE it is additional revenue for Sabre and represents lost business for a competitor. If a passenger books through EAASY SABRE and the reservation is sent to a Sabre agent for ticketing, EAASY SABRE collects $2 for saving the agent at least $2 worth of time. A new and potentially significant source of revenue is planned for the near future that will enable travel vendors to promote their wares by inserting advertisements on EAASY SABRE screens.

EAASY SABRE MAIN MENU

1 System Quick Tips	1 Helpful information
2 Travel Information and Reservations	2 Access to flights, cars and hotels
3 Weather Information	3 From National Weather Bureau
4 AAdvantage	4 Your personal AAdvantage account status
5 Applicaton to Use EAASY SABRE	5 Application for free membership with immediate booking privileges
6 Profile Review and Change	6 Contains personal data to expedite booking process
7 Travel Club	7 Access to discounted travel opportunities
8 Sign Off	8 Removes booking authorization

From any display . . . Type Help or ? for assistance, or
Res or /R to go to Reservations Menu
Top or /T to return to Main Menu or
Exit or /E to return to System Operator

Figure 14.9. A generic example of the main menu for EAASY SABRE. Courtesy of AMR Corporation.

14.4.3. Objectives

The specific reasons for launching the service are shown in Table 14.4. All of these objectives are bound together to form a strategic thrust designed to place the ability to find information and make reservations without going through a middleman in the hands of the traveler. The long-term impact is hoped to be more bookings, a chance to enter new businesses, additional sales, the distribution of information, increased market share, and improved constituent relations.

14.4.4. Technology

The AMR has two separate computer systems, one for supporting the internal needs of the corporation and another dedicated to operating the Sabre system that supports more than 24,000 travel agencies and 99,000 terminals in addition to those accessing the system through EAASY SABRE. It has nine IBM mainframes and 100 VAXes for handling reservations, baggage, and other travel-related functions. The interface to the communications network is a VAX computer cluster that is tied to the IBM computers on which the Sabre system resides. The interface between the VAX cluster and the network is the X.25 network standard.

The EAASY SABRE software acts as a protocol translator and resides on this VAX cluster. Its purpose is to pass requests from the users to the Sabre system, then pass the responses back to the user. This is an important point: Each of the more than twenty on-line services that offer EAASY SABRE are merely a conduit. A user's request for service is not handled by that service operator; instead the request is forwarded to the VAX cluster, which in turn forwards it to the Tulsa computer site hosting the Sabre system where the information resides.

Most of the on-line services use ASCII, a common transmission code. However each additional service that is added contributes to the complexity of the network and to the difficulty of interfacing with different service operators; Prodigy requires the most work. This is partly attributable to the graphics involved and the architecture of the service; a separate development team exists just to support Prodigy.

Table 14.4. The Objectives behind EAASY SABRE

Rank	Objectives
1.	Open new distribution channel for information and transactions supplied to different market
2.	Enter new business opportunities
3.	Sell products or services
4.	Distribute information
5.	Increase market share
6.	Improve public, customer, or stockholder relations

14.4.5. Designing to Meet User Needs and Foibles

SABRE was designed by company employees, and its contents are also maintained by AMR; changes in fares and schedules for the other airlines, which form a segment of the infobase, are provided by those carriers. Initially the group started with four people under a managing director; by the end of 1991, it had grown to include a marketing group of about a dozen people, a product development group, an applications development group, and a systems engineering group. At one point there were approximately 60 people working for EAASY SABRE but due to cutbacks there are presently 20; this does not include employees who maintain the Sabre system from which EAASY SABRE retrieves its information about fares and schedules.

A major factor of EAASY SABRE's success is the emphasis placed on user feedback: The service is continually being modified on the basis of what is learned from users or their surrogates. Studies have shown that the user population is slightly different from the one using commercial videotex services; for example, the medium income of videotex subscribers is in the high 50s, while EAASY SABRE subscribers are in the high 70s. Another important consideration is that EAASY SABRE appeals only to travelers among on-line users, which further limits the potential market. For such reasons, it was deemed crucial to understand EAASY SABRE subscribers.

Originally the entire user base of 11,000 subscribers was sent a questionnaire, but as the system grew, that proved impractical. Consequently a panel was established composed of the more enthusiastic members of EAASY SABRE. The panel is called the AAdvisory Board, and it currently has 700 active members. It is used to brainstorm and design questionnaires, test new features in the system, and comment on promotional campaigns. The membership changes slightly as members ask to leave and others are added. To be admitted to the board, a member must complete a questionnaire providing demographic, usage, and travel information. The questionnaire is the basis for creating focus groups for special studies.

14.4.6. Usability Laboratory

A facility called the Usability Laboratory has been built to take advantage of research opportunities available with the AAdvisory Board. The laboratory is also rented to other organizations wishing to do similar research. The laboratory, one of the most complete facilities of its type in the United States, is illustrated in Figure 14.10. There are two complete test rooms and support facilities. The person being studied sits in the test room, which contains a desk and PC. There are three ceiling-mounted cameras arranged to record facial expressions, the keyboard, and even material on the desk simultaneously. The PC is connected to a monitor in an adjoining control room to permit direct viewing and recording of what appears on the screen.

Facility Floor Plan

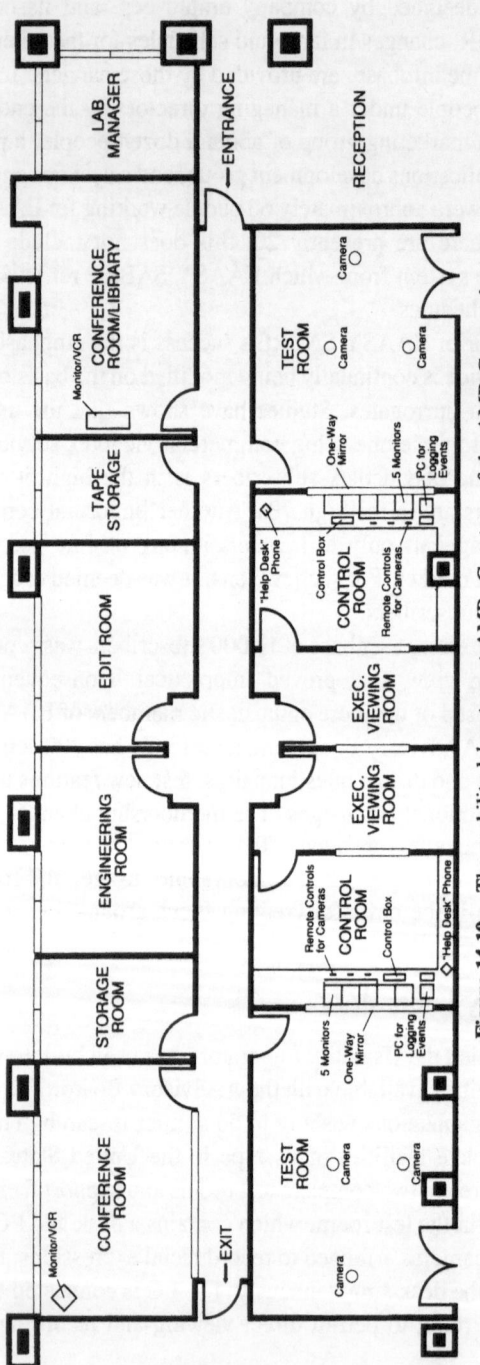

Figure 14.10. The usability laboratory at AMR. Courtesy of the AMR Corporation.

Adjoining the test room and separated by a one-way mirror is the control room. This soundproof room has five monitors, a video recorder, remote controls for the cameras in the test room, and a help telephone for assisting the test subject, five audio headsets with an internal and external intercom, and a PC for recording information about the evaluation. One of the five monitors displays what is being recorded on tape, which may be a mixture of several displays from the other monitors. A typical display might contain a view of the subject's screen with an inset showing the face or hands. Seating for five technicians is provided at a control console adjacent to the mirror. A third room in tandem is for viewing; it is elevated, and the side adjoining the control room has a large viewing window. Up to eight spectators can occupy this room to watch the test subject or observe activity on a monitor.

Several special purpose rooms are located near the test rooms. Next to the main entrance is the laboratory manager's office. There are two conference rooms, one of which also serves as a library. The extensive array of special equipment necessary to operate the facility is located in an engineering room. An edit room containing equipment for preparing tapes showing the research and two storage rooms complete the facility.

14.4.7. Probing the User's Psyche

To illustrate the extensive research that is done in support of EAASY SABRE, how the AAdvisory Board assists in the research effort, and how the Usability Laboratory is used, one of the research projects is described. This particular study involved four groups of six people. Each person spent between 3–4 hours in the laboratory. The first group consisted of EAASY SABRE bookers—regular users of the service who booked their own reservations. The second group consisted of subscribers who looked regularly but did not book their own reservations. The third group consisted of people who had never used a PC before. The last group consisted of highly experienced PC users who were not subscribers. The tasks given each subject involved four increasingly complex travel scenarios.

As expected the bookers experienced few problems and completed the four scenarios in the shortest time; the lookers finished second. Research uncovered some interesting facts about bookers and lookers. Bookers used many shortcuts and took advantage of some of the expert features, such as the bargain finder option; as a result, bookers never looked at the fare screens. In contrast lookers did not select that option, and none of the six ever selected it. Why did this behavior prevail? A later analysis of the demographics for a larger segment of the subscribers suggests the reason. Bookers outnumber lookers in every age category under 40, while lookers predominate in every age category over 40. Bookers take different kinds of vacations, more vacations, have different preferences for leisure products and services, and display other characteristics indicative of a different mindset. One of the emphases in the past has been to convert lookers into bookers.

Research may suggest that this is a lot more difficult than simply instructing someone how to select the correct keys; there may be other factors imbedded in the human psyche that must be considered.

The third and fourth groups in the study had their own set of hurdles to overcome. Individuals who had no PC experience varied in how well they did. Those who were frequent travelers had less difficulty figuring out instructions and eventually worked their way through the scenarios. The infrequent flyer had considerably more difficulty. Surprisingly the group with the most PC experience had the greatest difficulty. Presumably they expected system features to be slick and fast like the software they were accustomed to on their PCs. When function keys and wild card choices did not work, they were sometimes at a loss. They also spent considerable time studying fare rules and other technical portions of the content that are of little value to the typical user and make very little sense to the untrained. These findings reinforce the fact that research often uncovers unanticipated results.

14.4.8. Challenges and Plans

Understanding the needs of the consumer and clarifying mystifying industry data are the *sine qua non* for EAASY SABRE. The AMR has established the foundation for understanding the user psyche and devotes considerable resources to the effort. Such studies as the one just described reveal how great the challenge is. One of the most surprising results to come out of studies in the Usability Laboratory is that people simply do not read the screen in front of them, especially if they have a lot of experience with a PC. One of the scenarios in the study led users to a portion of the fare rules and restrictions. An option was a carriage return leading out of the morass and also a menu of options leading to highly technical information. Users almost universally opted for the menu choices that led them deeper into the incomprehensible details of the system and ignored the carriage return. The lesson appears to be that information on the screen must be arranged in a way that makes it absolutely clear what the user should do next. Designers of this particular screen now recognize the danger of putting a carriage return at the bottom of a screen containing a menu if the carriage return is the preferred choice.

One of the greatest challenges facing EAASY SABRE is making the complex fares and restrictions intelligible to the typical user. A small part of the solution lies in the user's self-education, but a major portion of the burden falls squarely on EAASY SABRE. Through its screen design and presentation, it must enable people to navigate successfully through the system. Future products under development, such as a *Windows* based interface to EAASY SABRE will continue to improve the dialogue between the system and the user.

EAASY SABRE will gradually become a larger and larger contributor to the company's revenue stream. The AMR recognizes the challenges it will face as this occurs and is moving aggressively to meet them by better understanding its customers and their needs.

Chapter 15

Cases:
Public Access Terminals

Many of the most innovative and successful systems studied were kiosks. A kiosk was defined in the questionnaire as an enhanced system placed in a public location, such as a shopping mall, hotel lobby, or conference facility, to be used by any passerby. Twenty-one organizations surveyed described a kiosk project; 15 were networked and six were stand-alone units. In addition to such advanced multimedia features as audio and video, many systems have a touch screen or a membrane keyboard, and some systems may incorporate a telephone handset, magnetic stripe or bar code reader, printer, or coupon dispenser. Fourteen of the projects in the survey were permanent and seven were considered trials. As Table 15.1 indicates, the majority of the systems were focused externally on strategic applications.

Systems distributing information outnumbered all other applications combined (see Figure 15.1). A few of the internal applications served the following purposes. One organization placed kiosks in various locations so that employees could obtain information about benefits and ask questions; employees could even perform transactions to change their plan enrollments. A university employed a kiosk system to provide campus information, and another organization implemented lobby systems to inform visitors and control access to certain areas.

Most of the applications involving information distribution were focused externally. One clothing manufacturer described a kiosk in retail stores that contained ads, size charts, a toll-free phone number for customer grievances, among other things. A half-dozen organizations described slightly less elaborate systems that contained multimedia features designed to promote either the organization or its products and services. Four organizations described how they used kiosks either for selling information they produced or selling a product or service.

Table 15.1. Form and Focus
of Kiosk Systems

	Focus	
Form	Internal	External
Operational	1	2
Strategic	4	14

Two units for selling products were developed by large retailers and contain both information about the products as well as the ability to place orders through the kiosk for delivery later. Four units were used for training and education. Three of the systems are installed in museums and integrated with the exhibits; the forth is more general in nature but directed at the museum industry for the time being.

The first case discussed in Chapter 15 describes the Unisystem at the Franklin Institute Science Museum in Philadelphia. When Unisystem was launched, it was undoubtedly the most sophisticated museum system of its kind in the world. The second case focuses on the Town Crier, a kiosk system installed throughout General Motors and one of the oldest and most successful information systems of its type in any company. The third case describes Heartfelt Advice at Merck & Co. This system is well designed, innovative, and one of the most attention-grabbing systems anywhere.

15.1. FRANKLIN INSTITUTE SCIENCE MUSEUM—UNISYSTEM

The Franklin Institute Science Museum in Philadelphia is developing a museumwide, visitor-oriented, multilevel computerized information system. A pilot study was conducted in December 1989, and the first phase of the system was implemented in May of the following year. When completed in 1991, the system had given birth to the world's first truly smart museum. The $3.4 million project is a joint effort of the museum and the Unisys Corporation, which is providing

Information distribution	13
Selling information	2
Selling a product other than information	2
Training and education	4

Figure 15.1. Major categories of applications for kiosk systems.

most of the hardware, software, and technical support. Additional funding is provided by the National Science Foundation.

15.1.1. A Dream That Came True

Like many science museums, the Franklin Institute has a long history of using technology in its exhibits; in fact the first computer was placed on the floor of the institute in 1958. Technology is integrated with exhibits in two ways: First technology itself is the focus of a display; second technology is a tool for interpreting or explaining an exhibit. Unisystem is an implementation of technology as a medium of interpretation.

In the mid-1980s, the museum staff began thinking about how to use information technology to facilitate a visitor's tour of the exhibits. Visitors to a museum form a diverse population: They are different ages, come from different socioethnic groups, have a wide array of educational backgrounds, and possess a unique set of interests. Traditionally this population was served by putting on each exhibit one or more labels tailored to appeal to the widest possible audience. Unfortunately there is a practical limit to how many labels can be placed on each exhibit. This problem is compounded when helping route the diverse population to different exhibits throughout the museum. Interpretation and routing problems were solved with Unisystem.

15.1.2. Unisystem

15.1.2.1. How the System Works

As visitors enter the museum, they are handed a bar-coded brochure explaining how to use the system and containing a unique bar code for that visitor; the brochure is shown in Figure 15.2. As the visitor tours the museum, the brochure is inserted into a reader attached to a computer terminal at each station. When the bar code is used, the visitor selects from among a series of options based on age level and interest. The available levels are adults, big kids, and little kids. As visitors tour the different exhibits, they insert the bar code in the reader at each exhibit and receive text and graphic information presented in full color tailored to their interests. If a user wishes, an option can be selected that is not consistent with the profile stored in the system; for example, adults often choose the kids' component because they are attracted by the lively graphics and the interactive format. Visitors can also use their bar-coded brochures to request a printed topic tour. Examples of tours are Franklin and His Work, History of Science, Transportation, and Black Achievers in Science.

The input at each station is through a touch screen. At a typical station, the opening screen has the selections shown in Figure 15.3. The exhibit overview describes the theme and topics of the exhibit and points out the many connections

Visitor's Guide to the Unisystem Computer Network

Slide this bar code through the bar code reader on any Unisystem terminal

About the Unisystem

With the Unisystem, The Franklin Institute is creating the world's first smart museum. The Museum staff has joined forces with engineers and programmers at Unisys to develop this sophisticated, yet easy-to-use visitor-operated computer system. Two years in development, it links exhibits in the Mandell Futures Center and the Science Center, provides science news, tours, games and printouts.

Teacher Access

Teachers who have access to a computer with a modem can hook up directly with The Franklin Institute's computer network to get information on exhibits, educational activities, references for study and experiment and ideas for cross-curriculum studies.

The access number is:

(215) 448-1105

24 hours a day; 7 days a week

300/1200/2400 bps

8 bits/1 stop/No parity

The Franklin Institute Science Museum

Benjamin Franklin Parkway
at 20th Street
Philadelphia, PA 19103-1194

0000966232

How to Get Your Printouts

Printouts may be picked up at your convenience. The Unisystem will keep a record of all the printouts you request. When you are ready, slide your bar code through the bar code reader at the printer station; your material will be printed as you wait.

Printer locations:

* Adjacent to the Preview Gallery, 2nd floor Science Center
* On the ramp in the Mandell Futures Center Mezzanine.

How to Use the Unisystem

You can use the Unisystem by touching the screen. However, if you want printed information to take home, you must use the bar code on the front of this brochure.

When you slide the bar code through the bar code reader on any Unisystem terminal, the screen will change, and you'll know that the system has opened a file just for you. *Be sure to choose the level of information that suits you or your group—adult, child, preschooler.*

Repeat each time you go to a new terminal. All the directions you need will appear on the screen.

Where to Start

The best place to start your visit is in the Preview Gallery, adjacent to the Benjamin Franklin National Memorial. Use the Unisystem terminals there to order free topic tours to enjoy as you visit the Museum. A printer station is just a few steps away.

Exhibit Information

Showtimes

Games

Topic Tours

Printouts

Take-home Activities

and more

The Unisystem is made possible by a generous grant from the Unisys Corporation, with additional support from the National Science Foundation.

UNISYS

Figure 15.2. Brochure with bar code for access to system. Courtesy of the Franklin Institute Science Museum.

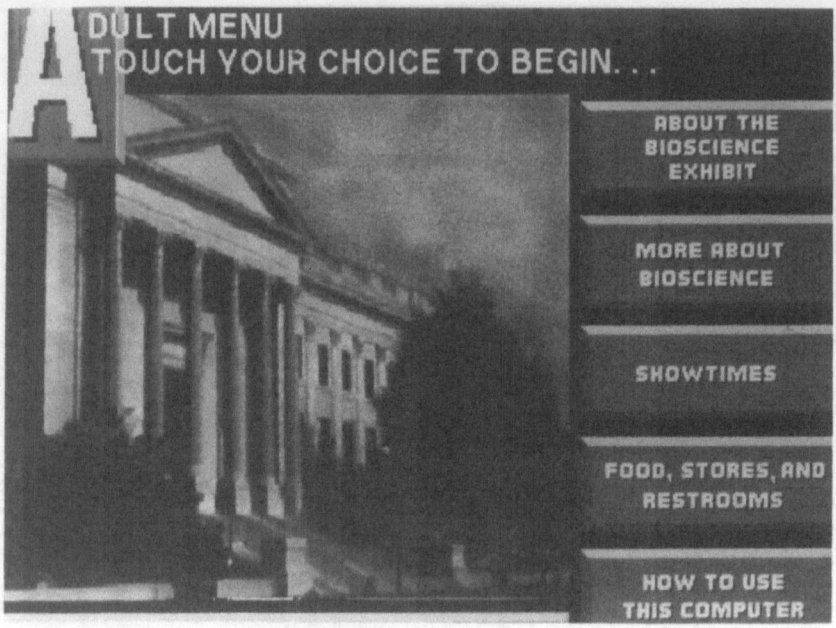

Figure 15.3. Opening screen to Unisystem. Courtesy of the Franklin Institute.

among exhibit topics. Other stations offer information about such things as bathroom and restaurant locations and the schedule of shows in various parts of the museum.

As visitors wend their way through exhibits, they are confronted with a variety of choices and a potpourri of information; they may be offered a chance to Choose the Future, which introduces futures-related issues and invites them to register their own opinion. The issues are raised by asking 5 to 10 thorny questions and accumulating the responses. The system then displays the cumulative responses to the question. Children can learn to Think Like a Futurist in Future and You or apply the ecological lessons of Future Earth and find out How to Help the Earth.

The bar codes can be used to create an electronic mailbox. As the visitor accesses information at various stations, copies of selected items can be requested. Copies can then be picked up at a central print station by using the bar code to request the contents of the mailbox to be printed. Printouts enable the learning process to continue after leaving the museum: Children can do experiments at home guided by the printouts, and adults can request information to conduct an energy audit of their home and learn how to use energy more efficiently. Or, they can carry away a list of environmental organizations or definitions of computer terms.

Another use of the museum computers is to help teachers. The institute is a popular destination for school trips: In 1989 more than 25% of the 726,000 visitors were school children on outings. The educational impact of the museum visit is vastly enhanced if the students prepare for the visit, then conduct focused activities during the visit, followed by a set of activities after returning to school. To accomplish this, two forms of special support are provided to the teachers. The first is an educator's pass for one year's unlimited free admission to the museum. The second is a limited text-only information base that can be accessed via modem from outside the museum. The information consists of topic-specific and age-appropriate activities for use before, during, and after a visit to the institute. Any part of the information may be downloaded and printed by the teacher. An average of 10 educators use this on-line Teacher Bulletin Board each day.

15.1.3. Objectives of the Project

There were a variety of objectives established for the project, and based on the preliminary information, it appears that they are being realized. One important objective was to improve information about each exhibit to the visitor; which entailed improving the form and timeliness of information. Using the computer medium eliminates the need for revising labels and graphic illustrations and speeds up the change process. Naturally it also integrated an interactive element into the exhibits that was not possible before and created an environment suitable for a science museum. Another objective was improving the accuracy of information, with frequent updates and revisions. Last the overall objective was to improve the public's enjoyment of the museum, which appears to have occurred.

15.1.4. Technology

The access terminals have touch screens rather than keyboards, and they do not require familiarity with computers. Presentations consist of a combination of text, color, graphics, and picture-quality high-resolution images. Every attempt has been made to create a colorful, playful graphics style that attracts visitors to the system and invites them to experiment with it, which required painstaking efforts to keep the interface as simple as possible. In the words of the director of the project, "Visitors don't care how clever the technology is or how elegant the operating system is. Visitors care what the experience is. After one or two interactions, the technology becomes transparent, and it is the experience that matters." Unlike many other systems, the high-quality images are not stored on a videodisc or CD-ROM; instead they are captured with an electronic video still camera, then digitized and stored in computer memory where they can be manipulated in visually exciting ways. Access terminals consist of a Unisys PW^2 300 microcomputer with an Intel i80286 processor and a 20-MB hard disk. Terminals are equipped with a touch-screen monitor with super VGA and an optical bar code reader.

15.1.5. Networking and Communications

More than 40 access terminals are networked into subgroups or LANs. Five Unisys 6000 minicomputers serve as hosts for the networks and at the present time can manage up to 10 access terminals; a sixth minicomputer, called the central host, keeps track of the five host computers. If any problems are detected with the LAN hosts or an access terminal, the central host can shut down that portion of the network until repairs can be made. The central host can also monitor accesses by a user as he/she accesses various terminals throughout the museum. The LANs are configured as a star and use Lattisnet. Networks are tied together in a ring configuration using ethernet. The software under which the system operates is called Mapper, a Unisys 4th-generation language.

Designing the network presented a series of challenges. During the pilot, response time was taking 8 or 9 seconds to change a screen; this was eventually reduced to below 3 seconds. A unique problem was the electrical environment in which the equipment had to function. Parts of the building were constructed more than 50 years ago, and the power supply experiences minor fluctuations. These problems are exacerbated by various exhibits, such as one containing a 300-ton locomotive that moves. Some redundancy is built into the network so that if one of the LAN hosts stops operating, the others can assume some of its tasks. If an access terminal stops working, it can be disconnected or turned off and on, and it will reboot automatically.

15.1.6. Design and Operating Issues

In addition to technical personnel supplied by Unisys, the museum's Special Projects Department supplied a staff consisting of a project director, a writer, a computer programmer, and a graphics designer. Supporting these four people were a consultant and the museum's content developers for the exhibits.

Since the system is designed to interpret the exhibits, it is necessary to design the content and interfaces to facilitate the interpretative process. The first step in creating the content was to ask exhibit developers and designers what the most important themes are and how those themes are carried out in physical components of this exhibit. After conferring for several hours, a script was developed; it was reviewed by the content experts and modified until it was in a suitable form. At this point, the graphics designer researched images, took pictures, and found or developed images. The results were again reviewed by content experts, and further modifications were made until everyone was satisfied.

Careful thought was given to the opening screen, but there were still surprises in store. One of the most frustrating issues was developing a set of navigation commands; these are the commands users employ to move through the information. For example, many of the screens have a hand with a finger pointing to the right and left; users were supposed to touch the appropriate hand to move forward or backward through the screens. It was quickly discovered however that a small sign with a pair of hands had to be placed at each access terminal to explain this means forward and this means backward. Another problem occurred when a user left without viewing all of the screens. If the next user began viewing before the access terminal timed out and returned to the first screen, the second viewer missed the opening screen with directions, including the choice of adult- or childlevel information.

Average attendance is 3000 people/day with peaks of 10,000. On the busiest days, every single station is in continuous use for up to 8 hours without a single time out. Time outs are set to occur after 45 seconds.

The whole question of what level of information to present was unpredictable. When the system was first set up, each visitor was issued a bar code as either an adult, a child, or a teacher. However it was discovered that individuals wanted to switch between levels based on their knowledge of particular subjects.

Once created the contents are maintained in-house. The actual rate of change is minimal. Exhibits containing a section called New Science may change as often as weekly, whereas information about a take-home experiment may change only seasonally.

15.1.7. Problems and Future Developments

A public access system designed to serve everyone from preschoolers to senior citizens has a vertical learning curve; stated differently, the interface between the users and the system must be designed to be mastered within a few seconds the first time it is encountered. If more than a few seconds are required, perhaps 5–10 seconds, the customer loses interest, and the younger the customers the more impatient they are and the less able to follow written directions.

The single greatest problem was refining the user interface, which communicates a wide range of options while remaining accessible to naive users. Although the process of tweaking the system will never be completed, it appears that the right combination has been found. The average number of accesses is 50–250 per hour depending on attendance, and the average length of access is 3–4 minutes. It is estimated that 40% of the nongroup visitors access the system during a visit; a much lower percentage of visitors with a group access it because of the way a group tour is conducted.

Another aspect of the interface problem was the bar code itself. Initially a credit-card-sized piece of paper with a bar code was given to each visitor, but research revealed that many visitors who accessed the system successfully were unaware of the full range of the bar code's functions. A brochure was later developed to alleviate the problem.

Topic tours will be enhanced by linking exhibits to the topic tour that was requested. Possibly large print will be added for the visually impaired and perhaps foreign language options. Additional technology will be incorporated, which may include artificial intelligence to deduce interests based on selection patterns, full-motion video footage, and maybe someday sound.

15.2. GENERAL MOTORS CORPORATION: THE TOWN CRIER

General Motors is a fully integrated manufacturer that makes everything from automobiles to locomotives. The Chevrolet-Pontiac-GM of Canada Group (C-P-C) is a major division of the company with headquarters in Warren, Michigan, and approximately 150,000 employees located in plants and offices throughout North America. In July 1985, C-P-C launched what is believed to be the first information service of its type. The Town Crier, nicknamed T.C., is an employee-centered interactive information service that contains fast-breaking news, employee opinion polls, special features, and event calendars. The service is housed in a series of 25 public access kiosks scattered throughout the company and connected to the company headquarters by the public telephone network.

15.2.1. Background

Large companies with employees located at multiple sites have always had problems communicating with their work force. Bulletin boards, newsletters, memos stuffed in paychecks, and special meetings have been the methods most organizations have used to overcome the communication barrier. However within the last several years, it has become increasingly practical to rely on computer-based information systems. Many organizations are establishing electronic bulletin boards that reside on the mainframe. This is an ideal form of communications for white-collar employees who have a computer on their desk. Unfortunately these systems fail to reach the multitude of workers who have neither a desk nor a computer.

The T.C. is a system intended to fill the breach. It is a well-designed and innovative application of technology conceived by the C-P-C Public Affairs staff and Electronic Data Systems (EDS) and begun as a pilot project at the headquarters. After proving its worth, it was expanded and distributed to factory sites. Impetus for the system came from valid complaints within the organization that employees heard job-related news from the public media first. A study of what other companies were doing to keep their employees informed was undertaken and some of these organizations were visited. Once conceived the system took shape rapidly; in fact from proposal to the introduction of a pilot system took only six months.

The T.C. has undergone several modifications since its introduction: The physical dimensions of the kiosk have been changed to permit handicapped accessibility: the screen resolution has been increased and graphics added. The latter were incorporated to permit plants to record information on the system, such as scrap rates. Formerly this type of information was displayed in wall charts at the plants. But in general, the system has withstood the test of time and remains much as it was when introduced in 1985

Other divisions within General Motors are using similar technology but for different applications. For example, one of the divisions has developed a show-room kiosk that permits the viewer to specify a car with certain accessories, then obtain a price. To share experiences, a company seminar was convened at one point in the late 1980s. Each unit brought its kiosks to demonstrate and ideas were shared. However there is currently no coordinated effort in the corporation to propagate systems of the type discussed in the following sections.

15.2.2. Town Crier

The kiosk that houses the T.C. is shown in Figure 15.4. The four major components of information are current news stories usually focusing on company news, opinion polls, special video features, and a calendar of events. There is room for eight news stories and a local weather report. These stories are written every morning and downloaded to the units via a telephone line. However the plants have the ability and the software to overwrite four of the stories to replace them with local information of greater interest to employees at that particular site. Local content is encouraged by headquarters to keep the system from being viewed as simply another vehicle for distributing information from the top down. Local content also brings credibility to the system by providing relevant information.

Opinion polls permit employees to vote on issues related to the division and the car industry. The polls are very nonscientific, and there is no way of prevent-ing an employee with a strong opinion from voting repeatedly. The purpose of the polling option was to generate interest, encourage use of the system and let employees voice opinions about particular issues. The polling segment, which changes regularly, deals with such questions as should smoking be banned in the building. Although the results are not statistically valid, they do provide input about issues that may be controversial. The answers to questions are automatically tallied and reported on the system. In one case, an allied component division posed a series of questions regarding new features on a car to gather some preliminary data about their desirability.

The feature selection showcases stories about C-P-C employees, programs, products, and events. It is updated twice a year, uses the kiosk's laser disc equipment, showing a 1-minute full-motion video story each week; 20 video features are contained on each disc. One of the features that garnered the most interest was about that invisible cadre of employees who clean the building after everyone goes home. Other features are stories about test labs and proving grounds, which is information employees do not normally receive firsthand.

The calendar, which primarily is business related, is updated continually. Local sites have the ability to insert events on the calendar. The main page of the kiosk contains closing stock prices from the previous day.

Figure 15.4. The Town Crier Information Center. Courtesy of the Chevrolet-Pontiac-Canada Group: General Motors Corp.

15.2.2.1. Obtaining Commitment

In 1990 a decision was made to move most of the units gradually out of headquarters and engineering locations and into the plants. This action was based on three separate evaluations that showed higher usage at plant locations than at office locations. The studies suggested that employees at headquarters and engineering locations have much better access to information then individuals working in plants as far away as in Van Nuys, California. Before the units are actually moved to a plant, the local management must agree to a specific level of support. The units are lent free of charge and left at the site as long as certain conditions are met. When a plant decides it wants a unit, two managers travel to the site to meet with the plant manager and the staff to secure a commitment to help maintain the hardware and provide local content. At that meeting, the importance of monitoring the system to make sure it functions properly is emphasized as is the importance of providing local content. Each site must agree to dedicate a portion of one person's time to writing local content and removing dated material. On several occasions, sites that made the commitment but then failed to honor it have had their kiosk removed.

15.2.2.2. Support Levels

At headquarters one person writes each day's news stories and transmits them to the kiosks; this is usually done early in the morning, and it takes several hours. Scanned images are a little more difficult to prepare, so these are done by a different person as needed, then transmitted. The same person also monitors the system and is responsible for modifications and other forms of day-to-day support, such as training those responsible for creating the local content. The combined time of these two people is thought to average between one and 1.5 person days. Video for the special features section is produced on tape by a company film crew and then sent outside the company for conversion to laser discs. Programming support is occasionally needed, and this is done internally by someone from another area of the company.

Many of the units are in very hostile environments, such as the factory floor, so they are subject to airborne particles and high temperatures. Despite this the units are surprisingly reliable. Although there are periods when multiple units are down and must be diagnosed, in a recent month, only one unit failed, and it took only five minutes to repair. The units are equipped with cooling fans and air filters to help combat environmental problems.

As part of their commitment to maintaining the system, plant management must dedicate some support for the system. For maintenance this is limited to making sure the system is working properly. On-site personnel receive training to diagnose minor problems, such as the kiosk becoming unplugged from the wall. If the problem is significant, such as hard drive failure, C-P-C has a contract with an outside company to perform equipment repairs. It is thought that the person maintaining the local content spends about two hours each day on that task.

15.2.2.3. Tracking and Monitoring Use

Each kiosk tracks every screen touch. This information is then transmitted back to the headquarters in an ASCII file where it is analyzed using dBase III and Lotus 1-2-3. Accesses to each area can then be examined to determine what is and is not popular; of course there is no way of determining if one person has accessed the same area more than once. The average number of daily touches is about 200, but it is not unusual to have the count exceed 700 a day. These statistics along with audience surveys have played an important role in determining where the units should be positioned and what changes should be made to the content.

15.2.3. Objectives

The purpose of the project is to disseminate information to employees; the reason for doing so electronically is the immediacy with which dissemination can occur. There were two reasons stated for placing so much emphasis on distributing information. First the more business-related information employees have, the better the decisions they make in their individual jobs, an issue that addresses the importance of understanding the company's competitive relationships to other firms and the forces at work in the marketplace. The second reason stems from a desire to provide information about the company or other industry-related information before employees read it in the public press. In a critical sense, the project provides a chance for the company to present its side whether the subject is the automobile business, what the competition is doing, or the organization's problems.

The specific objectives are listed separately in Table 15.2 in decreasing order of importance; all of these are regarded as being accomplished. A fifth objective has recently been added, which calls for integrating the Town Crier with existing communication services, such as newsletters, electronic display boards, and closed-circuit television systems, as well as using the Town Crier to support these various forms of communication.

15.2.4. Equipment

The T.C. combines text, color, audio, graphics, picture-quality images, and video into an integrated presentation medium. The information is created on a PC, a scanner, and a videotape camera. The text and graphics created on the PC and the scanned images are downloaded to the individual units from a PC via the telephone network. Each kiosk contains an IBM AT PC with 640-KB of memory, a

Table 15.2. Objectives behind the Town Crier

Rank	Objectives
1.	Distribute information sooner
2.	Improve communication
3.	Improve the organization's work climate
4.	Distribute information

video graphics interface board, a 30-MB hard drive, and a 2.4-Kbits/sec modem. The display device is a 19-inch Sony CGA monitor with a Visage, Inc. touch-screen membrane. The software is written in C by in-house programmers.

15.2.5. Design, Implementation, and Operation

Most of the initial design of the system was done by company employees, with some help from a contractor who had experience with touch-screen technology. The requirements were developed by the C-P-C Public Affairs staff, and the technical work was done by EDS. The initial development of text, graphics, picture-quality images, and a portion of the video contents were done internally. The audio and portions of the video were initially done outside the company. The content is maintained by the company, although the laser discs are prepared externally.

Portions of the text, graphics, and picture-quality images are updated daily. All of the audio is on the laser disc and it is recorded when stories are video-taped, or it is voiced over later; audio added later is professionally done. The video footage is produced in-house or adapted from existing videos and assembled on a laser disc manufactured twice a year. As stated previously, changes to the content are downloaded by telephone except for the video. For example, the eight stories that are changed daily are composed on a PC. When they have been completed, a communications program automatically dials a unit and transmits the eight stories to the proper location in each unit, and then proceeds to dial the next unit in sequence. The laser discs are mailed to each site and inserted by the designated support person.

Efforts are continually made to enliven screen presentations and render the content more appealing. At one point, statistical graphs were added in place of the opinion poll; charts were an assortment of pie, line, bar, and other popular charting forms tied to the news content. If for instance an article discussed market share, a chart would be developed to illustrate the story. However the audience evaluations indicated that the opinion poll had greater value. This combined with the fact that charts took more time than those assigned the work could afford to spend resulted in the charts being abandoned. Despite these experiences, attempts are continually being made to add interest to the screen displays. The recent addition of scanned images is one such attempt that has proven successful.

15.2.6. Problems and Future Developments

The single greatest problem encountered has been providing the right information to the right location from the user's perspective. This is a never ending challenge.

One of the discoveries made early in the life of the system was the nuisance the audio could create. At one point, T.C. was fitted with a proximity detector so that whenever anyone came within a few feet, music on the system started playing and information began cycling on the screen. If the unit were located in a hallway or lobby, it was constantly started by passersby, sometimes startling them. Even without the proximity detector, locating the unit in a lobby or near an area where someone, such as a receptionist or guard, is positioned can be very distracting. To deal with the disturbance, the sound volume was lowered. However when some of the units were moved into the plants, the sound was not loud enough, so amplifiers and speakers had to be added. This illustrates once again that sound is one of the trickiest media to deal with in systems of this type.

The most expensive portion of the content to create is the video. Although it adds pazzaz to the unit, it often requires a film crew to shoot a story. Depending on the amount of scripting necessary, this can be very labor intensive and time consuming. Attempts to use existing video have been successful to some extent but have not eliminated the need to produce new footage. Various solutions are being investigated to reduce the cost, including digitized video, but an economical alternative has not been found. Another problem related to the video is time versus cost trade-off. Videodiscs containing the information are produced twice a year, such a production schedule is much more economical than producing the discs monthly or quarterly, but it usually precludes using video for timely or late-breaking information.

The Town Crier is expected to change dramatically in the near future. Plans are underway to integrate the system with closed-circuit television and possibly to reconfigure the word-processing and story allocation system to facilitate updating. Other departments in the company have expressed interest in using kiosk systems for lobbies and providing information about employee benefits, but a proliferation of the technology is unlikely to occur. Whatever happens elsewhere in the organization, the Town Crier stands out as a successful vehicle for disseminating information to employees in a timely and interesting fashion.

15.3. MERCK & COMPANY, INC.—HEARTFELT ADVICE

Merck & Co., Inc., is the largest prescription pharmaceutical company in the world. It employs 37,000 people in more than two dozen nations and has annual sales of more than $8.7 billion. For an unprecedented six years, it emerged at the top of Fortune magazine's list of most admired corporations. Merck has achieved this recognition due to such corporate attributes as good management, products of high quality, social responsibility, and innovativeness. One example of the corporation's innovativeness is Heartfelt Advice, a creation of the marketing unit of the Merck Human Health Division, Merck & Co., Inc., Rahway, New Jersey. Heartfelt Advice is an interactive public education system designed to reduce heart risk factors. It became operational in December 1989, and it is used by employees, customers, and the general public.

15.3.1. Hold the Mayo and the Butter and the . . .

Work on Heartfelt Advice, begun in 1987, grew out of another project called the Merck Wellness Center, which was being developed at about the same time. Each year Merck attends medical conferences (congresses) all over the world. Just as many companies participate in trade shows to explain, demonstrate, and hopefully market their products, pharmaceutical companies display and explain their wares at congresses. As a means of drawing attention to its booth and establishing a presence in the exhibit area, Merck developed the Wellness Center. The Wellness Center is staffed by health care professionals, and its purpose is to assess the general health of those who visit and their risk of heart disease. Within a very short time, the Wellness Center has become a highly visible and popular feature of the Merck exhibit. It has become associated with the company, and doctors return to it at subsequent congresses. The success of the Wellness Center led to the question of how the concept could be expanded. Could a way be developed of assessing heart risk factors using information technology rather than health care professionals, and how could whatever is developed be deployed easily at a variety of sites?

15.3.2. "The Ability to Create What Has Not Existed before" Mozart

Heartfelt Advice is a unique interactive public education system for heart risk assessment. Designed to fit into a small kiosk, it incorporates a computer, touch-screen color monitor, laser disc, printer, and special software. The system consists of two programs: one designed for the general public and the other for access by doctors; both are resident on the system. The user selects the preferred version when the machine is booted. The focus of both versions is primarily educational.

Anecdotal evidence of the value of Heartfelt Advice is abundant. For example, after assessing and printing out her own risk of heart disease, an employee input information about her husband, who had high cholesterol and blood pressure levels, smoked, and was overweight. Interestingly the employee later reported that although her husband was aware of the risk he incurred, he was not motivated to correct the situation until seeing printouts of their respective vulnerability to heart disease. During her husband's new regimen, his wife provided updates showing how the risk of heart disease was changing, which was a stimulus to continue.

The company also gleans considerable health-related information from the system users, since the system keeps aggregate records of what takes place. This information assists in refining the Wellness Center, and it is also the basis of plans for some innovative programs that are germinating.

15.3.3. Objectives

The reasons for launching Heartfelt Advice are many and varied. The objectives from the survey listed in their order of importance are shown in Table 15.3. An overriding purpose behind the system was to enhance communication between the company and its customers, which is known as name recognition, brand identification, customer awareness, and a variety of other terms. The idea was to create a device that established a link between the company and its customers in a way that simply was not possible with print advertising and other promotional tools. The system focuses on educating the user about the risk of heart disease but in the process projects a positive image of Merck and reinforces the fact that the company is in the pharmaceutical industry. It is unlikely that anyone would spend as much time reading a brochure or remember the experience for as long as they will remember Heartfelt Advice.

In addition to providing information about heart disease, the system can calculate a patron's risk based on age, weight, and other user-supplied figures. Naturally the accuracy of the assessment depends on the honesty and knowledge of the person providing the data. Because of the myriad of variables required, the only other practical way of obtaining accurate information of this type would be with a staff of trained professionals.

Most of the objectives for the system are being realized. It is a positive statement of the company's involvement in the health industry. The information it supplies is relevant and if followed, it can improve a person's health. The value of the system is apparently recognized, because people return to it again and again. The system provides an efficient and entertaining way of alerting people to the dangers of heart disease, and it presents risk calculations based on current

Table 15.3. Principal Objectives
in Implementing the Application

Rank	Objectives
1.	Improve communication
2.	Improve accuracy of information
3.	Improve public, customer, or stockholder relations
4.	Distribute information sooner
5.	Enter new business opportunities
6.	Differentiate products or services from competitors'
7.	Increase market share

medical research. For example, if the level of cholesterol deemed safe declines, the system can be easily updated.

The system has clearly enabled Merck to differentiate itself from the competition. Whether or not it has been successful in increasing the company's market share can not be determined, but it certainly has contributed to making the public more aware of the company, especially those who work in the health care industry.

15.3.4. Technology

The system conveys its message through text, color, audio, graphics, picture-quality images, and video. Various types of equipment have been used without any significant reliability problems occurring. The system requires little maintenance and operates 24 hours a day. Twenty-seven units were built and distributed, and the only problem of any consequence has been the power supply in buildings where the machines have been located. On several occasions, power outages have de-activated the system, and it has had to be restarted. An important element in the system design is the touch screen, which serves as the only interface to the system. In a system designed to communicate information about the human body, this is a natural and comfortable extension of the technology employed.

15.3.5. Design and Operating Issues

Heartfelt Advice is one of the most effectively tailored systems found anywhere. It cleverly melds information with technology to create an eye- (and ear-) catching and appealing presentation. Figure 15.5a shows the screen that is displayed when the system is idle. Following instruction to touch the screen activates an audio and video sequence. Note that the initial display when the system is silent shows a French mime with a cartoonlike bubble emanating from its head and containing text. The opening video image is on a laser disc.

Figure 15.5b shows the main index for assessing heart risk. The icons and human caricature (not shown) reveal a clever use of graphics both to highlight the display as well as add interest to the screen. The typical alternative is a less interesting set of choices numbered 1, 2, 3, and so on. Anyone seeking in-depth information about the factors affecting the risk of heart disease can make a selection from the options shown in Figure 15.5c.

One segment of the system invites the user to feel his/her way through dinner by assessing the food value of each item on the plate shown in Figure 15.5d. By touching the items pictured for dinner, such as the asparagus, the calories, fat, cholesterol and sodium levels can be obtained.

A team consisting of Merck personnel, an outside company, a writer, and a consultant was assembled to design and create the contents. A company specializ-

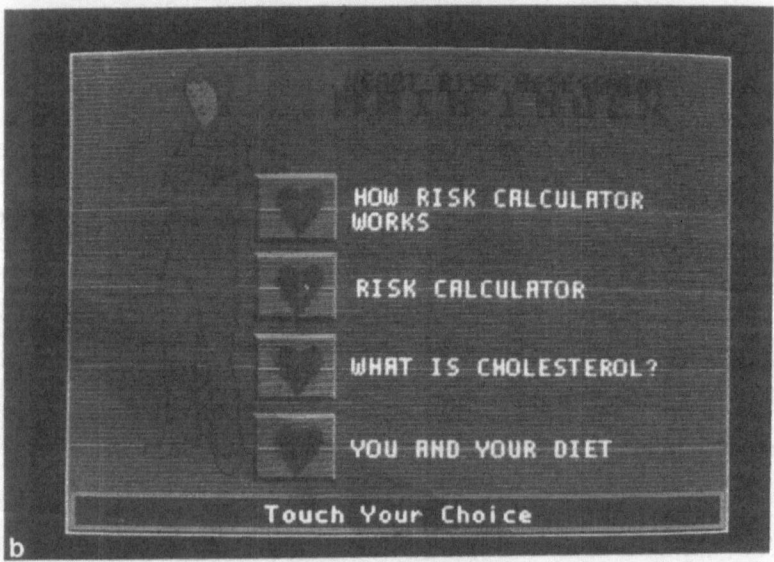

Figure 15.5. Healtfelt Advice. (a) Video display during idle state. (b) Heart risk assessment main index. (c) In-depth information. (d) Feeling your way through dinner to determine food value. (Courtesy of Merck & Co., Inc.)

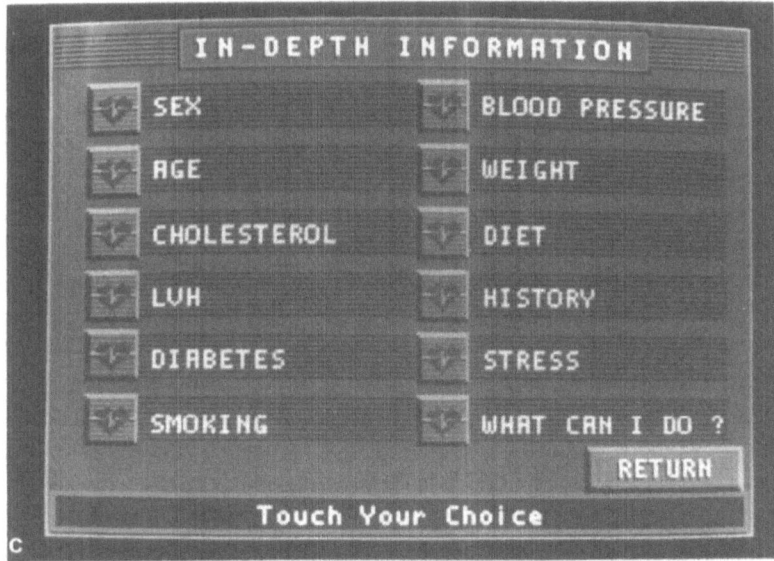

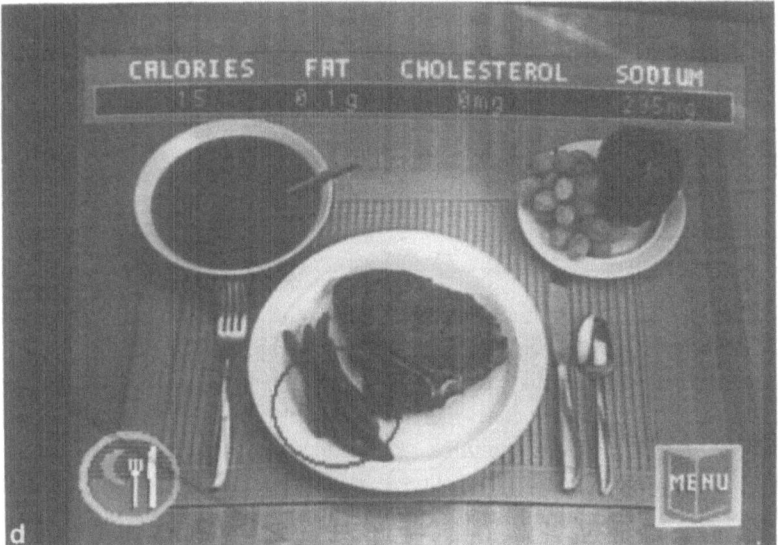

Figure 15.5. (*Continued*)

ing in designing and creating systems of this type was in charge. A writer who had done other work for Merck wrote the script, and a consultant was hired to provide guidance and support. The final text was created and is maintained in-house. The audio, graphics, picture-quality images, and video were created and are maintained by an outside company. There are plans to update the contents annually.

The number of accesses varies per location. A typical site may have 350 accesses per week, with a session lasting an average of 17 minutes. A unit located in a cafeteria at the Merck headquarters was accessed about 2500 times in the first three months it was installed.

Kiosks can be located wherever electrical power is available, although the physical safety of the unit must be considered. Kiosks are generally located in attended areas, such as lobbies and booths at medical congresses. Kiosks can be moved easily, and no difficulty has been encountered shifting them from location to location.

Several things have been learned about how people use Heartfelt Advice. Since the system keeps track of what areas are accessed, it has been observed that when people are at leisure, they will start to access the system but do not take time to complete the session as intended, which leads to more starts than finishes. Also some people apparently do not want to receive bad news. After responding to questions about age, weight, cholesterol, and other parameters necessary for calculating the risk of heart disease, some users terminate the session before answering all the questions and their level of risk can be computed. Audio is an important feature, but it can influence someone's willingness to access the system. When the unit is located in an area that is generally quiet, such as a hospital lobby, there is some reluctance to access it for fear of making too much noise.

The kiosks are not networked in any way. Statistics are accumulated on a floppy disk, and notes are kept in a paper log. When the unit is returned, the disks and log are also returned for analysis.

15.3.6. Problems and Future Developments

Perhaps the most difficult hurdle to overcome in creating the system was getting the necessary internal medical and legal approval. By the nature of its products, a pharmaceutical company is vulnerable to all sorts of legal suits. Consequently tremendous care was taken to ensure that there was nothing inaccurate or misleading in the infobase. Furthermore every effort was taken to promote the system as an educational medium, and not to establish a doctor–client type of relationship. In addition a disclaimer is incorporated into the system.

Many large companies have difficulty encouraging and fostering innovation. For example, at Merck there is no institutionally imbedded process for introducing innovative systems like Heartfelt Advice. Instead innovation often has to have a champion in the form of an individual or work unit that is willing to pursue an idea and fund enough front-end development to convince management to proceed. From small successes grow corporate acceptance and funding.

The Heartfelt Advice system has been very successful as an educational medium. As experience is acquired, it will be modified and improved; for example, the possibility of developing foreign language versions is being considered. The system has also been in part a catalyst or at least a model for other innovative projects. The wonderful thing about innovation is that no one can predict where one innovation will lead or what other ideas it may generate.

Survey of Enhanced Information Systems

Most of the information about the types of systems described in this book is anecdotal and is often provided by vendors and not practitioners. The survey was intended to replace anecdotes with facts. Organizations were identified that were known to have suitable projects and an introductory letter and survey were sent. This was followed by a reminder and second survey. All but one of the organizations were in North America. Fifty-one usable responses were received. The survey went directly to those most affected, the practitioners, and asked a set of questions formulated to answer five fundamental questions: (1) How are organizations using enhanced systems? (2) What are the objectives? (3) What technology and methodology is used? (4) What are the trials and tribulations of using enhanced systems? (5) How are these types of systems expected to evolve and what will be their likely impact on the respondent's organization over the next five years?

The following page contains the overview of definitons, survey parts, and guidelines followed by the eight-page survey.

Definitions

Enhanced System   An easy-to-use, interactive computer system that possesses some or all of the following features: text, graphics, color, picture quality displays, audio, and video.

Project   A work effort that you can think of as an entity.

Public Access Kiosk   An enhanced system which is put in a public location such as a shopping mall, hotel lobby, or conference facility to be used by any passerby.

Survey Parts

Part A.   Questions of general information.

Part B.   Questions about a specific project that meets portions of or all of the definition of an enhanced system as defined above.

Part C.   Questions about problems and future developments.

Guidelines

Please select a specific project from anywhere in your organization. While computers may be used throughout your organization select just one project that contains portions or all of the elements of an enhanced system as defined above.

The project may reside on a computer system operated by your organization or operated by another party; e.g., an on-line commercial service provider. Users may or may not be employees.

If a question applies and you are not absolutely sure of the answer, please make an educated guess.

Try to complete and return the survey within 7 working days using the pre-addressed and postage paid envelope. Thank you.

SURVEY OF ENHANCED INFORMATION SYSTEM PROJECTS

....
    A. QUESTIONS OF GENERAL INFORMATION
....

1. Name, address, and telephone number of person filling out survey. (optional)

Name _____   Phone _____
             (please print or attach business card)

Job Title _____

Organization _____

Address _____

         _____

City _____ State or Country _____ Zip _____

2. What is the principal product or service provided by the work unit that uses the enhanced system?

    _____

    _____

....
    B. QUESTIONS ABOUT A SPECIFIC PROJECT THAT MEETS PORTIONS OF OR ALL OF THE
        DEFINITION OF AN ENHANCED SYSTEM AS DEFINED ON THE COVER PAGE.
....

+-----------------------------+
| BACKGROUND INFORMATION |
+-----------------------------+

3. What is the name used in the organization to identify the project? _____

4. What is the approximate date the project became operational? _____

    Would you describe the project as a trial or as a "permanent" system?

                        Trial ☐   "Permanent" ☐

    What is the approximate date the project is scheduled to end or was ended?

                Not Applicable ☐   Termination Date _____

5. Who are the intended users? (check all that apply)

            Employees ☐   General Public ☐

            Customers ☐   Other ☐

6. Briefly describe the project. Include in your response the application (e.g. selling, dealer support, customer service, etc.) for which the system is used. If a description already exists in written form, please feel free to include that document instead.

    _____

    _____

    _____

    _____

    _____

    _____

    _____

    _____

                (Please continue on back, if additional space is needed.)

7. Please mark the original objectives of the project in the first column below. In the second column rank the objectives you checked. After each objective, please insert a mark if it was or is being accomplished. The next question gives you the opportunity to report about the actual benefits.

check all that apply	rank 1= highest	objective	accomplished
☐		Improve communication	☐
☐		Reduce cost of communication	☐
☐		Improve accuracy of information	☐
☐		Distribute information	☐
☐		Distribute information sooner	☐
☐		Sell products or services	☐
☐		Improve public, customer, or stockholder relations	☐
☐		Increase market share	☐
☐		Reduce operating cost	☐
☐		Improve the work climate of the organization	☐
☐		Differentiate products or services from competitor	☐
☐		Enter new business opportunities	☐
☐		Other	☐

8. What benefits have resulted in addition to the objectives marked in the right-most column in question 7 as being accomplished?

TECHNOLOGY

9. What forms of presentation are used? (check all that apply)

text ☐   graphics ☐
color ☐   picture quality images ☐
audio ☐   video ☐

other _____

10. Briefly describe the host computer hardware, including the manufacturer, model, memory size, and peripheral attachments or give the name of the on-line commercial service provider; e.g., Prodigy.

_____

_____

_____

Is the system you described in question 10 a public access kiosk? (A kiosk is defined on the cover page)

no ☐   Please go to question 11.

yes ☐   If the kiosk is not connected to the computer in question 10, please go to question 12.

11. Briefly describe the hardware through which the user accesses the system described in question 10, including any special requirements for memory, graphics capability, operating system, etc.

_____

_____

_____

_____

12. What is the name of the application software and was it developed in-house?

   Name of software _____

   Vendor ☐

   In-house ☐

NETWORKING AND COMMUNICATIONS

---

13. How is the user's access device or the kiosk networked to other computers? (check all that apply)

   It isn't linked ☐   (go to question 14)

   Telephone line ☐   Wide area network ☐

   Local area network ☐   Other _____

What is the transmission speed over the link in bits per second?

   Don't know ☐   2400 ☐

   300 ☐   9600 ☐

   1200 ☐   Other _____

How is the network configured?

   Don't know ☐   Bus ☐

   Star ☐   Mesh ☐

   Ring ☐   Other _____

What generic form of protocol is used?

   Don't know ☐   Token ☐

   CSMA/CD ☐   Other _____

DESIGN, IMPLEMENTATION, AND OPERATION

---

14. Who did most of the initial system design?

    Company employees ☐ ...... Name of Department _____

    External Sources ☐ ...... Name of Principal Source _____

15. Who creates the original or initial contents for a new set of screens? Please respond for only the forms of presentation you said were used in question 9.

Form	In-house	External Source	Other (specify)
text	☐	☐	_____
audio	☐	☐	_____
graphics	☐	☐	_____
picture quality images	☐	☐	_____
video	☐	☐	_____

16. Who maintains the contents that are displayed on the screen? Please respond for only the forms of presentation you said were used in question 9.

Form	In-house	External Source	Other (specify)
text	☐	☐	_____
audio	☐	☐	_____
graphics	☐	☐	_____
picture quality images	☐	☐	_____
video	☐	☐	_____

17. How frequently are the contents updated and what would you estimate is the percentage of the contents that is updated at that time? Please respond for only the forms of presentation you said were used in question 9.

Form	UPDATE FREQUENCY					percent (%) changed
	daily	weekly	monthly	yearly	never	
text						%
audio						%
graphics						%
picture quality images						%
video						%

18. How much is the system used as measured in the following ways?

How many users are there in a specified time period? _____ What is the time period? _____

  Unknown or proprietary information ☐

What is the average number of accesses per time period? _____ What is the time period? _____

  Unknown or proprietary information ☐

What is the average length of a session in minutes? _____ minutes

  Unknown or proprietary information ☐

C. QUESTIONS ABOUT PROBLEMS AND FUTURE DEVELOPMENTS

19. What was the single biggest problem that had to be overcome at any point in the system's design, implementation, or operation? Please use the reverse side if additional space is needed.

_____

_____

_____

_____

20. What do you think will be the evolution of systems of the type that you described during the next five years in your organization? Please use reverse side if additional space is needed.

_____

_____

_____

_____

21. What will be the impact on your company, if your prediction in question 20 is accurate? Please use reverse side if additional space is needed.

_____

_____

_____

_____

If the pre-addressed and postage paid envelope has become separated from the survey, please return the survey to Dr. Antone (Joe) F. Alber, College of Business Administration, Bradley University, Peoria, IL 61625. Please feel free to make any comments or suggestions about the subject of this survey in writing or by calling (309) 677-2253.   THANK YOU!

Appendix B

Acronyms and Abbreviations

ASCAP	American Society of Composers, Authors, and Publishers
ANI	Automatic Number Identification
ANSI	American National Standards Institute
ASCII	American Standard Code for Information Interchange
BIPs	billion instructions per second
BISDN	broadband integrated services digital network
BOC	Bell operating company
BOPs	billion operations per second
BSA	basic service arrangement
BSE	basic service elements
CATV	Community Antennae Television
CAV	constant angular velocity
CCITT	International Telegraph and Telephone Consultative Committee
CD-ROM	compact disc read-only memory
CEI	Comparably Efficient Interconnection
CEPT	Conference of European Post and Telecommunications Administrations
CGA	color graphics adapter
CLASS	Custom Local Area Signaling Service
CLV	constant linear velocity
CNS	Complementary Network Service
CO	central office
CO-LAN	central office—local area network
CPE	customer premises equipment
CPU	central processing unit
CRT	cathode ray tube
CUG	closed user group
CUI	character-based user interface

DAT	digital audio tape
DBS	direct broadcast satellite
DOV	data over voice
DTMF	dual-tone multifrequency
DVI	digital video interactive
EGA	enhanced graphics adapter
EMI	electromagnetic interference
ESP	enhanced service provider
FCC	Federal Communications Commission
FEP	front-end processor
FPS	frames per second
FT	fault tolerant
FT1	fractional T1
GKS	Graphical Kernel System
GUI	graphical user interface
H$_Z$	Hertz
IC	interexchange carrier
ISA	Interactive Services Association
ISDN	integrated services digital network
IVDT	integrated voice/data terminal
IVR	interactive voice response
JPEG	Joint Photographic Experts Group
K, k	1000
Kbits/sec	thousand bits per second
LADT	local area data transport
LAN	local area network
LATA	local access transport area
LEC	local exchange carrier
MAN	metropolitan area network
Mbits/sec	million bits per second
MDA	monochrome display adapter
MDS	multipoint distribution service
MFJ	Modified Final Judgment
MFLOPs	million floating point operations per second
MIDI	Musical Instrument Digital Interface
MIPs	million instructions per second
MME	Microsoft Multimedia Extensions (for Windows)
Modem	modulator—demodulator
MPC	multimedia personal computer
MPEG	Motion Photographic Experts Group
MTBF	mean time between failures
MTTR	mean time to repair
NAPLPS	North American Presentation Level Protocol Syntax
NTSC	National Television Standards Committee
OLTP	on-line transaction processing
ONA	open network architecture

PABX	private automatic branch exchange
PAT	public access terminal
PAV	public access videotex
PBX	private branch exchange
PC	personal computer
POS	point of sale
POTS	plain ordinary telephone service
PPSN	public packet switched network
PSN	public switched network
PSTN	public switched telephone network
PTT	Postal Telegraph and Telephone Administration
RBOC	regional Bell operating company
RGB	red, green, blue
SCA	Subsidiary Communications Authorization
SDVD	Simultaneous digital voice and data
sec	second or seconds
SIG	special interest group
SMDS	switched multimegabit data service
STV	subscriber TV
TPS	transactions per second
UART	Universal asynchronous receiver/transmitter
VAN	value-added network
VGA	video graphics array
VIA	Videotex Industry Association
VPEG	Video Pictures Expert Group
WAN	wide area network
WORM	write-once-read-many

Appendix C

Glossary

algorithm A series of prescribed steps for completing a task, *e.g.*, compressing images or audio in a prescribed manner

American National Standards Institute An organization serving as a coordinator for the development of standards and as a clearinghouse for the distribution of standards

analog signal Data represented in a continuous signal

architecture Organizational structure of an entity, such as a computer, a communications system, or an infobase

artificial intelligence The ability of a machine to perform functions normally associated with human beings, such as learning, reasoning, and adapting

ASCII (pronounced AS-KEE) The American Standard Code for Information Interchange is a widely used 7- or 8-bit code for representing characters in digital form.

audiotext A computer system that provides callers with access to information by dialing a telephone number and selecting options from a spoken set of choices; the system requires a pushbutton telephone.

authoring A structured approach for developing the contents of an infobase, often associated with interactive systems containing graphics or video

authoring language A computer program to facilitate the development of an infobase, often associated with interactive systems containing graphics or video

authoring system A combination that may include hardware, software, stored data and images, and documentation for designing and creating the content to be displayed

automatic number identification A feature of telephone service that allows a telephone subscriber to identify the source of an incoming call

availability A measure of the time a system is operable; the ratio of operating time to the sum of the operating time plus down time. The formula expressed in terms of mean time between failures (MTBF) and mean time to repair (MTTR) is Availability = MTBF/(MTBF + MTTR).

417

Appendix C

baseband A single data signal is transmitted in its unmodulated form and occupies the entire carrying capacity of the network.

basic service arrangement The switching and transport service purchased by the information service provider to communicate with its customers using the public telephone network; *see* **open network architecture.**

basic service element One of many elements comprising the public telephone network that can be combined with other basic service elements to create a new service; *see* **open network architecture.**

benchmark A group of programs and data run on different machines to establish a basis of comparison among the machines

bit A bit is the abbreviation for binary digit. In the binary numbering system, only two digits are used—0 and 1. These digits may be represented physically as two magnetic polarities, two frequencies, the existence or absence of an electric pulse, and other physical means.

bit rate The frequency with which binary digits, or pulses representing them, pass a given point in a communications medium

broadband A single cable that carries several logical channels of data at different frequencies

bus A network configuration in which units are connected to a single communications medium (*e.g.*, coaxial cable) that runs the length of the network

CD-I A specification for an interactive product that delivers still images, audio, and graphics

CD-ROM A format standard for storing digital data on a compact disc; it specifies how the data are to be arranged on the disc. The spelling of disc with a "c" is an accepted industry practice for optical storage media.

CGA Abbreviation for color graphics adapter, which has a 320-by-200-pixel resolution for color and a 640-by-200 pixel resolution in monochrome and a frequency of 15.75 KH_Z

central office Location of the telephone switching equipment for a particular geographic area

central office local area network A centrex-based LAN; normally the CO-LAN (pronounced *see* oh lan) equipment is located at the telephone company's CO.

centrex A business telephone service similar to a private branch exchange but usually (not always) originating in the telephone company's central office, hence the origin of the name central exchange. Centrex has two basic features—Direct Inward Dialing and Automatic Identified Outward Dialing.

CLASS A set of telephone call management features, such as displaying the calling number, distinctive ringing for preset numbers, selective call forwarding, and automatic callback

comparably efficient interconnection (CEI) An interim requirement for connecting to the telephone network established by the FCC until the telephone companies can establish their permanent interconnection arrangement under open network architecture; *see* **open network architecture**

complementary network services A service that resides on the user's side of the telephone line, such as distinctive ringing; *see* **open network architecture.**

compressed audio and video Audio and video that have been digitized, and then the amount of data needed to represent them has been algorithmically reduced

computer phone   A microcomputer with integrated voice and data components including an attached handset and special software with telephone-related functions

connectivity   The ability to interconnect equipment through a communications network

constant angular velocity (CAV)   The speed of the disc is constant, and the data near the center is packed more densely than data near the edge; contents are organized into sectors on each track.

constant linear velocity (CLV)   Contents are arranged in a single spiral track with a uniform density of bits. The closer the read head gets to the outer edge, the faster the disc spins.

Council on European Posts and Telegraphs   An organization made up of European postal telegraph and telephone administrations

cross talk   Interference created when a signal transmitted on one circuit or channel creates an undesirable effect on another circuit or channel

cursor   A position indicator marking the active location on a screen; the cursor may be a blinking line, a blinking white block, or any other symbol designed to attract the viewer's attention.

cyberspace   *See* **virtual reality**

DAT   Digital audio tape storage is an advanced form of helical scan recording with a capacity of 1.3 gigabytes on a credit-card-sized 4-mm cassette.

data   Raw material from which information is derived and consisting of facts, concepts, and instructions suitable for communication, interpretation, or processing by human or automatic means

decompression   Regeneration of an audio or video original that has been compressed.

desktop publishing   Producing professional-looking documents using a microcomputer and software with text-processing, graphics, and imaging capabilities

dedicated system (configuration)   A system that supports enhanced applications exclusively

digital signal   Data represented in a discrete signal

digital video interactive   A technology that makes possible the display of full-motion video, high-speed graphics, and high-quality audio.

digital video   Representing a video image by digitizing it; one common method is to describe the image with a two-dimensional array of samples called pixels. Each pixel is represented by a number that describes its color.

digitize   To convert an image, an audio signal, or even the dimensions of a physical object into a digital input that can be stored and manipulated by computer

disc   The conventional way of spelling the recording medium used for optical storage. A disc is usually made of a metal alloy capable of storing data sandwiched between a rigid substrate and a plastic protective coating. *See* **disk**.

disk   The conventional way of spelling the recording medium used for magnetic storage. *See* **disc**.

dot pitch   Distance between phosphors of the same color (dots) on the display screen; the smaller the dot pitch, the closer the dots are to each other, and the higher the resolution.

EGA   Abbreviation for enhanced graphics adapter, which has a 640-by-350 pixel resolution and a frequency of 21.85 KH_z; EGA offers 16 colors on screen from an 8-by-14 dot cell.

EGA/CGA monitor   A monitor with the capacity to switch between EGA or CGA mode

emulate   In information technology, the ability of one device to imitate another.

enhanced system   An easy-to-use, interactive computer system that possesses some or all of the following features: text, graphics, color, picture-quality displays, audio, and motion video

enhanced service provider   An organization or individual that offers such services as on-line transactions and alarm-monitoring primarily over the telephone network; *see* **open network architecture**.

ethernet   A LAN that enables two or more electronic devices to pass information back and forth

facsimile   A system that transmits scanned images consisting of text, graphics, and any other imagery that may be rendered on a paper document

fault tolerant   A system feature that ensures the integrity and continuous availability of data and programs despite hardware and software failures

floptical   A high-density magnetic recording floppy disk with a capacity of as many as 21 Mbytes; the capacity is the result of special optical-servo tracks that increase track densities almost 10 times the track densities of ordinary floppies.

footprint   Amount of space required by the base of a monitor or microcomputer

format   A predictable pattern of design consistently applied to the way in which information is entered and/or displayed

fractional T-1   A communication service that enables organizations to purchase portions of a T-1 line in 64-Kbits/sec increments

frame   Basic physical unit of information in a videotex system; corresponds to a single screen display; *see* **page**.

gateway   Combination of hardware and software through which a user can gain access to an information or transaction-based system

genlock   A device for synchronizing computer data and a video signal from another source, such as a VCR, to produce a recordable video picture

Hertz   Abbreviated H_Z; the number of cycles per second; a basic unit of frequency

hypercard   An individual frame of information in a hypertext application

hypermedia   *See* **multimedia**.

hypertext   Software in which information is stored in discrete nodes that can be given addresses or labels. Nodes can be displayed and direct software links can be created to connect files. *Alternative*: Information organized into chunks and linked together to support nonlinear access of the material

image capture   Digitizing an image and storing it in an electronic medium, such as on a floppy disk

image processing   Operating on an image that has been digitized; operating could include making it sharper, coloring it, changing its size, or analyzing it electronically for particular characteristics or features

infobase   A term used in this book to differentiate the repository of information from a repository of data; *see* **data** and **information**.

information   Data that have been transformed into a meaningful and useful form

information technology   Application of scientific and managerial principles to information resources for operational, tactical, and strategic purposes

integrated voice/data terminal   A device that combines the components needed for voice and data communications in the same unit

integrated services digital network (ISDN)   A digital network that carries voice, data, telemetry, slow-motion video, and other forms of communication separately or simultaneously over a transmission medium accessed by a limited set of multipurpose user network interfaces

interactive voice response   Computer-based voice information services that include voice messaging, call directing, and audiotext services

interexchange carrier   A carrier that provides intra- or interstate services, measured or flat rate, that cross a LATA

internet   A loose affiliation of private, academic, and government-supported networks operating in over 40 countries and providing electronic mail, file transfers, and computing resources

Joint Photographic Experts Group (JPEG)   A compression scheme for still-frame images primarily, although it is possible to compress motion video with it; devised by the Joint Photographic Experts Group

keyset   Set of pushbutton keys, *e.g.*, the 12 pushbuttons on a standard telephone handset

kiosk   A structure housing an easy-to-use interactive computer system that is usually intended for public access; often found in lobbies, airports and shopping centers

lands   Flat space between the pits on the surface of a compact disc

local access transport area (LATA)   An area as small as a single city or as large as a whole state that encompasses one or more contiguous local exchange areas and is serviced by a LEC

local area data transport (LADT)   A network service offering by a communications carrier for delivering low- to medium-speed data and information among users and infobases and for supporting high-speed exchanges among infobases

local area network (LAN)   A network encompassing a limited geographical area that links two or more electronic devices

local exchange carrier   A common carrier that provides telephone and subsidiary services within a local market

local loop   The telephone line or outside plant cable pair connecting a subscriber to a telephone company's central switching office

magneto-optical storage   A rewritable optical storage technology

MDA   Abbreviation for monochrome display adapter, which has a 720-by-348 pixel resolution and a frequency of 18 KH$_Z$; text is created from a 9-by-14 dot cell.

menu   A set of options presented to a viewer from which selections can be made, *e.g.*, a menu frame containing numbers (1, 2, etc.) to enter to select other frames

mixed media   *See* **multimedia**.

modem   A device for converting the digital signal produced by a computer into an analog signal for transmission over a network and vice versa; modem is the contraction of the terms modulator and demodulator.

Motion Photographic Experts Group (MPEG)   A compression scheme for full-motion video devised by the Motion Picture Experts Group

multicomputer   A combination of computers that can share a global memory and have a local memory and input/output capability; multicomputers can process single or multiple programs and can make use of closely coupled data transmission and concurrent programming.

multimedia   A term used to describe a computer system that combines a variety of the following features—text, graphics, color, picture-quality displays, audio, and motion video

multimedia system   An easy-to-use, interactive computer system that possesses some or all of the following features—text, graphics, color, picture-quality displays, audio, and motion video

multipoint distribution service   An omnidirectional microwave broadcast operating on the 2150–2162 MH$_Z$ portion of the frequency spectrum with a radius of 20–25 miles

multiprocessing   Simultaneous execution of two or more programs by two or more processors

multiprocessor   A computer with multiple processors for simultaneous use; processors may share memory and other system components.

multiprogramming   Concurrent execution of two or more programs by allowing the programs to share computer resources, *e.g.*, to execute one program while a second is outputting a result to a terminal

multitasking   Partitioning a job that has been submitted to the computer into several cooperating tasks that can be executed simultaneously

Musical Instrument Digital Interface (MIDI)   A digital protocol for transmitting musical data

NAPLPS   North American Presentation Level Protocol Syntax, a standard that specifies the coding scheme for videotex services

on-line transaction processing   A sequence of operations carried out under the direct control of a computer on a collection of related actions whose completion identifies a discrete unit of work

open network architecture (ONA)   Overall design of a carrier's basic network facilities and services to permit users to interconnect to specific network functions and interfaces on an unbundled and equal access basis

optical disc   A disc whose contents are read by a light beam, *i.e.*, laser

page   Basic logical unit of information in a videotex system consisting of combinations of letters, numbers, colors, and patterns; pages are sometimes referred to as frames, but strictly speaking, a frame is the basic physical unit, and the page is the logical unit consisting of one or more frames.

parallel processing   Simultaneous execution of two or more sequences of instructions, or one sequence of instructions operating on two or more sets of data, by a computer having multiple arithmetic and/or logic units

pel   Width of a line, dot, or predefined object, such as a cursor; it is sometimes referred to as a brush stroke; *i.e.*, a large pel creates a broad brush stroke; *see* **pixel**.

photovideotex   Combination of a still television picture, text, and graphics capability

picturephone   *See* **videophone**.

pixel   Short for picture element; a physical pixel is the smallest displayable unit on a screen. A logical pixel is a geometric construct associated with the drawing point. Its stroke width determines the width of a graphic primitive.

point-of-sale terminal   A device for automating retail sales operations, such as inventory control, credit authorization and verification, and electronic funds transfers

Prestel   National videotex service of Great Britain; it began commercial service in 1979.

PTT   An abbreviation for Postal Telegraph and Telephone administration; the branch of federal government that operates the communications services in most countries throughout the world. The United States vests this function in nongovernmental hands, *e.g.*, AT&T.

Private Branch Exchange (PBX) A telephone service operated by an organization serving the telephone extensions of its employees or residents in a building or campus. Access is also provided to the public network.

processor A device that carries out operations on data

Prodigy A graphics-based videotex service operating in the United States. It was created in 1984 and owned by IBM and Sears.

Project Victoria A trial videotex service started by Pacific Bell in August 1986, but discontinued in 1988, due to the regulatory uncertainty confronting the telephone industry at that time

protocol A set of formats, rules, and procedures governing the exchange of information between devices in a communications system

public access terminal A device used to interface with a public access videotex system

public access videotex An easy-to-use interactive electronic information and transaction service intended for public use

public packet switched network (PPSN) A switched network that transfers data in chunks called packets. The channel is occupied only for the duration of the transmission, which permits many users to share channels; also **PSN**—packet switched network.

public switched telephone network (PSTN) Public dial-up telephone system; also **PSN**—public switched network

public system (configuration) A computer system operated by a commercial service provider and made available to organizations and the general public for a fee

refresh rate Number of times the scan lines comprising a display are repeated; a television picture in the United States is refreshed 60 times a second.

regional Bell operating company (RBOC) A telephone company created by the divestiture of AT&T; the seven RBOCs are Ameritech, Bell Atlantic, Bellsouth, Nynex, Pacific Telesis, Southwestern Bell, and U.S. West.

resolution Number of distinguishable pixels per linear unit of measure

reliability Probability that the system will continue to function under given conditions for a specified period of time; the formula is: $R(t) = e^{-bt}$, where b is the inverse of the MTBF, and t is expressed in hours.

response time Total elapsed time between when a user completes an inquiry and the response begins to appear on the screen

ring A network configuration in which units are connected to a communications medium (*e.g.*, coaxial cable) that forms a continuous loop

service provider An entity responsible for providing part or all of one or more videotex services

session Period of time starting when a user establishes communications with another user or computer and ending when communications is broken off

shared system (configuration) An enhanced system that resides with other applications on a computer mainframe

soft keys Soft function keys correspond to keys on a keyboard or touch screen and appear on the top or bottom of the display screen; these keys change as the operation mode changes. For example, if a computerphone is being used as a computer, the soft keys may select programs or operating modes. When used as a terminal, soft keys may initiate or disconnect a call.

star   A network configuration in which each unit is connected to a central, traffic-routing hub

strategic management   Formulation, implementation, and evaluation of actions that will enable an organization to achieve its goals

subscriber TV (STV)   A little-used method for distributing pay-TV and special-service broadcasts over-the-air

Subsidiary Communications Authorization (SCA)   A digital data services broadcast using a subcarrier channel on an FM radio signal; also referred to as data radio and packet radio

system administrator   A person responsible for the configuration, integrity, and security of the infobase; the liaison between the frame creation activity and the viewers

system operator   Entity that operates a computer system; usually identified with the location of the system hardware

T-1   A 24-channel communication service operating at 1.544 Mbits/sec in the United States

telecommuting   Enabling employees to work at home or an intermediate location by providing an organizational infrastructure and the resources that support computer, office products, and E-mail

telemedia   *See* **multimedia**.

transaction   Collection of related actions whose completion identifies a discrete unit of work; all of the parts of a transaction must be completed or else the transaction has to be restarted.

tree   A network configuration in which units are connected to multiple central hubs, which in turn are connected to each other

value-added network (VAN)   A network that adds value to the common carrier's network services by adding such features as better error detection and control

VGA   Abbreviation for video graphics array, which has a 640-by-480 pixel resolution for color and 720-by-400 pixel resolution in monochrome and a frequency of 30.5 KH$_z$; VGA offers 256 colors from a palette of 262,144.

video; full-screen, full-motion   Receipt and display of 30 new images per second

videodisc   A storage media employing laser technology and normally containing video, audio, and perhaps text; the contents may be accessed directly rather than linearly.

videophone   A device with an integrated telephone, television camera, display screen, and frequently a computer, which permits users to be seen as well as heard

videotex   An easy-to-use interactive computer system

Videotex Industry Association   A trade association for the videotex industry; the organization changed its name to Interactive Services Association on January 1, 1992.

virtual reality   A state of the mind produced by having a computer-controlled environment respond to a user's sense of touch, hearing, and visual feedback.

visualization   Interactive visual exploration of large amounts of data

wide area network (WAN)   A communication network serving geographically separate areas, some of which may be served by LANs connected to the WAN.

workbench   Used in this book to represent the specific place where a job is performed and the equipment and communication services available at that location.

workstation   Resources, often computer in nature, and the location where used; these are dedicated to an individual

worm   A write-once-read-many times optical storage technology that ensures data cannot be altered once written by burning pits in the surface of the medium

Index

About the Author

Antone (Joe) F. Alber is Professor of Business Computer Systems at Bradley University where he teaches in the College of Business Administration and is currently president of the University Senate. He has degrees from Penn State (Ph.D., Operations Management), the University of Pittsburgh (MBA), and Lehigh University (B.A., Mathematics).

Dr. Alber has published in leading business and computer periodicals and is the author of *Videotex/Teletext: Principles and Practices*. He is currently working on a multimedia handbook for managers.

Dr. Alber consults with business on computer applications and interactive computer systems. He is a cofounder of Heartland Free-net, the first community supported computer information system in the U.S., and is one of the original members of the Interactive Services Association.